D1528518

ASTRONOMY AND ASTROPHYSICS LIBRARY

Series Editors:
G. Börner, Garching, Germany
A. Burkert, München, Germany
W. B. Burton, Charlottesville, VA, USA and
   Leiden, The Netherlands
A. Coustenis, Meudon Cedex, France
M. A. Dopita, Canberra, Australia
A. Eckart, Köln, Germany
E. K. Grebel, Heidelberg, Germany
B. Leibundgut, Garching, Germany
A. Maeder, Sauverny, Switzerland
V. Trimble, College Park, MD, and Irvine, CA, USA

For further volumes:
http://www.springer.com/series/848

M. Vázquez · E. Pallé · P. Montañés Rodríguez

The Earth as a Distant Planet

A Rosetta Stone for the Search of Earth-Like Worlds

M. Vázquez
Instituto de Astrofísica de
Canarias, Tenerife
Spain
mva@iac.es

P. Montañés Rodríguez
Instituto de Astrofísica de
Canarias, Tenerife
Spain
pmr@iac.es

E. Pallé
Instituto de Astrofísica de
Canarias, Tenerife
Spain
epalle@iac.es

ISSN 0941-7834
ISBN 978-1-4419-1683-9   e-ISBN 978-1-4419-1684-6
DOI 10.1007/978-1-4419-1684-6
Springer New York Dordrecht Heidelberg London

Library of Congress Control Number: 2010922052

© Springer Science+Business Media, LLC 2010
All rights reserved. This work may not be translated or copied in whole or in part without the written permission of the publisher (Springer Science+Business Media, LLC, 233 Spring Street, New York, NY 10013, USA), except for brief excerpts in connection with reviews or scholarly analysis. Use in connection with any form of information storage and retrieval, electronic adaptation, computer software, or by similar or dissimilar methodology now known or hereafter developed is forbidden.
The use in this publication of trade names, trademarks, service marks, and similar terms, even if they are not identified as such, is not to be taken as an expression of opinion as to whether or not they are subject to proprietary rights.

Cover illustration: G. Pérez (IAC)

Printed on acid-free paper

Springer is part of Springer Science+Business Media (www.springer.com)

Preface

Since the discovery of the first planet outside the Solar System (or exoplanet) in 1992, detection of the number of planets is increasing exponentially. This planet search is generating one of the most active and exciting fields in astrophysics for the next decades. Although we are not capable of detecting and exploring planets like our own yet, ambitious ground and space-based projects are already being planned for the next decades, and the discovery of Earth-like planets is only a matter of time.

The theory of stellar evolution has been tested and developed by observations of several stellar types at different times of their evolution. In the 1980s, the observations of 'The Sun as a Star' provided the role of our star as the Rosetta stone in interpreting the observations of sun-like stars with different mass, age and level of magnetic activity. This solar–stellar connection had a double avenue, because the stellar observations also contributed to a better understanding of the solar magnetism.

Although we are probably set for some surprises, the example of the Earth and the rest of the rocky planets of the solar system will be our guidance to classifying and understanding the multiplicity of planetary systems that might exist in our galaxy. In a similar way to that of the Sun as a star, it is reasonable to expect that the future observed population of planets in the galaxy will exhibit a wide range of planet types and evolutionary stages. Observations of 'The Earth as a Planet' will provide the key to understanding future observational spectra of such bodies. However, the *Earth-Exoplanets connection* will also work in both directions. When a substantial database of exoplanets becomes available, statistics of planetary formation and evolution will become possible. This will provide vital information in solving some of the questions about the formation and evolution of our own planet and the solar system, for which we still have no answers.

The current view on stellar evolution is very deterministic. The future and evolution of a star depend on two basic properties: its mass and its metallicity. If these two quantities are known, we can establish whether the star will explode as a supernova in a few million years or if it will end its days as a red giant. For planets, the picture is a little more complicated. At first instance, the mass of the planet, its composition and its distance from the parent star will determine its habitability and evolution. But other factors, such as the presence of gas giant planets, can also play a major role in its evolution. The parent star will also influence the evolution of the planet.

To establish the solar–stellar connection, we needed only to compare the stars; similarly, to establish the Earth–exoplanets parallelism, we need to compare not only the planets but also the physical properties and evolution of their planetary systems as a whole. For example, we may be able to determine how many of the 'rocky' planets that we detect have experienced a runaway greenhouse effect, such as Venus, or how many have lost their atmosphere as Mars has. By observing planets of different ages, we also learn about the state of our own planet in different epochs.

Undoubtedly, one of the main concerns for astrophysics in the coming years will be the search for life. If a planet has all the suitable original conditions to develop and sustain life, does life necessarily occur? And if it does, what are the average time scales for the development of bacteria, plants or intelligence? Are we alone? ... During its evolution, some of the most dramatic changes suffered by our planet affected the composition of the atmosphere. Extraterrestrial observers would obtain two different spectra of our planet depending on the epoch of the observation. Early on the Earth's history, the major atmospheric signatures were those of CO_2 and water vapour, but in recent times (in terms of millions of years), together with such spectral features, the bands of molecular oxygen (O_2) and ozone (O_3) are also present. This dramatic change, the rise of oxygen content, was triggered by the appearance of life. In the future, we may be able to infer whether life is common or not in the universe by observing the evolutionary stages of millions of planets. From the tiny bacteria to technological civilizations we can expect to see life signatures in the atmosphere of exoplanets.

In summary, this book will focus on observations of the Earth as a model for the search of exoplanets and on the information that we will be able to extract from their observation. We put ourselves in the position of an external observer looking at the solar system from an astronomical distance, and we try to answer how we could conclude that this particular planet, the third in distance to the central star, is essentially different from the others and capable of sustaining life. Then, we apply what we learn from this change of perspective to the search for exoplanets similar to Earth.

The first chapter of the book provides first a historical briefing on the progressive knowledge of our planet. Then, we start with a sort of space travel. Starting with first observations from the altitude, using balloons and rockets, we continue with the views of our blue planet from the Moon and the different planets of our solar System. One of the most important achievements of space research was the capability to observe the Earth from outside, floating in space.

This concept will be complemented with the second chapter, where we describe the main properties of our planet. A description of the present Earth from its interior to the atmosphere is given, followed by a review of the different periods of the Earth history.

The third chapter shows how the Earth should be observed from space as Sagans's blue dot. The photometric, spectroscopic and polarimetric properties of the globally integrated light reflected/emitted by Earth are discussed. Special emphasis is given to the Earthshine observations, the sunlight reflected by Earth toward the dark side of the Moon, as a proxy for such global observations.

The outer layers of the Earth are discussed in Chap. 4. Many interesting processes resulting from the interaction of the atmosphere with the high energy solar radiation, solar wind and cosmic rays occur at high altitude. Observed from space the Earth glows in discrete spectral lines (airglow), enhanced during transitory events such as the auroras. The UV and X-rays are excellent diagnostic tools for investigating these regions of the atmosphere.

The existence of life is one of the most relevant properties of our planet, and the Earth would look completely different without it. However, detecting it unambiguously from vast distances is no trivial matter and ingenious techniques should be used for this purpose. This matter will be handled in Chap. 5. Toward the end of the chapter, the main features of our technological society, as reflected in the Earth's electromagnetic spectrum, are described, providing a hint as to how to detect other extraterrestrial civilizations.

Detection of other Earths is the essential requirement to apply the Earth–Exoplanets connection. Chapter 6 is dedicated to a review of the current and future projects, explaining in detail the current limitations in detecting the less massive planets. A research field with a brief history, starting in the 1990s, but which has undergone rapid development and is becoming one of the most important fields in current Astrophysics.

The mass of an astronomical body, together with the chemical composition of the environment where it was formed, determines its future. Stars are massive enough to reach temperatures that permit thermonuclear reactions in their interiors. Less massive brown dwarfs can make this process only by burning deuterium. Planets were recently defined by an international committee and come mainly in two classes: Giants and terrestrial. For the latter, the Earth, Mercury, Mars and Venus are our references; however, the possibilities are larger. Lacking observations of terrestrial exoplanets, we can figure out theoretically how these exo-earths could be, changing some of the basic parameters. From Super-Earths and Super-Mercuries to Carbon planets, an ample diversity of possible worlds is presented in Chap. 7. Some selected exoplanets discovered already are studied in detail.

However, planets are not isolated bodies. They experience the influence from the parent star and the rest of planetary companions in the planetary system. The broad destiny of a planet is determined by its initial mass and chemistry, but the ultimate fate depends on how the planet is affected by the interactions with its companions. To be in the right place is a good recipe. In Chap. 8 we discuss these collective processes, along with the different theories on the formation of planetary systems. As usual we start with our own solar system, which can be confronted with the first observations of proto-planetary disks and multiple planetary systems.

The last chapter is a necessarily failed attempt to answer some fundamental questions regarding the position of our planet in the Universe: Is our Sun special? Is our Solar System common? and finally: Is our Earth unique? The complete answers can only be provided by observations, and search for bio-markers, of terrestrial exoplanets to be discovered in the future. In the meantime, we can, however, remark that we have a unique process of formation and an ample diversity of planetary systems. Birth and environmental factors determine together the structure and evolution of a planetary system and its components.

Many people have been involved at the Instituto de Astrofísica de Canarias (IAC), in different ways, in the preparation of this book. R. Castro elaborated and retouched a substantial number of the figures and the Library staff (M. Gómez, and L. Abellán) provided an excellent service in tracing old publications. The computer services maintained our informatics tools in operation, helping when a problem appeared.

We thank I. Gómez Leal and J.A. Belmonte for the task of reading critically the whole book. F. Anguita, V. Bejar, C. Esteban, A. García Muñoz, I. Ribas, J.A. Robles, J. Schneider and R.M. Zapatero-Osorio also have critically read different drafts of the book chapters and made valuable comments, advices and suggestions. Figures and data have also been kindly supplied by T. Bastian, C. Benn, E. Böhm-Vitense, C. Bounama, R. Buser, V. Courtillot, C. Esteban, M. Harris, P. Hoffman, B.W. Jones, L. Kaltenegger, E. Kokubo, M. Kuchner, G. Laughlin, M. Livio, N. Loeb, D. López, J.J. López Moreno, K. Menou, R. Neumeyer, A. Oscoz, M. Perryman, E. Pilat-Lohinger, I. Ribas, J. Robles, K. Shingareva, M. Sterzik and D. Valencia. We thank the NASA ADS service, which provides a wonderfully efficient service to the scientific community. Archives of NASA and ESA have also been used.

Anna Fagan helped to make this book readable in English. We alone, however, bear the responsibility for its contents. We thank the Springer staff, especially Dr. Harry Blom, for his confidence in our work. We also acknowledge the excellent work done by Ms A. Sridevi and her team (SPi) during the production of this book.

Finally, our families showed great patience and gave us their full support during the lengthy process of writing this book, which we dedicate to them.

La Laguna M. Vázquez
July 2009 E. Pallé
P. Montañés Rodríguez

Acronyms

AIM	Aeronomy of ice in the mesosphere
AU	Astronomical unit
BCE	Before common era
COROT	Convection rotation and planetary transits
ELT	Extremely large telescope
E-ELT	European extremely large telescope
ESA	European Space Agency
ESO	European Southern Observatory
EUV	Extreme ultraviolet
GOES	Geostationary Operational Environmental Satellite
HST	Hubble Space Telescope
IAU	International Astronomical Union
IMF	Initial mass function
INGRID	Isaac Newton Group red imaging device
IPMO	Isolated planetary-mass objects
LHB	Late heavy bombardment
MAHRSI	Middle High Resolution Spectrograph Investigation
NAOS	Nasmyth adaptive optics system
NASA	National Aeronautics and Space Agency
NEAR	Near earth asteroid rendezvous
OWL	Overwhelmingly large telescope
PAL	Present atmospheric level
SeaWiFS	Sea-viewing wide field-of-view sensor
SOHO	Solar and Heliospheric Observatory
TIMED	Thermosphere ionosphere mesosphere energetics and dynamics
TIROS	Television infra-red observation Satellite
TP	Terrestrial planet
TPF	Terrestrial planet finder
UV	Ultraviolet
VIRTIS	Visible and InfraRed Thermal Imaging Spectrometer
VLT	Very large telescope, ESO
WHT	William Herschel telescope

Units

$1\,\mu m = 1{,}000\,nm = 10{,}000\,\text{Å}$

1 Astronomical unit (AU) = Mean Sun–Earth distance = 149.60×10^6 km
1 Light Year (ly) = $9{,}461 \times 10^{12}$ km = 63,241 km
1 Parsec (pc) = 3,2616 ly

Ga (Gigayears) = 10^9 years
Ma (Million of years) = 10^6 years
Earth Mass (M_E) = 5.9736×10^{24} kg
Jupiter Mass (M_J) = $1{,}8996 \times 10^{27}$ kg
Solar Mass (M_S) = $1{,}989 \times 10^{30}$ kg

1 Joule = 10^7 ergs

Contents

1 Observing the Earth 1
  1.1 The Exploration of Our Planet 1
  1.2 First Observations of Our Planet from the Air 7
      1.2.1 Early Balloon Pictures 7
      1.2.2 The Space Research 9
  1.3 The Earth–Moon System 20
  1.4 The Solar System 21
      1.4.1 General Characteristics 22
      1.4.2 A View from the Edge 23
      1.4.3 Our Environment 25
  References 31

2 The Earth in Time 35
  2.1 The Earth at the Present Time 38
      2.1.1 The Interior 39
      2.1.2 Plate Tectonics 44
      2.1.3 The Atmosphere 47
      2.1.4 Energy Balance of the Atmosphere 50
  2.2 The Precambrian Era (4,500–4,550 Ma BP) 54
      2.2.1 The Formation of the Earth: The Hadean Era 55
      2.2.2 The Archaean and Proterozoic Times 63
  2.3 The Phanerozoic Era 78
      2.3.1 The Drift, Breakup and Assembly of the Continents 80
      2.3.2 Supereruptions and Hot Spots 81
      2.3.3 The Connection Temperature-Greenhouse Gases 82
      2.3.4 Temporal Variations of the Magnetic Field 84
      2.3.5 Mass Extinctions in the Fossil Record 85
  2.4 The Quaternary 89
      2.4.1 The Ice Ages 90
      2.4.2 The Present Warming: The Anthropocene 91

	2.5	The Future of Earth	94
		2.5.1 The End of Life	94
		2.5.2 The End of the Earth	95
	References		96
3	**The Pale Blue Dot**		**107**
	3.1	Globally Integrated Observations of the Earth	107
		3.1.1 Earth Orbiting Satellites	108
		3.1.2 Observations from Long-range Spacecrafts	111
		3.1.3 An Indirect View of the Earth: Earthshine	112
	3.2	The Earth's Photometric Variability in Reflected Light	116
		3.2.1 Observational Data	117
		3.2.2 Reflectance Models	121
		3.2.3 The Earth's Light Curves	123
		3.2.4 The Rotational Period	124
		3.2.5 Cloudiness and Apparent Rotation	127
		3.2.6 Glint Scattering	128
	3.3	Earth's Infrared Photometry	131
	3.4	Spectroscopy of Planet Earth	134
		3.4.1 The Visible Spectrum	135
		3.4.2 The Infrared Spectrum	137
		3.4.3 The Earth's Transmission Spectrum	139
	3.5	Polarimetry of Planet Earth	143
		3.5.1 Linear Polarization	143
		3.5.2 Circular Polarization	145
	References		146
4	**The Outer Layers of the Earth**		**151**
	4.1	Temperature Profile and the Energy Balance	151
	4.2	Stratosphere: The Ozone Layer	156
		4.2.1 Natural Processes of Ozone Formation and Destruction	158
	4.3	Mesosphere	160
	4.4	The Thermosphere	161
	4.5	The Exosphere: Geocorona	162
	4.6	Airglow	164
		4.6.1 Nightglow	166
		4.6.2 Dayglow	169
		4.6.3 Twilight Airglow	170
	4.7	The Ionosphere	171
		4.7.1 General Structure	172
		4.7.2 Ionosphere Indicators	176
		4.7.3 Lightnings	178

4.8	The Magnetosphere	179
	4.8.1 Description	179
	4.8.2 Radiation Belts	180
	4.8.3 Aurorae	182
4.9	Radio Emission of the Earth and Other Planets	183
4.10	The Earth in X-Rays	186
4.11	The Earth's Gamma Ray Emission	187
4.12	The Outer Layers of the Early Earth	188
References		190

5 Biosignatures and the Search for Life on Earth … 197

5.1	The Physical Concept of Life	197
5.2	Astrobiology: New Perspectives for an Old Question	200
5.3	Requirements for Life	201
	5.3.1 Biogenic Elements	201
	5.3.2 A Solvent: Water	202
	5.3.3 Energy Source	205
5.4	Biosignatures on Present Earth	209
	5.4.1 Spectral Biosignatures in the Atmosphere	209
	5.4.2 Chlorophyll and Other Spectral Biosignatures of the Planetary Surface: The Red Edge	213
	5.4.3 Chirality and Polarization as Biosignatures	222
5.5	Biosignatures on Early-Earth	223
5.6	Life in the Universe	225
	5.6.1 Circumstellar Habitable Zone	225
	5.6.2 Additional Constraints for Habitability	232
	5.6.3 Galactic Habitable Zone	234
5.7	Signatures of Technological Civilizations	235
	5.7.1 Night Lights	236
	5.7.2 Spectral Features	238
	5.7.3 Artificial Radioemission	239
	5.7.4 Nuclear Explosions	241
	5.7.5 Extraterrestrial Pulses	243
References		243

6 Detecting Extrasolar Earth-like Planets … 251

6.1	First Attempts to Discover Exoplanets	251
6.2	The Mass Limit: From Brown Dwarfs to Giant Planets	253
	6.2.1 The Brown Dwarf Desert	258
6.3	The Detection of Earth-like Planets: A Complex Problem	258
	6.3.1 Brightness Ratio	258
	6.3.2 Angular Distance	260

	6.4	Methods for the Detection of Exoplanets	261
		6.4.1 Indirect Detection of Exoplanets	263
		6.4.2 Direct Observations of Exoplanets	275
	6.5	The Next 20 Years	280
	References		281

7 The Worlds Out There 289
- 7.1 Definition of a Planet 289
- 7.2 Our Solar System 291
  - 7.2.1 General Facts 291
  - 7.2.2 Chemical Abundances in the Solar System 292
  - 7.2.3 Giant Planets 293
  - 7.2.4 Terrestrial Planets 294
  - 7.2.5 Dwarf Planets and Other Minor Bodies 295
- 7.3 Planetary Atmospheres 299
- 7.4 Statistical Properties of the Extrasolar Giant Planets 302
  - 7.4.1 Mass Distribution 302
  - 7.4.2 Hot Jupiters 302
  - 7.4.3 Eccentric Planets 305
  - 7.4.4 Role of the Metallicity 305
  - 7.4.5 Stellar Masses 306
- 7.5 Types of Terrestrial Planets 307
  - 7.5.1 Rocky Planets 308
  - 7.5.2 Super-Earths 309
  - 7.5.3 Carbon–Oxygen Ratio: The Carbon Planets 314
  - 7.5.4 Super-Mercuries 315
  - 7.5.5 Planets Around Pulsars in Metal-Poor Environments 317
  - 7.5.6 Terrestrial Planets Around Giant Planets: The Rocky Moons 317
  - 7.5.7 Free-Floating Planets 318
- 7.6 Characterization of Exoplanets 319
  - 7.6.1 Mass–Radius Relationships 319
  - 7.6.2 Atmospheres of Exoplanets 322
  - 7.6.3 Radio Emission of Exoplanets 326
- 7.7 Terraformed Planets 326
- 7.8 Expect the Unexpected 327
- References 328

8 Extrasolar Planetary Systems 337
- 8.1 The Origin of the Solar System: Early Attempts 337
  - 8.1.1 Nebular Theory 337
  - 8.1.2 Catastrophic Theories 339

	8.2	Formation of Planetary Systems	340
		8.2.1 Stellar Formation	340
		8.2.2 The Early Accretion Phase	342
		8.2.3 The Protoplanetary and Debris Disks	343
		8.2.4 Formation of Giant Planets	346
		8.2.5 Formation of Terrestrial Planets	348
	8.3	Planetary Orbits	350
		8.3.1 Basic Orbital Elements	350
		8.3.2 Keplerian Orbits	352
		8.3.3 Harmony and Chaos	355
		8.3.4 Relevant Parameters of Dynamical Stability	358
		8.3.5 Resonances in Planetary Systems	360
		8.3.6 Lagrangian Points	363
	8.4	The Dynamically Habitable Zone	364
	8.5	Architecture of Planetary Systems	367
		8.5.1 Systems with Hot Jupiters: The Planetary Migration	369
		8.5.2 Binary Systems	376
		8.5.3 Multiple Planetary Systems	377
	8.6	Violence and Harmony	383
	References		383
9	**Is Our Environment Special?**		**391**
	9.1	Is the Sun Anomalous?	392
		9.1.1 Singularity	392
		9.1.2 Mass	393
		9.1.3 Location	394
		9.1.4 Age	395
		9.1.5 Chemical Composition: Metallicity	396
		9.1.6 Magnetic Activity	397
		9.1.7 Solar Analogs	399
	9.2	Is the Solar System Unique?	400
		9.2.1 Nature vs. Nurture	400
		9.2.2 Debris Disks	405
		9.2.3 The Energetic Environment	406
		9.2.4 Solar System Analogs	407
	9.3	Is the Earth Something Special?	408
		9.3.1 Habitability	408
		9.3.2 Variations of Orbital Parameters	409
		9.3.3 Presence of a Large Satellite	410
	9.4	The Ultimate Factor: Life	412
	References		413
Index			**419**

Chapter 1
Observing the Earth

The planet Earth plays the leading role in this book. The detailed knowledge we have of it at present has been accumulated mainly over the last decades. Over the centuries, the task of producing the first maps of our planet's surface has slowly and painstakingly been carried out. From our towns and villages, humanity has mapped the observable Earth, with the depths of the oceans remaining as the last obstacle. Simultaneously, human ingenuity has devised methods to measure its basic parameters.

At present, we have already detected around 300 exoplanets and we are progressing in the characterization of their physical properties, including global maps of their surfaces. We are proceeding inversely to the way in which we observed the Earth, making first global observations and, step by step, increasing the observed level of detail. In the past, the Sun was our guide for interpreting the observations of other stars; now our planet will be the Rosetta Stone used to decipher the data from other planetary worlds.

The proposed *Earth–Exoplanets connection* is based on the universality of processes leading to the formation of planets around stars. Thus, what we can learn by observing the Earth as a distant planet can be used to interpret the observations of future terrestrial planets. However, we must keep in mind that our reference is 4.6 Ga old, and has only one Earth mass and radius, while certainly other ranges of these parameters will be observed on exoplanets.

1.1 The Exploration of Our Planet

For many centuries, the Earth was considered as something different from the rest of the Universe. This was largely a final consequence of a debate in classical Greece between two different models.

Leucippus (first half of fifth century BCE) was the founder of atomism. His ideas are mainly known from Democritus (460–370 BCE),[1] who suggested that the

---

[1] No word written by Leucippus has survived. From Democritus we have only some fragments of his books, but a good description of his principles are included in the writings of Aristotle and Diogenes Laertius.

Universe is composed of 'atoms' and 'voids'. Both are infinite and constitute the primary elements of everything. Each atom is uniform, homogeneous and indivisible. To him is attributed the following sentence:

> There are innumerable worlds of different sizes. In some there is neither sun or moon, in others they are larger than in ours and others have more than one. These worlds are at irregular distances, more in one direction and less in another, and some are flourishing, others declining. Here they come into being, there they die, and they are destroyed by collision with one another. Some of the worlds have no animal or vegetable life nor any water. (Guthrie 1979, p. 405).

This theory was spread throughout Europe by Lucretius (99–55 BCE), who in his *De Rerum Natura* (On the Nature of the things) supported the existence of other worlds different from the Earth (cf. Dick 1982, 1996), presenting a primitive version of the plenitude principle: *when abundant matter is ready, when space is to hand, and no thing and no cause hinders, things must assuredly be done and completed.* However, this philosophy remained in darkness with respect to that presented by the Socratic philosophers.

Aristotle (384–322 BCE) was probably the first philosopher interested in knowing the nature of things through observations, but when he came to discuss the structure of the Universe, he was strongly influenced by religious prejudices and the ideas of Ionic Philosophy. According this school, our planet was composed only of four basic elements: earth, water, air and fire and had physical imperfections.[2] The heavens and objects in the heavens were composed of ether, the fifth element, the essence of the divine, perfect in their circular shape and movement (Fig. 1.1). In this view Earth and the rest of the Universe were completely different things.

Medieval scholastic philosophers granted ether changes in density. The bodies of the planets were considered to be denser than the medium that filled the rest of the universe. The five elements of nature were still commonly referenced in the sixteenth and seventeenth centuries; however, the diverse chemical properties of earth were already being systematically examined and analysed (and eventually reduced to the preparation of the Periodic Table of Atomic Elements), and the physical properties of air and water were systematically researched.

Giordano Bruno (1548–1600) emerges as an essential reference for this change of paradigm. He published three books in the form of dialogues. In the *De l'infinito Universo e Mondi* (On the infinite Universe and Worlds), he assumed that the Universe was infinite and that the Sun was a normal star. He said, 'Innumerable suns exist; innumerable earths revolve around these suns in a manner similar to the way the seven planets revolve around our Sun'. Convinced of the universality of the laws ruling the Universe, he used the Earth as a model, declaring that there is life everywhere. See Gatti (2002) for a monograph on the life of Bruno.

The determination of Earth's size and distance to the Sun were critical to place our planet in an universal context. However, its central position remained until the break of the paradigm with the publication of the N. Copernicus (1473–1543) *De*

---

[2] For the Chinese, the basic elements were five: earth, wood, metal, water and fire.

1.1 The Exploration of Our Planet

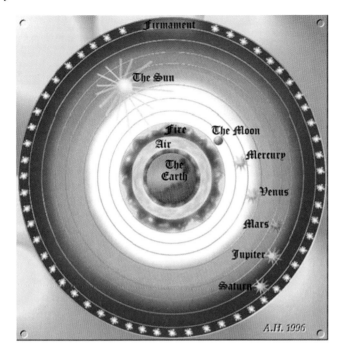

Fig. 1.1 The Greek Cosmos following the model of Aristotle

revolutionibus orbium coelestium (On the Revolutions of the Celestial Spheres).[3] The atomists speculated that laws of nature are observed to operate universally. The verification of such assumption became possible in the Renaissance through the development of instruments (telescope and microscope) to observe the nature in an objective way. The humans entered for the first time in the sphere of the unknown, the fixed stars, (Fig. 1.2) and step by step it was confirmed that the outside world obeys the same rules and physical laws as those that rule our planet.

These discoveries were complemented with the improved knowledge of our own planet. Anaximander (610–546 BCE) is credited with having created the first map of the world, circular in form and showing the known parts of the world around the Aegean Sea at the centre, all of this surrounded by a global ocean (Fig. 1.3). Eratosthenes (276–194 BC) drew an improved map (Fig. 1.4), incorporating parallels and meridians, and was the first to estimate the size of our planet.[4]

The period from the early fifteenth century until the seventeenth century is known as the Age of Discovery. At that time European ships travelled around the world mainly to search for new trading routes, but also driven by curiosity. The main

---

[3] It was published in 1543, though he had arrived at his theory some time earlier.

[4] Unfortunately, his work *On the measurement of the Earth* was lost. We know about his method from indirect sources and the value obtained for the Earth diameter is estimated between 39,690 and 46,620 km.

Fig. 1.2 The Renaissance. Figure adapted from C. Flammarion (1888) *L'Atmosphère: Meteorologie Populaire*, Librairie Hachette, Paris

Fig. 1.3 World Map by Anaximander. Courtesy: Wikipedia

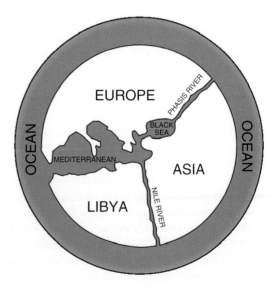

events were the discovery of America by Christopher Columbus (1451–1506) on 12 October 1492 and the circumnavigation of the globe by Ferdinand Magellan (1480–1521) and Juan Sebastián Elcano (1476–1526) from 10 August 1519 to 6 September 1522, events that forever settled the debate about the roundness of the

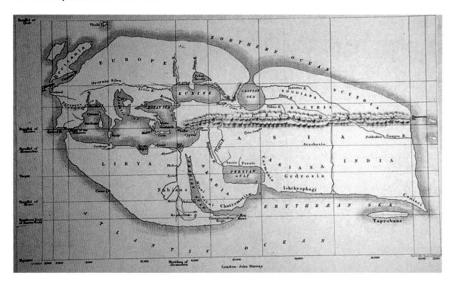

Fig. 1.4 World map by Eratosthenes, around 194 B.C.E. Credit: Heritage History

Earth. Juan de la Cosa (1460–1510) made maps of the new American continent, of which the only survivor is the Mappa Mundi of 1500, preserved at the Naval Museum of Madrid.

Some of the new information was included in the map of Martin Waldseemüller (1470–1521), published in 1507 accompanying his *Universalis Cosmographia* (Fig. 1.5).[5] He was the first to use the name America and to show the Pacific Sea.[6]

G. Mercator (1512–1594) developed a cylindrical projection that allowed planispheres to be produced (Fig. 1.6), where parallels and meridians were straight and perpendicular to each other (see Crane 2003). The projections trying to fit a curved space onto a flat sheet distort the true layout of the Earth's surface, and exaggerate the sizes of the areas near the equator.

The determination of the first stellar parallax by F. Bessel (1784–1846) in 1839 allowed him to measure the distance to a fixed star (61 Cygni, 10.3 ly). A few years later, the development of spectroscopic techniques made it possible to characterize the chemical composition of the stars and to verify that they were similar to our Sun.

In the near future, the application of sophisticated methods will give us the opportunity to detect and study planetary bodies similar to our Earth. The day we arrive at another Earth-like planet, we will already have in our possession detailed cartographic maps of it, taking advantage of the knowledge we have acquired in finding out about our own planet.

---

[5] The map was kept in the castle of Prince Johannes Waldburg in Germany until it was acquired, in 2001, by the Library of the US Congress. It is now on display in its Thomas Jefferson Building.

[6] Curiously, the Pacific Sea was discovered only in 1513 by Vasco Nuñez de Balboa.

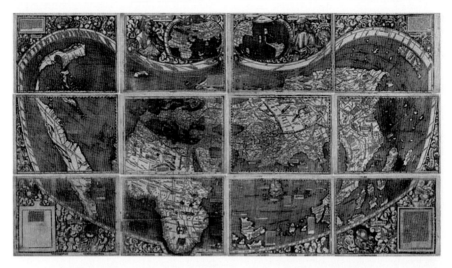

Fig. 1.5 World map by Martin Waldseemüller, produced in 1507. Courtesy: Geography and Map Division, Library of US Congress. Available as a Wikimedia Commons file

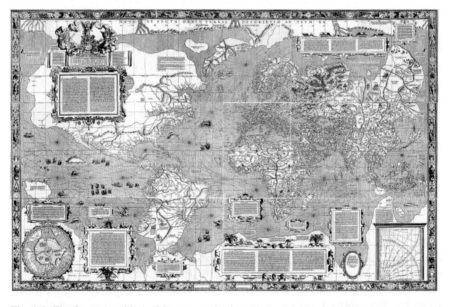

Fig. 1.6 The first map with the Mercator projection: *Nova et Aucta Orbis Terrae Descriptio ad Usum Navigatium Emendate*

Our planet is currently being studied in detail using different techniques. Space research has clearly shown what was already expected. The best procedure to see something as a whole is to go some distance away and contemplate it from afar. This was a consequence of human curiosity to reach new frontiers. Schellnhuber (1999)

1.2 First Observations of Our Planet from the Air

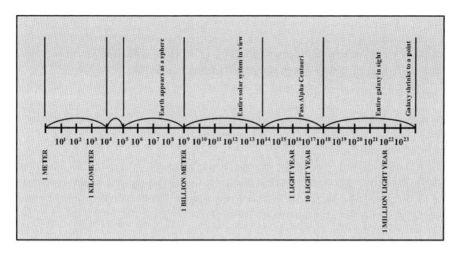

Fig. 1.7 The distance scales to observe our planet. Adapted from Morrison (1985)

suggests the bird's eye principle to get a panoramic view of our planet, observing it from the distance. A less expansive method is simulation modelling, where components and processes of the Earth are replaced by mathematical proxies. In this book we follow both approaches.

Let us imagine starting a journey from the Earth's surface to outer space progressively increasing our distance (Fig. 1.7). For this purpose we use different spatial scales, namely, metres, kilometres (10^3 m), light-years (9.461×10^{15} m) and parsecs (3.262 ly, 3.086×10^{16} m).

By increasing the distance d to our target of dimension D, we observe smaller angles of sight θ. At a certain threshold, the eye, or our observing instrument, can no longer observe such small angles and the target will disappear from our view.

1.2 First Observations of Our Planet from the Air

The history of mankind was clearly marked by the importance of a better knowledge of our planet. However, observations from the ground have a very limited scope. It was necessary to ascend in the air to have an adequate perspective (Hildebrandt 1908; Hannavy 2007) .

1.2.1 Early Balloon Pictures

In October 1858, G.F. Tournachon (1820–1910), known as Nadar, took the first picture from a captive balloon, 'The Giant', above the Earth's surface at an altitude of

100 m.[7] It was carried out in a French valley (Bievre, near Paris) and unfortunately the photograph has been lost.[8] On 13 October 1860, J. Wallace Black (1825–1896) ascended to an altitude of 400 m in the 'Queen of the Air' balloon to photograph parts of Boston.[9] A cable was used to hold the balloon in place.[10]

The first free flight photo mission was carried out by J.N. Truchelut, sometimes called Triboulet, in 1879. Meanwhile, an alternate approach, consisting of mounting cameras on kites, became popular in the last two decades of the nineteenth century. The British meteorologist E.D. Archibald began this method in 1882, with the kites carrying scientific instruments.

In 1888, the first well-preserved aerial photographs were taken by A. Batut (1846–1919) over Labruguiere (France) using a kite. The camera, attached directly to the kite, had an altimeter that encoded the exposure altitude on the film, allowing scaling of the image. A slow burning fuse, responding to a rubber band-driven device, activated the shutter within a few minutes after the kite was launched. A small flag dropped once the shutter was released to indicate that it was time to bring down the kite (Batut 1890; Gernsheim and Gernsheim 1969).

Alfred Maul (1864–1941) succeeded in photographing the landscape at a height of 600 m during the launch of a rocket in 1904 (Fig. 1.8). At this height the nose of the rocket separated, the camera was released into the air with a parachute and finally the image was taken.

In 1903, J. Neubronner (1852–1932) designed and patented a breast-mounted aerial camera for carrier pigeons. Mostly used for military purposes, the birds were introduced at the 1909 Dresden International Photographic Exhibition, where postcards of aerial photographs taken above the exhibition were very popular with the public. In 1906 George R. Lawrence (1869–1938) photographed the aftermath of the San Francisco earthquake using a string of 17 kites to lift a handmade panoramic camera aloft (Fig. 1.9).

It appears that Wilbur Wright (1867–1912) – the co-developer of the first aeroplane to leave the ground in free flight – was also the first to take pictures from an airplane, in Centocelli (Italy) in 1909. On 10 November 1935, A.W. Stevens (1886–1949) and O. Anderson (1895–1961) took the first photograph[11] of the curvature of the Earth from a free helium balloon, the Explorer II, at an altitude of 22 km. They reported that at this altitude *the sky appears very dark indeed, but it can still be called blue.*

---

[7] For this photograph to be made, the camera, light-sensitive material and a development system had to be taken on the balloon in order to develop the picture instantly after it was exposed.

[8] A caricature of the event remains, prepared by Honorè Daumier (1808–1879) for the 25 May 1862 issue of *Le Boulevard*.

[9] The photograph is preserved at the Boston Public Library.

[10] Balloons were explored as observation platforms during the American Civil War, with Wallace urging aerial photography as a technique for reconnaissance.

[11] Published in National Geographic, May 1936.

1.2 First Observations of Our Planet from the Air

Fig. 1.8 Photograph of the landscape from a small rocket by A. Maul in 1904

Fig. 1.9 Photograph of San Francisco from an altitude of 600 m by G. Lawrence on 28 May, 1906, 6 weeks after the famous earthquake

1.2.2 The Space Research

The advances in space research have made possible the observation of our planet as a distant cosmic body (see Poole 2008), enabling the application of techniques of remote sensing used in Astrophysics.

1.2.2.1 The First Attempts

After the World War II, the US Naval Research Laboratory began experimenting with German-designed V-2 rockets. On 24 October 1946, a grainy black-and-white picture was taken from a V-2, launched from White Sands (New Mexico), at an altitude of 90 km (Fig. 1.10). Clyde Holliday, an engineer of the project, wrote in National Geographic in 1950: *the V2 photos showed for the first time, how our Earth would look to visitors from another planet coming in a spaceship.* Smaller sounding rockets, such as the Wac Corporal, and the Viking and Aerobee series, were developed and launched by the military in the late 1940s and 1950s. These rockets, although not attaining orbit, contained automated still or movie cameras that took pictures as the vehicle ascended.

In April 1960, NASA launched the first Television and Infrared Observation Satellite (TIROS), followed by Nimbus. Together, they collected and beamed back thousands of images of cloud cover as well as images of different weather patterns (see Fig. 1.11).

Kilston et al. (1966) tried to detect signs of intelligent life by analyzing thousands of pictures obtained by the Tiros and Nimbus meteorological satellites, with a spatial resolution of 1 km. Only 0.1% of the photographs were indicative of some type of technological civilization: an interstate highway in the US and an orthogonal grid produced by Canadian loggers. Both shared the property of being linear structures. However, we must keep in mind that the human eye tends to connect disconnected features into rectilinear ones, as was the case with the Martian canals after the announcement by G. Schiapparelli in 1877 (see Evans and Maunder 1903; Antoniadi 1908).

Sagan and Wallace (1971) used for the same purpose images with a better resolution (100 m) obtained by the manned spacecrafts Gemini and Apollo. The fraction of images showing signs of intelligent life was ∼1.5%. Allowing for astronaut selection effects, the fraction was reduced to 1%.

Fig. 1.10 The curvature of the Earth visible from a V2 rocket at 90 km altitude on 24 October 1946. Credit: White Sands Missile Range/Applied Physics laboratory

Fig. 1.11 The Television Infrared Observational Satellite (TIROS) was the first weather satellite. Above is the first television image taken from an altitude of about 700 km on 1 April 1960. Credit: TIROS program and NASA

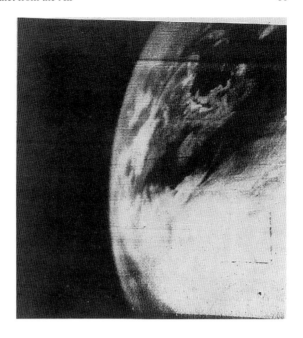

In 1950, the British astronomer Fred Hoyle (1915–2001) had looked ahead saying: *Once a photograph of the Earth, taken from outside, is available [...] once the sheer isolation of the Earth becomes plain to every man whatever his nationality or creed, and a new idea as powerful as any in history will be let loose* (Hoyle 1950). It took some time to get these global images.

The ATS-3 spacecraft was launched into a synchronous orbit on 5 November 1967. It was equipped with an improved spin scan camera that could take pictures in colour. On 10 November, the satellite transmitted the first colour pictures of Earth (Fig. 1.12) taken from synchronous altitude (36,000 km). However, this image failed to have a major impact in the media. It was necessary to go farther from Earth for Hoyle's idea to become a fact.

Automatic spacecrafts were launched to the lunar orbit to study our satellite in detail and prepare manned landings. On 23 August 1966 the Lunar Orbiter 1 took the first image of the Earth floating in the space (Fig. 1.13). Although heralded by some journalists as the Image of the Century, it remained practically unknown until recently when the original photograph was restored using modern technology in the framework of the Lunar Orbiter Image Recovery Project (LORP).[12]

Probably less known is that the Soviet spacecraft Zond 6 obtained a similar image on 14 November 1968 but in black and white and with much less spatial resolution (Fig. 1.14 and Stooke 2008).[13]

---

[12] This project was started 20 years ago by Nancy Evans and completed recently by Dennis Wingo and Keith Cowing.

[13] A crash landing on Earth flattened and broke open the film canister, but 52 photographs were recovered with some degree of laceration and fogging.

Fig. 1.12 The whole Earth from 36,000 km. South America and West Africa are clearly visible. Courtesy: NASA

Fig. 1.13 Image of the Earth obtained by the Lunar Orbit 1 spacecraft in August 1966. Courtesy: NASA/LORP

1.2.2.2 The Manned Flights

The purpose of the Apollo program was to land men on the lunar surface and to return them safely to the Earth. Moreover, it constituted an excellent observatory to see our planet as a whole. Images obtained from the Apollo VIII over Christmas 1968 allowed us to see, for the first time, the Earth as an isolated body in space

1.2 First Observations of Our Planet from the Air

Fig. 1.14 Image of the Earth obtained by the Soviet Zond 6 spacecraft in November 1968 from the lunar orbit with a 400 mm camera shooting 13 × 18 cm photographs. Credit: Kira Shingareva and Zond-6 Russian Mission. A catalogue of Earth images obtained by Soviet lunar missions is available at http://www.mentallandscape.com/C_CatalogMoon.htm

(Fig. 1.15). Three photographs were taken, one in black and white and two in colour. The black and white shot was taken first by F. Borman, and the two colour shots were taken moments later by William Anders (Zimmerman 1999; Poole 2008).

We are now at the scale of 10^8 m from our home planet.

The astronauts of the Apollo VII had different feelings about the spectacle they were contemplating from the lunar orbit. For James Lovell, seeing the distant Earth strengthened his conviction that we existed for a purpose.[14] On the other hand, for his crewmate William Anders the view suggested a lonely purposelessness. *We are like ants on a log* he commented later (Zimmerman 1999). For F. Borman, the Earth *was the only thing in space that had any colour to it. Everything else was either black or white, but not the Earth* (Borman 1988).

The day after the image was taken, the writer Archibald MacLeish (1892–1982) wrote[15]: *To see the Earth as it truly is, small and blue and beautiful in that eternal silence where it floats, is to see ourselves as riders on the Earth together, brothers on that bright loveliness in the eternal cold-brothers who know now that they are truly brothers.*

---

[14] He also commented: *The Earth from here is a grand ovation to the big vastness of space.*
[15] Article on The New York Times, 25 December 1968.

Fig. 1.15 The Earth seen as a planet: Image obtained from the Moon taken by the Apollo VIII on 24 December 1968, known as the 'Earth Rise'. Credit: NASA and Apollo 8 crew (Frank Borman, James Lowell and William Anders)

Remembering his space travels, Edwin Aldrin[16] describes the colour of the oceans as composed of different tones, from greenish to a deep blue. The continents were bronze-coloured, as an olive, and it was difficult to distinguish any green masses. The whiteness of the polar caps was one of the prominent features of our planet. From the Moon the Earth seemed a bright jewel in a dark and velvety sky.

The Apollo 14 astronaut Edgar Mitchell also transmitted his feeling after contemplating our planet from the Moon's orbit: *Suddenly, from behind the rim of the moon, in long, slow-motion moments of immense majesty, there emerges a sparkling blue and white jewel, a light, delicate sky-blue sphere laced with slowly swirling veils of white, rising gradually like a small pearl in a thick sea of black mystery. It takes more than a moment to fully realize this is Earth . . . home.*

The Apollo 17 crew took in December 1972 the only photograph of the Earth where the Sun was directly behind the spacecraft. Therefore, instead of being partly shrouded in darkness, the planet appears fully illuminated (Fig. 1.16).

Very recently, in September 2007, the Japanese spacecraft Kaguya (Selene) was launched to the Moon, where it was placed in an orbit at 100 km altitude above the lunar surface. Transmitted images allowed a new perspective of Earth-rise and Earth setting (Fig. 1.17).

---

[16] He was a member of the Apollo 11 crew and the second man on the Moon. His feelings while observing the Earth from the outer space have been taken from a interview appearing recently (23 April 2008) in the Spanish newspaper *El País*.

1.2 First Observations of Our Planet from the Air

Fig. 1.16 The Earth seen as a planet: Image taken by the Apollo 17 astronaut Harrison Schmidt from a point halfway between the Earth and the Moon (December 7, 1972). Credit: NASA

Fig. 1.17 The composite shows the Earth setting on the horizon near the Moon's South Pole. It took about 70 s from the left image to the right image (complete setting). Credit: Japan Aerospace Exploration Agency (JAXA) and NHK (Japan Broadcasting Corporation)

Since 1982, many astronauts have observed the Earth from space, most of them using the International Space Station, the Mir Station and the Space Shuttle.[17] As a

---

[17] Thousands of images of our planet, taken by astronauts, are archived at the NASA-Johnson Space Center (http://eol.jsc.nasa.gov), including around 800 images of the Whole Earth.

final reflection, we can mention a comment of the Saudi Arabia astronaut Salman Abdaluziz al-Saud[18] that reflects a promising perspective, probably a little utopic, for human progress: *The first day or so we all pointed to our countries. The third or fourth day we were pointing to our continents. By the fifth day, we were aware of only one Earth.* Soviet astronauts had similar feelings. Vladimir Kovalyonok[19] commented: *After an orange cloud – formed as a result of a dust storm over the Sahara and caught up by air currents – reached the Philippines and settled there with rain, I understood that we are all sailing in the same boat.* The book of White (1998) summarizes the comments from astronauts about how viewing Earth from space affected perceptions of themselves, their planet of origin, and their own place in space and time.

1.2.2.3 The Earth Observatory

Space is a privileged site not only for studying the Universe but also for observing our planet. Many satellites control different aspects of it, such as the ozone layer, the weather, etc. In particular, the NASA Earth Observatory provides an excellent platform to free access Earth images.

In 2002, NASA produced the first *Blue Marble* picture, the most detailed true-colour image ever produced. The maximum resolution was of 1 km per pixel. It was composed from data obtained with the Terra satellite. It was followed by the *Blue Marble: Next Generation*, a mosaic of satellite data taken mostly from the MODIS sensor on board the Terra and Aqua satellites. They consist of monthly composites at a spatial resolution of 500 m, revealing seasonal changes to the land surface (see Fig. 1.18). The Earth viewed from space is a nearly perfect sphere with the equatorial diameter being a little larger. Seventy percent of its surface is covered by oceans and this is why the planet appears blue.

1.2.2.4 Infrared Images

While it is relatively easy to distinguish clouds from land areas in the visible range, there is more detailed information about the clouds themselves when observing in the infrared. Darker clouds are warmer, while lighter clouds are cooler. A full image of the Earth at the wavelength of $11.2\,\mu$m obtained by the GOES 6 satellite on 21 September 1986 is shown in Fig. 1.19. A temperature threshold was used to isolate the clouds. The land and sea were separated and then the clouds, land and sea separately coloured and combined back together.

Water vapour absorbs and re-radiates electromagnetic radiation, especially in the infrared 6–7 μm band. Such infrared radiation, emitted by the Earth's surface/

---

[18] He flew on the mission STS-51-G in the shuttle Discovery from 17–24 June 1985, together with a French astronaut and five US astronauts.

[19] He was commander of three missions: Soyuz 25, Soyuz 29 and Soyuz T-4.

1.2 First Observations of Our Planet from the Air

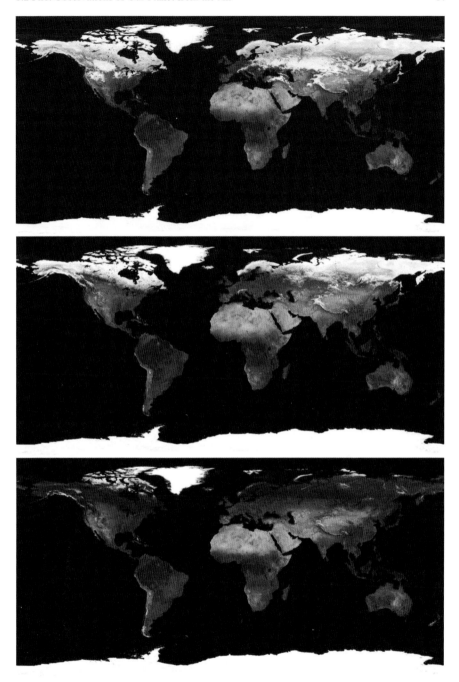

Fig. 1.18 The Blue Marble Next Generation seasonal composites. From *top* to *bottom*: January, April and August compositions are shown. The original resolution has been degraded to 8 km/pixel. Credit: NASA

Fig. 1.19 Earth observed in the infrared. Credit: Richard Kohrs, University of Wisconsin-Madison

atmosphere and intercepted by satellites, is the basis for remote sensing of tropospheric water vapour. On a water vapour image, each pixel is assigned a gray shade according to its measured brightness temperature. Typically, white indicates a very cold brightness temperature (radiation from a moist layer or cloud in the upper troposphere), and black indicates a warm brightness temperature (radiation from the ground or a dry layer in the middle troposphere). The first TIROS and Nimbus satellites obtained the pioneering images of this kind. Figure 1.20 shows one of these images, obtained with hourly resolution, by the GOES 8 satellite.

On its way to the comet 67P Churyumov-Gerasimenko, the spacecraft Rosetta made an Earth fly-by[20] on 4th and 5th March 2005. The on-board spectrometer VIRTIS obtained more than 850 monochromatic images of our planet, ranging from the UV to the thermal infrared (5,000 nm). After the closest approach to Earth and at a distance of 250,000 km from our planet, VIRTIS took two high spatial (62 km per pixel) and spectral resolution images of the Earth (Coradini et al. 2005). On 20 November 2007, the spacecraft performed a second approach to our planet, obtaining also different images in the visible and infrared ranges.

The MESSENGER spacecraft is now in orbit around Mercury, but during its 2 August 2005 Earth fly-by it took images in different spectral bands. Figure 1.21 illustrates the differences between visible and near-infrared images (see caption for details). Blue light is easily scattered in the Earth atmosphere, producing the blue

---

[20] Fly-bys make use of the gravitational attraction of planets to modify a spacecraft's trajectory and to gain the orbital energy needed to reach the final target.

1.2 First Observations of Our Planet from the Air

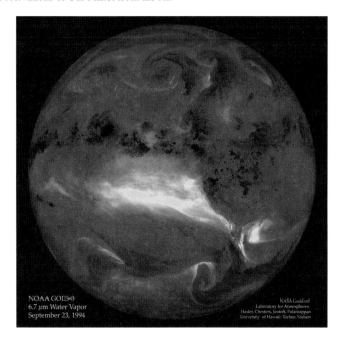

Fig. 1.20 This image, taken by GOES-8 (Geostationary Operational Environmental Satellite) on 23 September 1994, shows atmospheric water vapour observed at a wavelength of 6.7 μm. Image produced by F. Hasler, D. Chesters, M. Jentoft-Nilsen, and K. Palaniappan (NASA/Goddard) and T. Nielsen (Univ. of Hawaii)

Fig. 1.21 Earth images taken by the wide-angle camera of MESSENGER at 102,000 km above the Earth. (*Left*) Composite made from combining filters with peak sensitivities at 480, 560 and 630 nm (B, G and R filters, respectively). (*Right*) Composite made from combining the filters at 560, 630 and 750 nm. Credit: NASA/Johns Hopkins University Applied Physics Laboratory/Carnegie Institution of Washington. Image PIA10122 of the Planetary Photojournal

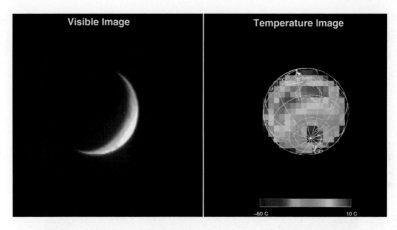

Fig. 1.22 Earth images taken by Thermal Emission Imaging System (THEMIS) onboard the 2001 Mars Odyssey. (*Left*) Image in the visible range showing the planet as a thin crescent. (*Right*) Infrared image showing the light emitted by all regions of the Earth, according to their temperature. Credit: NASA/JPL/Arizona State University. Image PIA00558 of the Planetary Photojournal

skies. Infrared light is not easily scattered, and so images of Earth remain sharp. Continental areas are mostly red due to the high reflectance of vegetation in the near-infrared (see Chap. 5 for more details). The red colouring of the image is a reflection of the Brazilian rain forests and other vegetation in South America.

Similar images were also acquired on 19 April 2001 at a distance of 3,563,735 km when the Mars Odyssey spacecraft left the Earth for Mars. Here, what stands out is the capability of the infrared for 'night-vision' and the possibility to produce temperature maps (Fig. 1.22).

1.3 The Earth–Moon System

In its journey around the Sun, the Earth is accompanied by a relatively large satellite, the Moon. The history of our planet cannot be dissociated from the evolution of our satellite. Gravitational tides are the most remarkable interactions between them. Figure 1.23 shows a mosaic of some of the images of the Earth–Moon system from space,[21] where the reflective clouds and atmosphere of Earth contrast strongly with the dark tones of the lunar surface. NEAR images view both bodies from above their south poles. The Mars Global Surveyor obtained the first image ever taken from another planet showing our Earth as a whole.

---

[21] The Galileo (1992), Rosetta (2005) and Venus Express (2005) spacecrafts also took images of the Earth–Moon system from the vicinity of the Earth.

1.4 The Solar System

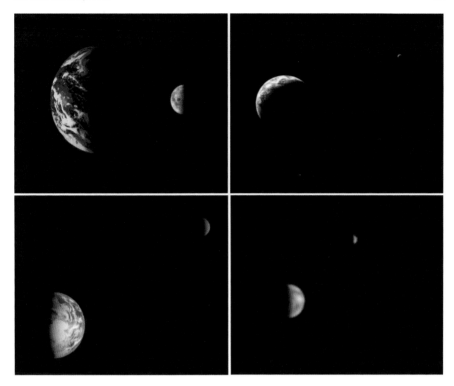

Fig. 1.23 The Earth–Moon system viewed from different satellites. From *left* to *right* and from *top* to *bottom*: Mariner X (November 1973), Mars Global Surveyor (2003), NEAR (23 January 1998) and Nozomi (1998). Credit: NASA and Japanese Space Agency

The relatively large size of the Moon with respect to the Earth is the most remarkable parameter to be pointed out. This value together with its present distance to our planet give to the Moon an angular size equivalent to that of the Sun. This coincidence is unique in the Solar System and allows us to contemplate the impressive total solar eclipses.

1.4 The Solar System

In nature, isolated systems do not exist. This principle can also be applied in our context. The Earth cannot be considered, in the study of exoplanets, as an isolated object. It needs to be studied together with its surroundings.

Fig. 1.24 The Solar System containing the Sun, eight planets and three dwarf planets. Courtesy: Wikipedia

We can classify the components of the Solar System in the following groups (Fig. 1.24)[22]:

- One star: The Sun
- Eight Planets with their 162 known satellites
  - Four rocky or terrestrial planets
  - Two gas giants
  - Two ice giants
- Three dwarf planets with their four known satellites
- Thousands of small bodies mainly located at
  - Asteroid Belt
  - Kuiper Belt
  - Oort Cloud

1.4.1 General Characteristics

As the Sun orbits around an axis perpendicular to the galactic disk, the planets, asteroids and comets of the Solar System turn around the centre of masses of the system, located inside the solar's volume near its geometrical centre. Similar

---

[22] After a decision made by the International Astronomical Union on 24 August 2006. See Chap. 7 for more details.

Table 1.1 Orbital parameters of the planets

Planet	Average distance to the Sun (AU)	Eccentricity	Inclination angle with respect to the ecliptic
Mercury	0.39	0.21	7.00
Venus	0.72	<0.01	3.40
Earth	1.00	0.02	
Mars	1.52	0.09	1.85
Jupiter	5.20	0.05	1.30
Saturn	9.55	0.06	2.48
Uranus	19.22	0.05	0.77
Neptune	30.11	<0.01	1.77

1 AU = 1.5×10^{11} m (average Sun–Earth distance)

orbital dynamics are exhibited by satellites orbiting around the planets. The features of planetary orbits arise from the well known laws of orbital motion deduced by Johannes Kepler (1571–1630), mostly based on detailed observational data from Tycho Brahe (1546–1601). These are the following:

- Planets orbit following elliptical trajectories with the Sun located at one focus.
- The radio-vector of a planet sweeps out equal areas at equal intervals.
- The square of the orbital period is proportional to the cube of the average distance from the Sun, the semi-major axis of the ellipse. Each orbital system has a characteristic and unique proportionality constant that differs from others of different orbital systems with different centre of masses.

These laws based on observational data (Table 1.1) did not take into account the dynamical aspects of planetary motion until Isaac Newton (1642–1727) solved this problem by analyzing the motion of two bodies moving together under an inverse square law of attraction.

All planets move around the Sun along direct orbits (counter clockwise from a north perspective). However, some moons travel around their planets following retrograde orbits.

The last two chapters will be dedicated to the peculiarities of the planets, in our Solar System and beyond.

At 10^{10} km of distance from the Earth, we have the entire Solar System in sight.

1.4.2 A View from the Edge

On 14 February 1990, NASA commanded the Voyager 1 spacecraft to turn the narrow-angle camera (1,500 focal length) around to photograph the planets it had visited. NASA compiled 60 images from this unique event into a mosaic of the Solar

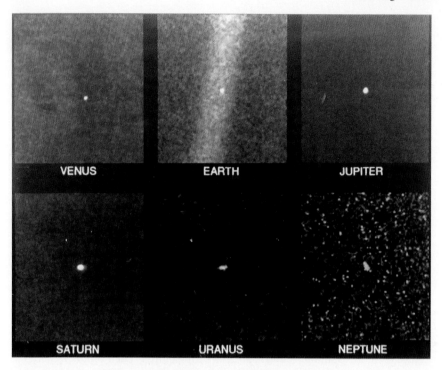

Fig. 1.25 Images of six planets of the Solar System from the edge of the Solar System taken by the Voyager I spacecraft. They were taken through three colour filters (V, R, G) and recombined to produce the colour images. The size of the Earth was 0.12 pixels. Credit: NASA

System (Fig. 1.25). Earth appears to be in a band of light because it coincidentally lies right in the centre of a beam of scattered rays resulting from taking the image so close to the Sun.[23]

Sagan said the following in an address delivered on 11 May 1996 about what he felt the photo demonstrated:

> We succeeded in taking that picture from deep space, and if you look at it, you see a dot. That's here. That's home. That's us. On it everyone you love, everyone you know, everyone you ever heard of, every human being who ever was, lived out their lives. The aggregate of our joy and suffering, thousands of confident religions, ideologies, and economic doctrines, every hunter and forager, every hero and coward, every creator and destroyer of civilization, every king and peasant, every young couple in love, every mother and father, hopeful child, inventor and explorer, every teacher of morals, every corrupt politician, every 'superstar', every 'supreme leader', every saint and sinner in the history of our species lived there – on a mote of dust suspended in a sunbeam

---

[23] Mercury was too close to the Sun to be seen. Mars was not detectable by the cameras due to scattered sunlight in the optics.

1.4 The Solar System

Fig. 1.26 The Earth seen as a planet from the orbit of Saturn on 15 September 2006 by the Cassini spacecraft. The bump at the *left* is the Moon. Credit: NASA

This image inspired his book *Pale Blue Dot*, where the philosophy about humankind's place in the Universe and the description of what was known about the Solar System at that time are mixed.

Recently (15 September 2006), the Cassini spacecraft, in orbit around Saturn, took with its wide-field camera an image of the Earth. At that time, Cassini was looking down on the Atlantic Ocean and the western coast of north Africa. The phase angle of Earth seen from Cassini is about 30°. Figure 1.26 shows an isolated view of our planet, where the Moon is seen as a dim protrusion at the upper left.

1.4.3 Our Environment

As was the case for Earth, the Solar system is not isolated from the rest of the Universe and, therefore, we must consider its position with respect to the other structures where it is embedded. For this purpose, we place ourselves at 20 light years distance from the Sun.

1.4.3.1 Nearby Stars

Around 131 stellar and sub-stellar objects are located within 20 light-years from the Sun, most of them M red stars and brown dwarfs (see Fig. 1.27). The closest ones to the Sun form the α Centauri system. Proxima Century is only 4.22 light years away, whereas the other two (α Cen A and α Cen B) form a bounded binary system, are brighter and are a little farther away (4.36 light years). Thebault et al. (2008) have studied the probability of habitable Earth-like planets in α Centauri B.

Fig. 1.27 Stars in the vicinity of the Sun (20 light years)

Two solar-like stars are of special interest in the solar neighbourhood.

Epsilon Eridani is located about 10.5 light-years away in the northeastern part of constellation Eridanus and has an age in the range of 0.85–1 Ga and a spectral type K2V. Somewhat smaller and cooler than our own Sun, Epsilon Eridani is also less luminous. It shows an infrared excess indicative of cool dusty material rotating around the star (Gillett 1986). This debris disk was first imaged at 850 μm by Greaves et al. (1998). Radial velocity measurements have discovered a Jovian class planet (M sin i = $0.86M_J$) with an estimated semi-major axis of 3.4 AU and an eccentricity of 0.6 (Hatzes et al. 2000) orbiting the star with a period of 7 years. No further giant companions seem to exist (Janson et al. 2008). Recently, Backman et al. (2009) have discovered two rocky asteroid belts and an outer icy ring, making it a triple-ring system. The inner asteroid belt is a virtual twin of the belt in our solar system, while the outer asteroid belt holds 20 times more material. Moreover, the presence of these three rings of material implies that unseen planets confine and shape them.

The G8V star Tau Ceti is the nearest sun-like star. It is 'metal-deficient' (see Chap. 7 for details) and therefore is thought to be less likely to have rocky planets around it. However, observations have detected over 10 times as much dust surrounding Tau Ceti as is present in the Solar System (Greaves et al. 2004). As was expected, no companions have yet been detected. Along with Epsilon Eridani, it was searched (unsuccessfully) for any sign of intelligent life in 1960 (Project Ozma).

Going farther away by an order of magnitude, we see that the Sun is located within three superclusters of stars. The youngest (70 Ma old) and closest (410 ly) is centred on the Pleiades star cluster. In the middle, we have the Sirius supercluster (Ursa Major Group, 300 Ma old)) containing stars such as Sirius and the central stars of the Big Dipper. Finally, the oldest (formed 630 Ma ago) was born along the Hyades, with all their stars heading away from us, in the direction marked by Orion. For data about these stars see Hoffleit and Jascheck (1982).

1.4.3.2 The Gaseous and Dusty Neighbourhood

The space between the stars is not empty. The 'interstellar medium' is the name the astronomers give to the gas and dust that pervade the interstellar space. The densest of these clouds, the molecular clouds, are the cradle for the formation of new stars. To see how they are distributed in our vicinity, we need to move a little farther away and place our observing viewpoint at 1,500 light-years from the Earth.

The morphology of this region, the *Local Interstellar Medium*, was configured 4×10^7 years ago by the passage of an expanding shell (see Ferlet 1999). As star formation was triggered in the high-density gas ahead of the shell, several OB stellar associations were formed, which now delineate the structure of what is known as Gould's Belt. Most stars in the solar neighbourhood younger than about 60 Ma are located in this flattened structure a few hundred parsecs in size, with the Sun inside it. It contains many young low-mass stars and interstellar gas (Stothers and Frogel 1974; Grenier 2000) and hosts 432 ± 15 Supernova progenitors with masses $>8\,M_S$ and its estimated lifetime implies a minimum of 40 supernova events $Ma^{-1}\,kpc^{-2}$ in a few tens of Ma (Comeron et al. 1994; Grenier 2000). The Belt is evident in the night sky as a band of very bright stars inclined about $20°$ relative to the Galactic plane. The map in Fig. 1.28 shows the surrounding 1,500 light-years and the position of nearby high-density molecular clouds (Frisch 2000), where the process of star formation takes place.

The Local Bubble (Breitschwerdt et al. 1998; Maíz-Apellániz 2001; Frisch 2006, 2007) is about 300 light years long. It is almost completely empty, being 1,000 times less dense (0.001 atoms per cm^3) and 100–100,000 times hotter (T $\sim 10^6$ K) than ordinary interstellar material and flows through the Solar System with a relative Sun–cloud velocity of $\sim 26\,km\,s^{-1}$ (Zank and Frisch 1999; Génova and Beckman 2003). According to Welsh et al. (2004), it is shaped more like a tube and should be called the Local Chimney. It was probably caused, a few million years ago, by a nearby supernova explosion or a strong stellar wind from hot stars.

Inside the Local Bubble we have some cloudlets such as The Local Interstellar Cloud, also known as the Local Fluff (Fig. 1.29), a region of relatively higher density (~ 0.3 atoms per cm^3) and lower temperatures (T $\sim 7,000$ K) into which the Sun has (relatively) recently entered (2,000–8,000 years ago; see Frisch 1996). Using the

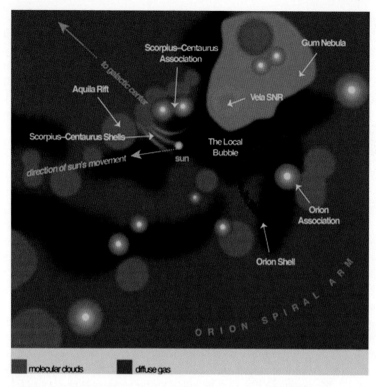

Fig. 1.28 The Galactic environment Within 1,500 light-years of the Sun contains gas clouds of various densities and temperatures. The Sun has passed through a hot, very low-density region – the Local Bubble (*black*) – over the course of several million years and it is now embedded in a shell of warm, partly ionized material flowing from the Scorpius–Centaurus star-forming region. Adapted from Frisch (2000), Fig. 3. Reproduced by permission of American Scientist

UV instrumentation of the HST, Linsky et al. (2000) have produced detailed maps of this 20 light-years long region. Based on radio-surveys of our Galaxy, Dickey (2004) found that structures like the Local Bubble and the local clouds within it, such as the Local Fluff, are very common.

The Sun itself is not at rest in the local reference frame, but has its own motion of about $15\,\mathrm{km\,s^{-1}}$ in the direction of the constellation Hercules (approximately in the direction of the star Vega, see previous images), so that it will plow through the Local Fluff in less than 3,000 years time, and eventually come out through the left wall of the Local Bubble.

The consequences of the passage of our Solar System through these kinds of discontinuities are not very well known. Especially relevant is its location with respect to nearby molecular clouds, where massive stars are born and therefore frequent supernova explosions take place. The Scorpius–Centaurus OB association (see Fig. 1.28) is the closest star-forming region on the outskirts of the Local Bubble. Currently located at a distance of 450 light-years, it is receding in a direction

1.4 The Solar System

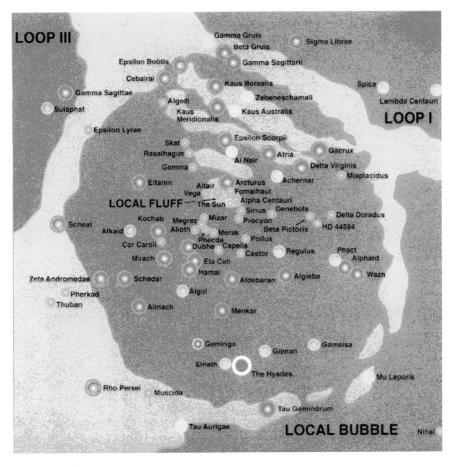

Fig. 1.29 The Local Bubble, in which the Sun is presently embedded (scale: 25 light years). Adapted from N. Henbest and H. Couper (1994) 'The Guide to the Galaxy' Chap. 6 Fig. C. Reproduced by permission of Cambridge University Press

towards the above mentioned constellations. Maíz Apellániz (2001) has shown that this association has generated 20 supernova explosions during the last 11 Ma, and that it was at its closest approach to the Earth, at a distance of 130 light-years, 5 Ma ago. One of the supernovae exploded 2 Ma ago, close enough to Earth to provoke, or at least contribute, to the Pliocene–Pleistocene boundary marine extinction, where the plankton and bivalve mollusks, all UV sensitive, were the species most affected (Benítez et al. 2002).[24] The event seems to be confirmed by the finding of an excess of ^{60}Fe atoms in the corresponding layers of deep ocean (Knie et al. 1999).

---

[24] Conventional explanations for this event are based on the emergence of the Panama isthmus or climate cooling. See Chap. 2.

1.4.3.3 The Galaxy

The home of our Solar System is the Milky Way Galaxy. Like other spiral galaxies, it consists of a thin disk of gas, dust and bright young stars, a swarm of older stars forming a central bulge, and a faint surrounding halo, composed of very old stars. Its main disk is about 80,000–100,000 light-years in diameter, about 250,000–300,000 light-years in circumference and, outside the Galactic core, about 1,000 light-years in thickness. To see the Galaxy as a whole, we need to travel $\sim 10^{19}$ km away.

Looking at the Galaxy from a face-on perspective (Fig. 1.30), we could easily detect the spiral arms on the disk, formed as a consequence of density waves, which trigger star formation. The spiral arms are not actually solid groups of stars. Instead, they represent areas where there are more stars, gas and dust. The arms are about 0.5 kpc wide and the spacing between the spiral arms is about 1.2–1.6 kpc.

The Sun (and therefore the Earth and the Solar System) is at present located close to the inner rim of the small Orion Arm at a distance of $\sim$7.8 kparsec from the Galactic Center (see Reid 1993; Eisenhauer et al. 2003, 2005; Nishiyama et al. 2006; Groenewegen et al. 2008 for different measurements of this parameter). The

Fig. 1.30 Map of Milky Way as seen from far Galactic North showing the spiral arms. The Orion arm contains our Sun and the Solar System. Credit: Richard Powell, The Atlas of the Universe (http://www.atlasoftheuniverse.com)

clockwise orbital speed of the Solar System relative to the centre of the Galaxy is $\sim$220 km s^{-1}, completing a revolution every 225–250 Ma. The Sun has circled the Galaxy more than 18 times during its 4.6 billion year lifetime.

Molecular clouds are located preferentially in the spiral arms. Their masses range from the Giant Molecular Clouds (M$\sim$10^5–10^6 solar masses) to small ones. Their mass spectrum follows a power law. Massive stars are born in these clouds, which ionize the hydrogen gas in their vicinity, thus forming the HII regions (Russeil 2003).

Stars have different velocities as a function of their age. Low-mass, older stars, like the Sun, have relatively high random velocities and as a result can move farther out of the galactic plane.

An important aspect to consider is the motion of the Sun with respect to the spiral arms of our Galaxy. The co-rotation line is defined as the zone where the angular velocity of the Galaxy, individual stars, and that of the pattern of spiral arms are similar. Marochnik (1983) and Mishurov et al. (1997) studied the position of our Sun close to this co-rotation line, suggesting that this placement is exceptional in the sense that it minimizes the frequencies of passage through large molecular clouds contained in the spiral arms.

At a distance of 10^{22} km our Galaxy is reduced to a point, and we can no longer distinguish the position of our Sun. Our space travel has come to an end. In the next chapter, we change the dimension, and we start a time travel through the History of the Earth.

References

Antoniadi, E.M.: Mars, note on photographs of, taken by Percival Lowell in 1907. Mon. Not. Roy. Astron. Soc. **69**, 110–114 (1908)
Backman, D., Marengo, M., Stapelfeldt, K., Su, K., Wilner, D., Dowell, C.D., Watson, D., Stansberry, J., Rieke, G., Megeath, T., Fazio, G., Werner, M.: Epsilon Eridani's Planetary Debris Disk: Structure and Dynamics based on Spitzer and CSO Observations. Strophys. J. (2009)
Batut, A.: La Photographie aérienne par cerf-volant. Editions Gauthier-Villars (1890)
Benítez, N., Maíz-Apellániz, J., Cañelles, M.: Evidence for Nearby Supernova Explosions. Phys. Rev. Lett. **88**(8), 081,101–081,105 (2002)
Bessel, F.W.: Bestimmung der Entfernung des 61sten Sterns des Schwans. Astron. Nachr. **16**, 65–96 (1839)
Borman, F.: Countdown: An Autobiography. William Morrow and Company (1988)
Breitschwerdt, D., Freyberg, M.J., Truemper, J. (eds.): The Local Bubble and Beyond (1998)
Comerón, F., Torra, J., Gómez, A.E.: On the characteristics and origin of the expansion of the local system of young objects. Astron. Astrophys. **286**, 789–798 (1994)
Coradini, A., Capaccioni, F., Drossart, P., Adriani, A., Capria, M.T., De Sanctis, M.C., Filacchione, G., Piccioni, G., Rosetta-VIRTIS Science Team: VIRTIS Rosetta Earth-Moon imaging spectroscopy. In: Bulletin of the American Astronomical Society, Bulletin of the American Astronomical Society, vol. 37, p. 650 (2005)
Crane, N.: Mercator: The Man Who Mapped the Planet. Henry Holt and Co (2003)
Dick, S.J.: Plurality of Worlds. Cambridge University Press, Cambridge (1982)
Dick, S.J.: The biological Universe. Cambridge University Press, Cambridge (1996)
Dickey, J.M.: Is the Local Fluff typical? Adv. Space Res. **34**, 14–19 (2004)

Eisenhauer, F., Genzel, R., Alexander, T., Abuter, R., Paumard, T., Ott, T., Gilbert, A., Gillessen, S., Horrobin, M., Trippe, S., Bonnet, H., Dumas, C., Hubin, N., Kaufer, A., Kissler-Patig, M., Monnet, G., Ströbele, S., Szeifert, T., Eckart, A., Schödel, R., Zucker, S.: SINFONI in the Galactic Center: Young Stars and Infrared Flares in the Central Light-Month. Astrophys. J. **628**, 246–259 (2005)

Eisenhauer, F., Schödel, R., Genzel, R., Ott, T., Tecza, M., Abuter, R., Eckart, A., Alexander, T.: A Geometric Determination of the Distance to the Galactic Center. Astrophys. J. **597**, L121–L124 (2003)

Evans, J.E., Maunder, E.W.: Experiments as to the actuality of the "Canals" observed on Mars. Mon. Not. Roy. Astron. Soc. **63**, 488–499 (1903)

Ferlet, R.: The Local Interstellar Medium. Astron. Astrophys. Rev. **9**, 153–169 (1999)

Frisch, P.C.: LISM Structure - Fragmented Superbubble Shell? Space Sci. Rev. **78**, 213–222 (1996)

Frisch, P.C.: The Galactic Environment of the Sun. Am. Sci. **88**, 52–59 (2000)

Frisch, P.C.: Solar Journey: The significance of our Galactic Environment for the Heliosphere and the Earth. Springer (2006)

Frisch, P.C.: The Local Bubble and Interstellar Material Near the Sun. Space Sci. Rev. **130**, 355–365 (2007)

Génova, R., Beckman, J.E.: Kinematical Structure of the Local Interstellar Medium: The Galactic Anticenter Hemisphere. Astrophys. J. Suppl. **145**, 355–412 (2003)

Gatti, H.: Giordano Bruno and Renaissance Science. Cornell University Press, New York (2002)

Gernsheim, H., Gernsheim, A.: The History of Photography. Mc Graw-Hill Co., New York (1969)

Gillett, F.C.: IRAS observations of cool excess around main sequence stars. In: F.P. Israel (ed.) ASSL Vol. 124: Light on Dark Matter, pp. 61–69 (1986)

Greaves, J.S., Holland, W.S., Moriarty-Schieven, G., Jenness, T., Dent, W.R.F., Zuckerman, B., McCarthy, C., Webb, R.A., Butner, H.M., Gear, W.K., Walker, H.J.: A Dust Ring around epsilon Eridani: Analog to the Young Solar System. Astrophys. J. Lett. **506**, L133–L137 (1998)

Greaves, J.S., Wyatt, M.C., Holland, W.S., Dent, W.R.F.: The debris disc around τ Ceti: a massive analogue to the Kuiper Belt. Mon. Not. Roy. Astron. Soc. **351**, L54–L58 (2004)

Grenier, I.A.: Gamma-ray sources as relics of recent supernovae in the nearby Gould Belt. Astron. Astrophys. **364**, L93–L96 (2000)

Groenewegen, M.A.T., Udalski, A., Bono, G.: The distance to the Galactic centre based on Population II Cepheids and RR Lyrae stars. Astron. Astrophys. **481**, 441–448 (2008)

Guthrie, W.K.: A History of Greek Philosophy, Vol. 2: The Presocratic Tradition from Parmenides to Democritus. Cambridge University Press, Cambridge (1979)

Hannavy, J. (ed.): Encyclopedia of Nineteenth-Century Photography. Routledge (2007)

Hatzes, A.P., Cochran, W.D., McArthur, B., Baliunas, S.L., Walker, G.A.H., Campbell, B., Irwin, A.W., Yang, S., Kürster, M., Endl, M., Els, S., Butler, R.P., Marcy, G.W.: Evidence for a Long-Period Planet Orbiting ϵ Eridani. Astrophys. J. Lett. **544**, L145–L148 (2000)

Hildebrandt, A.: Airships, Past and Present. Archibald Constable, London (1908)

Hoffleit, D., Jaschek, C.: The Bright Star Catalogue. New Haven: Yale University Observatory, 4th edn (1982)

Hoyle, F.: The Nature of the Universe. Harper and Brothers (1950)

Janson, M., Reffert, S., Brandner, W., Henning, T., Lenzen, R., Hippler, S.: A comprehensive examination of the ϵ Eridani system. Verification of a 4 micron narrow-band high-contrast imaging approach for planet searches. Astron. Astrophys. **488**, 771–780 (2008)

Kilston, S.D., Drummond, R.R., Sagan, C.: A search for life on Earth at kilometer resolution. Icarus **5**, 79–98 (1966)

Knie, K., Korschinek, G., Faestermann, T., Wallner, C., Scholten, J., Hillebrandt, W.: Indication for Supernova Produced 60Fe Activity on Earth. Phys. Rev. Lett. **83**, 18–21 (1999)

Linsky, J.L., Redfield, S., Wood, B.E., Piskunov, N.: The Three-dimensional Structure of the Warm Local Interstellar Medium. I. Methodology. Astrophys. J. **528**, 756–766 (2000)

Maíz-Apellániz, J.: The Origin of the Local Bubble. Astrophys. J. **560**, L83–L86 (2001)

Marochnik, L.S.: On the origin of the solar system and the exceptional position of the sun in the galaxy. Astrophys. Space Sci. **89**, 61–75 (1983)

References

Mishurov, Y.N., Zenina, I.A., Dambis, A.K., Mel'Nik, A.M., Rastorguev, A.S.: Is the Sun located near the corotation circle? Astron. Astrophys. **323**, 775–780 (1997)

Morrison, P.: Powers of Ten. Freeman, W.H. (1985)

Nishiyama, S., Nagata, T., Sato, S., Kato, D., Nagayama, T., Kusakabe, N., Matsunaga, N., Naoi, T., Sugitani, K., Tamura, M.: The Distance to the Galactic Center Derived from Infrared Photometry of Bulge Red Clump Stars. Astrophys. J. **647**, 1093–1098 (2006)

Poole, R.: Earthrise: How Man First saw the Earth. Yale University Press, London (2008)

Reid, M.J.: The distance to the center of the Galaxy. Annu. Rev. Astron. Astrophys. **31**, 345–372 (1993)

Russeil, D.: Star-forming complexes and the spiral structure of our Galaxy. Astron. Astrophys. **397**, 133–146 (2003)

Sagan, C.: Pale Ble Dot: A vision of the Human Future in Space. Random House, New York (1994)

Sagan, C., Wallace, D.: A Search for Life on Earth at 100 Meters Resolution. Icarus **15**, 515–554 (1971)

Schellnhuber, H.J.: Earth system Analysis and the second Copernican Revolution. Nature **402**, C19–C23 (1999)

Stooke, P.: The International Atlas of Lunar Exploration. Cambridge University Press, Cambridge (2007)

Stothers, R., Frogel, J.A.: The local complex of O and B stars. I. Distribution of stars and interstellar dust. Astron. J. **79**, 456–471 (1974)

Thebault, P., Marzari, F., Scholl, H.: Planet formation in the habitable zone of alpha Centauri B. Mon. Not. Roy. Astron. Soc. (2008)

Welsh, B.Y., Sallmen, S., Lallement, R.: Probing the inner halo and IVC gas through the Local Interstellar Chimney. Astron. Astrophys. **414**, 261–274 (2004)

White, F.: The Overview Effect: Space Exploration and Human Evolution. American Institute of Aeronautics, 2nd edn (1998)

Zank, G.P., Frisch, P.C.: Consequences of a Change in the Galactic Environment of the Sun. Astrophys. J. **518**, 965–973 (1999)

Zimmerman, R.: Genesis: the story of Apollo 8. Dell Publishing, New York (1999)

Chapter 2
The Earth in Time

The Earth is now characterized by a set of physical parameters and features observed on its surface. However, the Earth in its present state represents just one stage in an evolutionary process starting from the moment the planet formed to the present day. In our future search for Earth-like planets, we may encounter a planet at any stage in its evolution, and therefore, we can feasibly use the history of the Earth as a guide to characterize these planets. In this chapter we study the different components of the Earth system over time, starting with an introduction to the Earth as it is at present.

Most of the observed changes in the surface of the Earth have come about as a consequence of the dissipation of the energy stored in its interior. Variations in the gaseous envelope, the atmosphere, are mainly driven by changes in the solar output and in the concentration of greenhouse gases, together with a strong internal variability. Thus, we can divide all the changes suffered by our planet into two main classes: abiotic (solar and geological) and biotic.

A holistic view of our planet is based on the assumption that all its properties cannot be determined only by the sum of its components. This constitutes the core of the Earth System Sciences (Kump et al. 2004; Steffen et al. 2005).

We can consider our planet as constituted by four vast reservoirs of material with flows of matter and energy between them: atmosphere, hydrosphere, biosphere and geosphere (Fig. 2.1). The entire system evolves as a result of positive and negative feedbacks between constituent parts.

For Schellnhuber (1999), the Earth System, E, can be represented by the equation

$$E = (N, H),$$

where $N = (a,b,c,...)$ is the ecosphere and consists in a set of linked planetary sub-spheres: a (atmosphere), b (biosphere), c (cryosphere), and so on; $H = (A,S)$ embraces the antroposphere A, and the component S reflects the emergence of a global subject, manifested, for instance, by adopting international laws for climate protection.

Table 2.1 and Fig. 2.2 show the main periods of the Earth history and the main geological and biological events, respectively, which have taken place throughout the evolution of our planet.

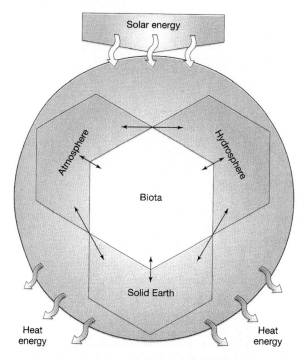

Fig. 2.1 Diagram of the Earth system, showing interactions among the components. From R.W. Christopherson, *Geosystems: an introduction to physical geography*, 1997. Copyright: Prentice-Hall

Table 2.1 Geologic periods

Precambrian	Hadean (4500–3800)		Moon formation
			Origin of the oceans
	Archaean (3800–2500)		Origin of life
	Proterozoic (2500–550)		Rise in atmospheric oxygen
			First cells with nucleus
Phanerozoic	Paleozoic	Cambrian (550–490)	Complex Life
		Ordovician (490–443)	
		Silurian (443–417)	
		Devonian (417–354)	
		Carboniferous (354–290)	
		Permian (290–248)	Massive extinction
	Mesozoic	Triassic (248–206)	
		Jurassic (206–144)	
		Cretaceous (144–65)	Massive extinction
	Cenozoic	Tertiary (65–1.8)	
		Quaternary (1.8–Today)	Humans

The time in brackets is in Ma

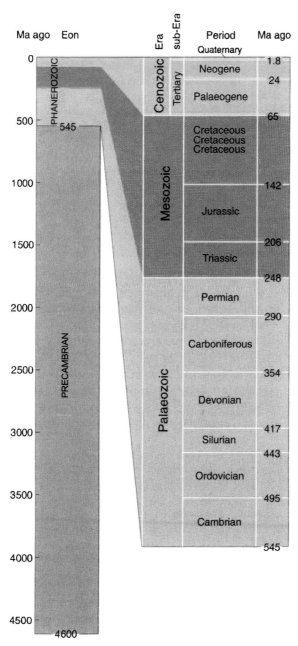

Fig. 2.2 Main geological periods

Different textbooks and monographs have handled the topic of this chapter. We can indicate only some of them (Stanley 1992, 1999; Lunine 1999; Lane 2002; Anguita 2002; Knoll 2003 and Zahnle et al. 2007).[1]

2.1 The Earth at the Present Time

The mass of our planet is approximately 5.98×10^{24} kg. It is composed mostly of iron (32.1%), oxygen (30.1%), silicon (15.1%), magnesium (13.9%), sulphur (2.9%), nickel (1.8%), calcium (1.5%) and aluminium (1.4%), with the remaining 1.2% consisting of trace amounts of other elements.

To understand the past and to predict the future, it is essential to know what is the present structure of our 4.6 Ga old planet. We follow a classical approach describing the different layers (Fig. 2.3) and the processes taking place there. Then we begin our time travel back to its past and into its future.

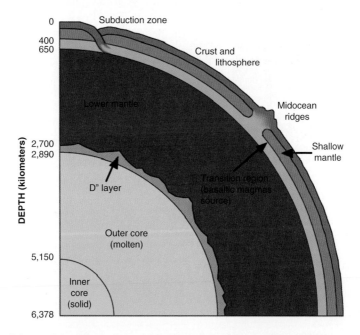

Fig. 2.3 Scheme of the different layers of the Earth, from the core to the surface

---

[1] See also the 2008 report of the National Research Council on *Origin and Evolution of Earth: Research Questions for a changing Planet*.

2.1.1 The Interior

The interior of the Earth is chemically divided into layers as a result of its molten state, early in its formation. Together with theoretical work, the main sources of information about this region comes from the recordings of free oscillations excited by large earthquakes.

The simplest model of the Earth is based on the assumption of an average density, $\bar{\rho} = 5.52 \,\text{g cm}^{-3}$. It is obvious that this value is higher than the average density (2.7–3.3) of the rocks of the Earth's surface. This clearly points to a concentration of mass near the centre of the Earth. Assuming that the planet is in hydrostatic equilibrium and is spherically symmetrical

$$dP = -\rho g\, dr.$$

The gravity is given by

$$g = \frac{Gm}{r^2} = \frac{4\pi G}{r^2} \int_0^r \rho r^2 dr,$$

where G is the gravitational constant and $m = \frac{4\pi}{3} r^3 \rho$ is the mass enclosed within the sphere of radius r and density ρ. Hence,

$$\frac{dP}{dr} = -\frac{4\pi G \rho}{r^2} \int_0^r \rho r^2\, dr.$$

Now, by splitting the density variation in two parts

$$\frac{d\rho}{dr} = \frac{d\rho}{dp} \frac{dP}{dr},$$

we obtain

$$\frac{d\rho}{dp} = \frac{\rho}{K},$$

where K is the compressibility module. Therefore,

$$\frac{d\rho}{dr} = -\frac{\rho}{K} \frac{Gm\rho}{r^2}.$$

From the theory of propagation of waves in the interior of the Earth, we know that

$$\frac{K}{\rho} = \alpha^2 - \frac{4}{3}\beta^2,$$

where α and β are the velocities of the P and S waves,[2] respectively, given by

$$\alpha = \left[\left(K + \frac{4}{3}\mu\right) / \rho\right]^{1/2}$$
$$\beta = (K/\rho)^{1/2},$$

where ρ is the density and μ is the rigidity modulus

$$\longrightarrow \frac{d\rho}{dr} = -\frac{Gm\rho}{r^2\left(\alpha^2 - \frac{4}{3}\beta^2\right)}$$

Moreover, we must also know the equation of state of the material in the interior that establishes a relation of density with temperature and pressure at all the depths.

The Preliminary Earth Reference Model constitutes a good framework to study the processes taking place in the Earth's Interior (Dziewonski and Anderson 1981). Figure 2.4 shows the variations with the radius of the most relevant parameters. For monographs on the Earth's internal structure see Zharkov (1986) and Poirier (1991).

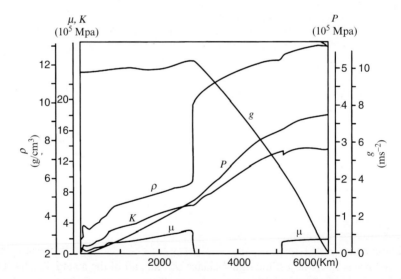

Fig. 2.4 The preliminary Earth reference model. The radial variation of the parameters cited in the text is shown. The surface is indicated by the level 0. Data available from http://solid_earth.ou.edu/prem.html

---

[2] Pressure (P) waves travel at the greatest velocities within solids and the particle motion is parallel to the direction of wave propagation. Shear (S) waves are transverse waves.

2.1.1.1 Inner Core

The core was the first internal structural element to be identified. Oldham (1906), by studying earthquake records, first suggested that the Earth must have a molten interior. Temperatures at the Earth's centre can reach to 5,000–6,000°C. However, the inner core remains solid due to the extremely high pressure overcoming the effect of high temperatures. It is composed of mainly nickel–iron alloy and some lighter elements (probably sulphur, carbon, oxygen, silicon and potassium). The inner core seems to be rotating faster than the Earth's surface (Glatzmaier and Roberts 1996).

2.1.1.2 Outer Core

The outer core is in liquid state, with convection as the main form of energy transport. The heat necessary to drive convection is derived from the gradual growth of the inner core. As the deeper outer core cools, liquid iron slowly solidifies, releasing heat in the process. The newly formed particles of solid iron also heat the outer core frictionally as they settle down to join the inner core.

The rotating fluid in this layer is responsible for producing the Earth's magnetic field via a dynamo process. Our planet possesses one of the strongest magnetic fields of the Solar System.[3] This is due to the combination of a relatively high rotation rate (24 h) and the thickness of the outer core, where we have fluids in motion. The magnetic field is practically a dipole, with the axis inclined approximately 11.3° from the planetary axis of rotation.

2.1.1.3 Mantle

The mantle lies roughly between 30 and 2,900 km below the Earth's surface and occupies about 70% of the Earth's volume (see Fig. 2.3). Temperatures range between 1,000°C at upper boundary and over 4,000°C at the boundary with the core. Although these temperatures far exceed the melting points of the mantle rocks at the surface, particularly in deeper layers, the materials are almost exclusively solid. The enormous lithostatic pressure exerted on the mantle prevents them from melting, because the melting temperature increases with pressure.

In the upper mantle there is a major area, the asthenosphere, where the temperature and pressure are at just the right balance so that part of the material melts. The rocks become soft plastic and flow like warm tar. This layer is 100–200 km thick and its top is about 100 km below the Earth's surface. It is mainly composed of oxygen, silicon, iron, aluminium and magnesium. The amount of water in the mantle is still a matter of debate (Bolfan-Casanova 2005).

---

[3] The strength of the field at the Earth's surface ranges from less than 30 μT (0.3 gauss) in an area including most of South America and South Africa to over 60 μT (0.6 gauss) around the magnetic poles in northern Canada and southern Australia, and in parts of Siberia.

Most of the heat generated is transported through the mantle via convection, providing an effective way to transport material and heat from the deep interior.[4] Evidence for the descent of crustal slabs into the mantle indicates that materials travel both ways. The Rayleigh number for the convecting mantle is

$$\mathrm{Ra} = \frac{g\alpha(T_m - T_s)(R_m - R_c)^3}{\kappa \nu},$$

where g is the acceleration due to gravity, α the coefficient of thermal expansion, ν the viscosity, κ the thermal diffusivity, R_m and R_c the outer and the inner radii of the mantle, respectively and T_m and T_s the temperatures of the mantle and surface, respectively. Convection exists for Rayleigh numbers greater than 10^3. For the mantle, values ranges between 10^6 and 10^8.

The mantle heat flow is parametrized in terms of this Rayleigh number Ra.

$$Q_m = \frac{k(T_m - T_s)}{(R_m - R_c)} \left(\frac{\mathrm{Ra}}{\mathrm{Ra}_{cr}}\right)^\beta,$$

where k is the thermal conductivity, Ra_{cr} the critical value for the onset of convection and β an empirical constant.

2.1.1.4 Lithosphere

The lithosphere is the outermost shell of the planet. It includes the crust and the uppermost mantle, which are joined across the Mohorovic layer. The solid crust of the Earth floats on top of the upper mantle and has two main components: oceanic and continental, the latter being thicker. Figure 2.5 illustrates the relative abundance of chemical elements in the upper continental crust. Rock-forming elements are the most abundant. Oxygen and silicon compose approximately 72% of the rocks, with the rest being aluminium, iron, calcium, magnesium and sodium.

2.1.1.5 Energy Budget

The internal energy of our planet has two main sources: (1) the potential energy acquired during the accretion process and the energy added by impacts during the Earth's initial growth and (2) the energy released by the radioactive decay of elements such as ^{40}K, ^{238}U and ^{232}Th through the following reactions:

$$^{238}\mathrm{U} \longrightarrow {}^{206}\mathrm{Pb} + 8{}^4\mathrm{He} + 6e^- + 6\bar{\nu} + 51.7\,\mathrm{Mev}$$
$$^{232}\mathrm{Th} \longrightarrow {}^{208}\mathrm{Pb} + 6{}^4\mathrm{He} + 4e^- + 4\bar{\nu} + 42.8\,\mathrm{Mev}$$
$$^{40}\mathrm{K} + e^- \longrightarrow {}^{40}\mathrm{Ar} + \bar{\nu} + 1.513\,\mathrm{Mev},$$

where $\bar{\nu}$ are the antineutrinos.

---

[4] Mantle materials are poor conductors of heat.

2.1 The Earth at the Present Time

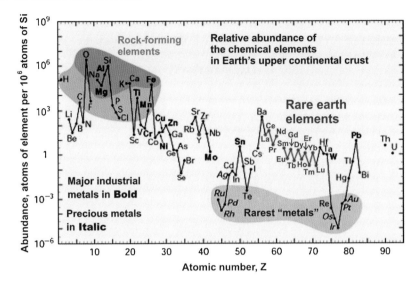

Fig. 2.5 Abundance (atom fraction) of chemical elements in the upper lithosphere as a function of the atomic number

The total power dissipated by the Earth's interior is estimated to be between 30 and 44 TW (Pollack et al. 1993; Garzón and Garzón 2001).[5]

Both types of energy, accretion and radiogenic, decline with time and are proportional to the mass of the planet.[6] They are dissipated, gradually and abruptly, toward the outer layers, giving rise to processes that have configured the structure of the Earth's crust (Condie 2004b). These values are small compared with the 174 PW ($\sim$340 W m^{-2}) received from the Sun at the top of the atmosphere.

The fundamental energy equation for the transfer of energy per unit volume is

$$C\frac{\partial T}{\partial t} = Q_c + Q_H - Q_{conv},$$

where C is the heat capacity of the Earth and Q_c is the heat transfer by conduction, which can be expressed by the Fourier law

$$Q_c = -\kappa \nabla^2 T,$$

κ being the thermal conductivity. Taking into account only Fourier's conductive cooling, W. Thompson (Lord Kelvin) estimated in 1863 the age of the Earth in 100 Ma (Richter 1986; Burchfield 1990). We clearly need other energy sources and cooling mechanisms.

---

[5] The heat flow is larger in the oceans than in the continents.

[6] The amount of generated energy is proportional to the planetary volume ($\propto R^3$), and the amount of dissipated energy depends on the planetary surface ($\propto R^2$). Therefore, it takes a certain time to cool.

The mantle is a viscous fluid and the heat advented by mass transfer with velocity V, convective heat flux, across the mantle is given by

$$Q_{conv} = c_p V \nabla T.$$

In 1895, John Perry (1850–1920) produced an age of Earth estimate of 2–3 billion years old using a model of a convective mantle and thin crust. However, the main event was the discovery of radioactivity by Henry Becquerel (1852–1908), in 1896, leading to the possibility of longer ages not only for the Earth but also for the Sun and the totality of the Solar System.

Q_H is the local heat generation by radioactivity decaying with time. It depends on the mass, m, of the main radioactive isotopes (expressed in units of 10^{17} kg.)

$$Q_H(TW) = 9.5m(^{238}U) + 2.7m(^{232}Th) + 3.6m(^{40}K)$$

From theoretical estimates we know that $m(^{232}Th):m(U):m(^{40}K) = 4:1:1$. Therefore,

$$Q_H(TW) = 24m(^{238}U).$$

The Urey ratio, U, is defined as

$$U = \frac{Q_H}{Q_{conv}}.$$

Mantle convection models typically assume U values from 0.4 to almost 1 (implicating a surface heat flux of 46 TW), whereas geochemical models (e.g. Korenaga 2008) predict $0.3 < U < 0.5$, implicating smaller heat fluxes ($\approx$30 TW). Latter models assume that U and Th are mainly in the lithosphere and mantle is in the form of oxides, leading to smaller values for the radioactive heat. Other models assume that potassium is alloyed with iron in the interior (Lee and Jeanloz 2003), providing an important source of radioactive energy to generate the magnetic field (Herndon 1996).

Araki et al. (2005) have measured antineutrinos produced by radioactive beta-decay at the heart of the Earth. The results obtained from these so-called geoneutrinos, 19 TW for radiogenic heat, are consistent with compositional models of the planet (Palme and O'Neill 2003; McDonough 2003), and provide a new way of determining where unstable isotopes are stored inside the planet and in what concentrations.

2.1.2 Plate Tectonics

Observed over time, the Earth shows clear changes in the aspect of its surface, which can be easily interpreted as the consequence of internal energy release from its interior. Figure 2.6 shows a computer generated view of the crust relief.

2.1 The Earth at the Present Time 45

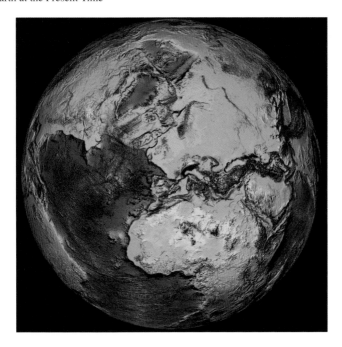

Fig. 2.6 A view of the crust relief showing land and undersea topography. Courtesy of National Geophysical Data Center

During the nineteenth and the early twentieth century, geologists explored the idea that the continents may have moved across the surface.[7] They were inspired by the remarkable fit between Atlantic coasts of Africa and South America, already noted by Francis Bacon.[8] This hypothesis was first developed by Alfred Wegener (1880–1930), who also studied the distribution of animals and fossils to help him in his interpretations.

After World War II, large chains of underwater volcanoes were discovered, known as mid-ocean ridges. The Mid-Atlantic Ridge was mapped first in some detail by M. Ewing (1906–1974) and B. Heezen (1924–1977). These rifts were later identified as newly formed sea floor that is extruded along the ridges. Once emerged, the new floor expands to the sides, process known as sea-floor spreading (Hess 1962).

In principle, there are three primary modes for releasing the heat contained in the Earth's interior, cooling the planet: magma ocean (see Sect. 2.3.1), stagnant lid[9] and

---

[7] In a letter written on September 1782 to the Abbe Soularie, Benjamin Franklin (1706–1790) recognized that the crust was a shell, which could be broken and parts moved about.

[8] In 1620 he said, *if the fit between South America and Africa is not genetic, surely it is a device of Satan for our confusion*.

[9] In stagnant lid convection, heat is transported by conduction in most of the top layer. The convectively unstable bottom is restricted to this region, which is significantly colder than the interior and not so cold that it is too stiff to participate in the convection.

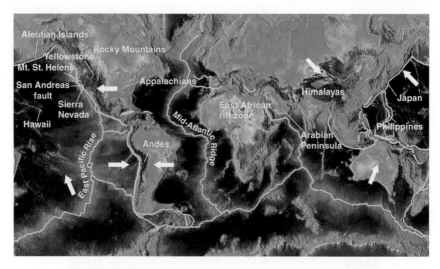

Fig. 2.7 Global plates of the planet

plate tectonics (Schubert et al. 2001). The latter process has been observed only in our planet[10] (see Martin et al. 2008 for a didactic summary).

The lithosphere essentially floats on the top of the mantle, the plastic asthenosphere, and is broken up into what are called tectonic plates, which are moving in relation to one another, continuously changing shape and size. The lithosphere is divided into about 20 rigid plates (Fig. 2.7). Dissipation of heat from the mantle is the original source of energy driving plate tectonics. Earth activity (earthquakes and volcanoes) is concentrated at the plate boundaries.

Three types of plate boundaries exist, differentiated by the way the plates move relative to each other. They are also associated with different types of surface phenomena. The different types of plate boundaries are the following:

Divergent boundaries: Here two plates slide apart from each other (examples of which can be seen at mid-ocean ridges and active zones of rifting). Material from the mantle rises from beneath a mid-ocean and partially melts, forming magma and creating new ocean crust.

Convergent boundaries: Two plates move together forming either a subduction zone (if one plate moves underneath the other) or a continental collision (if both plates contain continental crust). Deep marine trenches are typically associated with subduction zones. Because of friction and heating of the subducting slab, volcanism and earthquakes are almost always closely linked to convergent boundaries. The sinking of lithosphere in subduction zones provides most of the force needed to drive the plates and cause mid-ocean ridges to spread.

---

[10] Venus has a thick lithosphere and mantle plumes (stagnant lid convection) and although plate tectonics may have existed in the past in Mars, it now has become a 'single-plate' planet dominated by hot-spot volcanism (see also Sect. 2.3.2).

Transform boundaries: These occur where plates slide or, perhaps more accurately, grind past each other along transform faults. The relative motion of the two plates is either sinistral (left side toward the observer) or dextral (right side toward the observer).

The destruction (recycling) of crust takes place along convergent boundaries where plates are moving toward each other, and sometimes one plate sinks (is subducted) under another.

2.1.3 The Atmosphere

The atmosphere is the gas layer surrounding the planet (Fig. 2.8). The average pressure at the ground is 1,013 mbar, with a column atmospheric mass of 1 kg cm^{-2} and a density of 2.7×10^9 molecules cm^{-3}. Table 2.2 gives the best estimates about its main constituents.

Oxygen in Earth's atmosphere is very abundant. However, it is relatively rare at the cosmic scale ($\sim$0.06%) and therefore also in the protosolar nebula. Moreover, the noble gases (Kr, Ne) are thousands of times more abundant in the Sun than in our atmosphere. Some mechanism has blown up the primordial atmosphere, leaving the inert nitrogen as the only residual of those early times. We speak in Sect. 2.3.1 on the origin of carbon dioxide and water.

Fig. 2.8 Image of the Earth atmosphere taken on 20 July 2006 from the International Space Station by the astronaut Jeffrey Williams using a digital camera equipped with a 400 mm lens. Astronaut photograph ISS013-E-54329. Courtesy: NASA JSC

Table 2.2 Principal constituents of the atmosphere

Constituent		Composition by volume (%)
Nitrogen	N_2	78.08
Oxygen	O_2	20.95
Argon	Ar	0.93
Carbon dioxide	CO_2	0.0375
Water vapour	H_2O	0.001-4
Neon	Ne	0.0018
Helium	He	0.0005
Methane	CH_4	0.00017
Nitrous oxide	N_2O	0.00003
Hydrogen	H_2	0.00005
Xenon	Xe	0.000009
Ozone	O_3	0.000004

Measurements from balloon soundings and aircraft flights, together with theoretical developments, brought a clearer understanding of the physical structure of the terrestrial atmosphere, which was divided in different layers according to the different processes taking place in them (see Fig. 2.9).

Troposphere: Named after the Greek word for overturning, it extends from the terrestrial surface up to approximately 11 km. The heat source is infrared radiation emitted from the surface, warmed by visible solar radiation. The heat is transferred from the surface to the troposphere by the following processes:

- The evaporation of water and the release of latent heat[11] through the formation of clouds.
- Infrared emission and absorption by greenhouse gases, such as water vapour, CO_2 and CH_4.
- Sensible heat flux, the heat absorbed or transmitted by a substance during a change of temperature that is not accompanied by a change of state.

Assuming hydrostatic equilibrium $(dP(z)/dz = -\rho(z)g(z))$ and the equation of state for an ideal gas $(P = n\,k\,T)$, we obtain the following expressions:

$$\frac{dP(z)}{dz} = -\frac{P(z)\mu(z)g(z)}{kT(z)}$$

$$P(z) = P_0 e^{-mgz/kT}; \quad n(z) = n_0 e^{-z/H},$$

where H (z) is the pressure scale height given by

$$H = \frac{kT(z)}{\mu(z)g(z)}$$

---

[11] Heat absorbed during a change of state.

2.1 The Earth at the Present Time

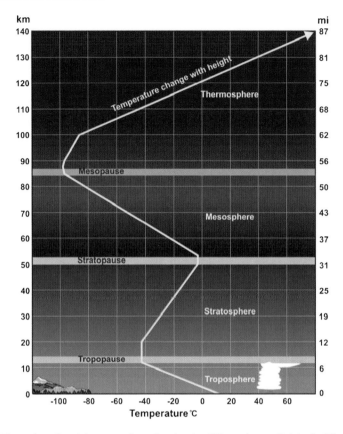

Fig. 2.9 Thermal profile of the atmosphere showing the different layers. Height (in kilometres and miles) is indicated along each side. Credit: National Weather Service

T being the temperature, z the geometrical altitude, p the pressure, g the gravitational acceleration, $\rho = nm$ the mass density, k the Boltzmann constant and μ the mean molecular weight.

Convection is the dominant mechanism for energy transport in the troposphere. Assuming adiabatic motion of the convective cells, the temperature gradient, dT/dz, can be derived by applying the first law of thermodynamics and the consideration of latent heat

$$dT/dz = -(g(z)/c_p)/[1 + (L/c_p)(dW/dT)] \approx 6.5°C/km,$$

where z is the vertical coordinate, g the gravitational acceleration, c_p the specific heat at constant pressure, W the mass of saturated air and L the latent heat of vapourization. Moisture can decrease dT/dz by releasing latent heat. James P. Espy (1785–1860) first derived this parameter empirically for dry and saturated conditions, and some years later verified theoretically by W. Thompson (Lord Kelvin).

The troposphere contains 99% of the water vapour in the atmosphere and its content decreases rapidly with height in this layer. Water vapour plays a major role in regulating air temperature because it absorbs solar energy and thermal radiation from the planet's surface.

The tropopause is highest in the tropics ($\sim$16 km) and lowest in the polar regions ($\sim$8 km), and also undergoes seasonal changes. Here, radiative processes start to dominate. Meteorological processes take place in the lower atmosphere (the tropo- and stratosphere).

Stratosphere: This layer lies between 10 and 50 km altitude and it is mainly a radiation-driven environment. The temperature increases with altitude due to the absorption of UV radiation by ozone, a topic that will be dealt with later (Fig. 2.26). Other major absorbing and emitting gases in this region are carbon dioxide and water vapour (see, e.g. Taylor 2003 for a summary).

Mesosphere: This is the coldest of the atmospheric layers and is created by the emission of radiation from carbon dioxide, CO_2. The temperature decreases with the altitude and reaches low enough values to freeze water vapour producing ice clouds, also called noctilucent clouds. Because of oxidation processes and the penetration of UV radiation, which dissociates polyatomic molecules, this layer is more complex than those below.

Thermosphere: The temperature rises in this layer again because of the heat released from the dissociation of molecular oxygen by UV light and photoionization by X-rays. Here, conduction is the main mode of energy transport. In this layer the absorption of solar energy is less than 1% of that in the stratosphere, but the air is so thin that a small increase in deposited energy can cause a large increase in temperature.

Exosphere: A region where most of the particles have enough kinetic energy to escape from the terrestrial atmosphere. The minimum escape velocity from the Earth, the critical escape velocity, is about $11.3 \,\mathrm{km\,s^{-1}}$.

The outer layers of our planet will be described in more detail in Chap. 4. See Chamberlain (1987) and Houghton (2002) for basic concepts on planetary atmospheres.

2.1.4 Energy Balance of the Atmosphere

2.1.4.1 Albedo

The unit-less quantity albedo (Latin for white) is a measure of the overall reflection coefficient of an object. The geometric albedo, p, is defined as the amount of radiation relative to that from a flat Lambertian surface, which is an ideal reflector at all wavelengths. The bond albedo, a, is the total radiation reflected from an object compared to the total incident radiation. For Earth, the bond albedo (the fraction of incoming sunlight that our planet reflects back to space) is 0.29 while the geometrical albedo is 0.37 (de Pater and Lissauer 2001). Unless otherwise specified in the following, we refer to the bond albedo (see next chapter for more details).

There are many factors in Earth's climate system that influence how much sunlight our world reflects back to space vs. how much it catches and stores in the form of heat. Potential parameters affecting the albedo are volcanic eruptions, changes in surface vegetation and/or desertification (Betts 2000), variations in snow and ice coverage (Randall et al. 1994), and atmospheric constituents such as aerosols, water vapour and clouds (Cess et al. 1996; Ramanathan et al. 1989; Charlson et al. 1992). Albedo changes will be determined by the total effect of the variations in all these parameters. However, these changing parameters will bring along multiple climate feedbacks, which make assessing the exact implications in the albedo a hard task.

2.1.4.2 The Planet's Mean Temperature

Let us assume a purely radiative balance of the Earth's atmosphere. Climate changes are produced by any perturbation to this balance. Figure 2.10 shows a scheme of the different external factors playing a role in the climate system.

Solar radiation is the primary energy source for the Earth's climate. Its flux at the Earth is determined by

$$F_S = \frac{L}{4\pi d^2},$$

where L is the solar luminosity (energy per time unit) and d the distance Earth–Sun.

The Earth's surface, aerosols in the atmosphere and clouds all reflect some of the incoming solar short-wavelength radiation back to space, preventing that energy from warming the planet. Furthermore, about 13% of the solar radiation incident in the atmosphere is Rayleigh scattered, half of this reaching the Earth's surface as diffuse radiation and the other half being returned to space (Houghton 2002). Short-wavelength radiation, usually defined as having wavelengths between 0.15

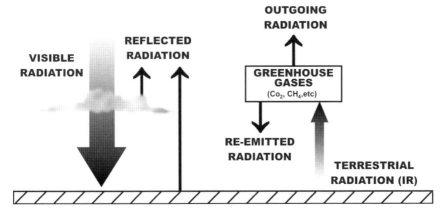

Fig. 2.10 The various factors related to changes in the Earth's surface temperatures

and 4.0 μm, includes about 99% of the sun's radiation; of this energy, 46% is infrared (>0.74 μm), 9% is ultraviolet (<0.4 μm) and the remaining 45% is visible, with wavelengths between 0.4 and 0.74 μm (Liou 2002). A significant portion of the solar energy is absorbed by the Earth (~70%), where it drives terrestrial phenomena before being radiated back into space through the atmospheric window as infrared radiation, peaking at about 10 μm.

The temperature of the Earth is determined by the balance between the received shortwave (visible) flux, F_{in}, and the emitted infrared (long-wave) radiation, F_{out}, so that

$$\left(\pi R_E^2\right) F_S (1-a) = 4\pi R_E^2 F_E.$$

Assuming that the Earth radiates as a black body $F_E = \sigma T^4$, we can define an equilibrium temperature as

$$T_{eq} = \left[\frac{F_S(1-a)}{4\sigma}\right]^{1/4} = \left[\frac{L(1-a)}{4\pi\sigma d^2}\right]^{1/4},$$

where σ is the Stefan–Boltzmann constant and T_{eq} (~255 K) is the equilibrium temperature of the Earth, a physical averaged long-wave emission temperature at about 5.5 km height in the atmosphere (depending on wavelength and cloud cover, altitudes from 0 to 30 km contribute to this emission).

Simple calculations indicate that T_{eq} is much less than the present surface temperature $T_s = 288$ K. Therefore, one must introduce a greenhouse forcing parameter $G[W/m^2]$ defined as the difference between the emission at the top of the atmosphere and the surface. The forcing G increases with increasing concentration of greenhouse gases (see Harries 1996 for an overview on the global energy balance of the Earth).

After Raval and Ramanathan (1989), we can define the normalized greenhouse effect g, with $g = G/\sigma T_s^4$. Then the outgoing power can be written as

$$F_{out} = 4\pi R_e^2 \sigma (1-g) T_s^4.$$

If the planet is in radiative equilibrium, $F_{in} = F_{out}$, then we have

$$T_s^4 = \frac{F_S}{4\sigma(1-g)}(1-a).$$

This means that the albedo, together with solar irradiance and the greenhouse effect, directly controls the Earth's temperature.

2.1.4.3 Greenhouse Gases

Greenhouse gases are the gases present in the atmosphere, which reduce the loss of heat (infrared long-wave radiation) into space by absorbing it and therefore contribute to global temperatures through the greenhouse effect. Two pioneers must be mentioned in this context: Joseph Fourier (1768–1830) established the concept of planetary energy balance (Fourier 1824), and J. Tyndall (1820–1893) began to study the capacities of various gases to absorb or transmit the heat emitted by the Earth (infrared radiation). He showed that the main atmospheric gases, nitrogen and oxygen, are almost transparent to radiant heat, while water vapour, carbon dioxide and ozone are such good absorbers that, even in small quantities, these gases absorb heat radiation much more strongly than the rest of the atmosphere (Tyndall 1863).

The major atmospheric constituents (nitrogen and oxygen) are not greenhouse gases. This is because homonuclear diatomic molecules[12] such as N_2 and O_2 neither absorb nor emit infrared radiation, as there is no net change in the dipole moment of these molecules when they vibrate. Molecular vibrations occur at energies that are of the same magnitude as the energy of the photons on infrared light. The most important greenhouse gases are diatomic heteronuclear molecules such as water vapour and carbon dioxide (Fig. 2.11); methane, nitrous oxide and other trace gases contribute as well. A simple relation between the surface temperature, T_s, and the partial pressure of carbon dioxide has been given by Walker et al. (1981).

$$T_s = 2T_e + 1.6 \left(P_t/P^0_{CO2}\right)^{0.364} - 226.4,$$

where t is time in billions of years and

$$T_e = 255/(1 + 0.087t)^{0.25}.$$

2.1.4.4 2D Models

In zonally averaged climate models, the rate of solar energy input to a latitude belt is locally balanced by the sum of the energy leaving the latitude belt as infrared radiation to space and the net heat transport to other latitude belts. This may be expressed by the relation (North et al. 1981)

$$-\frac{d}{dx}D(1-x^2)\frac{dT(x)}{dx} + I(x) = S(x)(1-a(x)),$$

where D is the meridional heat diffusion coefficient, x the sine of latitude, $T(x)$ is the zonally averaged temperature in a given latitude band, $I(x)$ is the outgoing infrared radiation, $S(x)$ is the annual solar radiation reaching the top of the atmosphere and $a(x)$ is the zonally averaged top-of-atmosphere albedo. Caldeira and Kasting (1992) give polynomial fits for these parameters.

---

[12] Diatomic molecules are molecules made only of two atoms. If a diatomic molecule consists of two atoms of the same element, then it is said to be homonuclear, otherwise it is said to be heteronuclear.

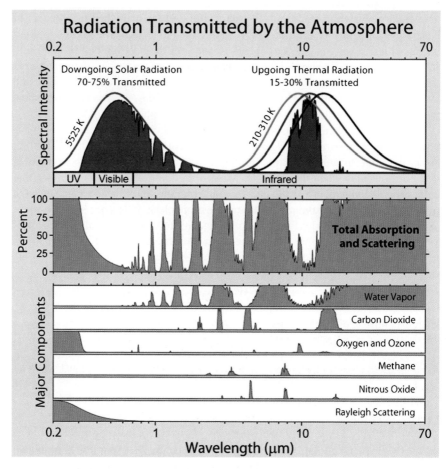

Fig. 2.11 Pattern of absorption bands generated by various greenhouse gases and their impact on both solar radiation and upgoing thermal radiation from the Earth's surface. Note that a greater quantity of upgoing radiation is absorbed, which contributes to the greenhouse effect

2.2 The Precambrian Era (4,500–4,550 Ma BP)

We have now the tools to describe the main epochs of the evolution of our planet. The debate over the age of the Earth and the Sun has been ongoing for over 2,000 years. The discovery of radioactivity near the end of nineteenth century made it possible to clarify the discussions providing a long-standing source for the solar energy and a technique of isotope geochronology. Measurement of the decay of radioactive elements has been applied in meteorites and the Moon, permitting the age of our planet to be estimated at 4,550 millions of years (Patterson 1956; Allegre et al. 1995 and Zhang 2002). The Pb and Hf-W isotopic systems have been widely used for this purpose. It is the beginning of our countdown.

2.2 The Precambrian Era (4,500–4,550 Ma BP)   55

Fig. 2.12 The main events during the Precambrian

HADEAN (4.500 - 3.800)	FORMATION OF CORE MOON FORMATION FAINT YOUNG SUN LATE HEAVY BOMBARDMENT
ARCHAEAN (3800 - 2500)	TECTONIC ACTIVITY ORIGIN OF LIFE (PROCARIOTS) METHANE
PROTEROZOIC (2500 – 550)	RISE ATMOSPHERIC OXYGEN CARBON DIOXIDE CYCLE EUCARIOTS FIRST SNOWBALL EPISODE

Figure 2.12 shows the main periods of the Precambrian era, covering most of the history of the Earth, and the most relevant events occurring during this time period.

2.2.1 The Formation of the Earth: The Hadean Era

Three main phases can be considered in the formation of the Earth (Goldreich et al. 2004). It started with a quick runaway accretion in a disk of small bodies, a process lasting less than one million years. During several hundred million years, the embryos grew at the expense of smaller bodies until finally the orbits of the embryos began to cross, colliding and coalescing in a protoearth (see Chap. 8 for more details).

The Earth's core began to grow after the formation of the protoearth as the temperature increased to the point where dense, liquid iron began to sink toward the centre of the planet. According to a value given by isotopic signatures of radiogenic element pairs ^{182}Hf/^{182}W (Kleine et al. 2002), the core formation was completed 30 Ma years after the formation of the Earth. Sometime during this period the surface of the Earth became solid and the first rocks were formed (Wood et al. 2006).

No more than 100 Ma after accretion, the Earth had already reached its present size. Temperatures in the interior were high enough to partially melt the mixed solids of silicate and iron. The differentiation process released a considerable amount of energy. The melting of hot dry mantle at ocean ridges and plumes resulted in a crust about 30 km thick, overlaid in places by extensive volcanic plateaus. The continental crust, in contrast, was relatively thin and mostly submarine.

Figure 2.13 shows a scheme of the evolution of temperature, water and carbon dioxide during the Hadean era. Substantial greenhouse and tidal heating were able to maintain a magma ocean for a few million years.

The Moon would have been formed by a grazing collision with a Mars-mass object during the late stage of Earth accretion (Hartmann and Davis 1975; Stevenson 1987; Canup and Asphaug 2001 and Canup 2004), but with our planet already differentiated into mantle and core (Toboul et al. 2007). The collision formed a dense atmosphere of gaseous silicates that rapidly cooled down and precipitated. The residual atmosphere was constituted by water vapour and carbon dioxide.

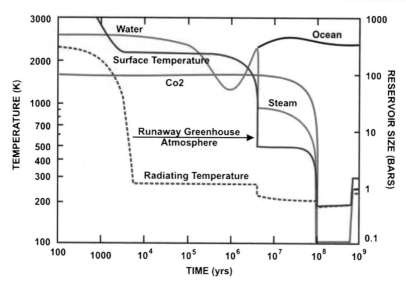

Fig. 2.13 Evolution of temperature, water and carbon dioxide during the Hadean. Source: Zahnle et al. (2007). Copyright: Springer

The partial pressure of CO_2 was between 40 and 200 bars[13] and the temperature around 1,300 K. In a few million years the atmosphere cooled down, the water vapour condensed and precipitated, and at 4.4 Ga BP had already formed a stable ocean (see Wilde et al. 2001 and Pinti 2005).

The building blocks of the Earth were likely dry because temperatures were too high for water to condense or form from hydrated materials. The best alternative source are the bodies of the asteroid belt. Carbonaceous chondrite meteorites originating from C-type asteroids in the outer asteroid belt have Deuterium/Hydrogen ratios similar to our oceans. Models have reproduced the current volatiles inventory via these minor bodies formed in the outer Solar System and gravitationally shifted inward (Morbidelli et al. 2000; Raymond et al. 2004). Comets are not able to fit the D/H ratio of the oceans.[14]

2.2.1.1 The Moon and the Earth Rotation

We will discuss later (Chap. 9) the importance of the Moon for the climate stability of the Earth, and therefore for its habitability. For the moment it is important to understand the processes that originated a satellite so large compared to the mass of the host planet.

---

[13] Calculated according to the current amount in carbonates, continental sediments and the biosphere.

[14] This parameter has been measured only in three comets: Halley, Hale-Bopp and Hyakutake. The values are factor two larger than the Earth water.

2.2 The Precambrian Era (4,500–4,550 Ma BP)

The principle of conservation of angular momentum requires that changes in the Earth's rotation must be produced by the following: (1) torques acting on the solid Earth and (2) changes in the mass distribution within the solid Earth. Here we are interested in the first process produced by the Moon.

The tides are the most important manifestation of the interaction between the Moon and the Earth, producing two ocean bulges, which are instantaneously directly beneath the moon and directly opposite the moon. The Earth's rotation carries the Earth's tidal bulges slightly ahead of the point directly beneath the Moon. This means that the force between the Earth and the Moon is not exactly along the line between their centres, producing a torque on the Earth and an accelerating force on the Moon. This causes a net transfer of rotational energy from the Earth to the Moon, slowing down the Earth's rotation by about 1.5 ms/century and raising the Moon into a higher orbit by about 3.8 cm per year.

Conservation of the angular momentum of the Earth–Moon system implies

$$L + S = L_{tot} = \text{constant},$$

where S is the spin angular momentum of the Earth and L the orbital angular momentum of the Moon. At present, $L_0/S_0 = 4.83$, but it is important to know the previous history (Arbab 2003).

The Earth's rotational energy is given by

$$E_R = \frac{1}{2}C\omega^2$$

and the Moon's orbital energy by

$$E_M = -\frac{GM_M M_E}{2r},$$

where ω, r, M_M and M_E are the Earth's angular speed, Earth–Moon distance, Moon mass and Earth mass, respectively. The moment of inertia, C, is given by

$$C = \frac{L - L_{tot}}{\omega}.$$

The rate of total tidal energy, $E = E_R + E_M$, dissipated in the system is

$$P = -\frac{dE}{dt}.$$

The Moon retreats at a rate proportional to its semi-major axis to the $-11/2$ power (as it gets further away, it retreats more slowly). The present rate of lunar recession of 3.82 ± 0.07 cm per year was obtained by lunar laser ranging (Dickey et al. 1994). Paleontological evidence comes from tidally laminated sediments. Williams (1990) reported that 650 Ma ago, the lunar rate of retreat was 1.95 ± 0.29 cm per year, and that over the period 2.5–0.65 Ga ago, the recession rate was smaller (1.27 cm per year). Williams (1997) reanalysed the same data set later, showing a mean recession rate of 2.16 cm per year in the period between now and 650 million years ago.

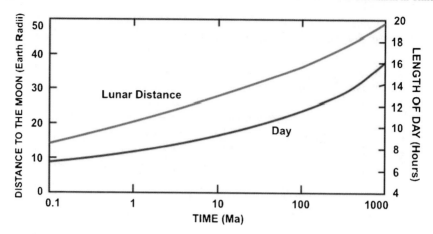

Fig. 2.14 Variation of the distance to the Moon and the length of the Earth's day during the first eon (1 Ga) after the Moon-forming impact. Source: Zahnle et al. (2007) Fig. 9. Copyright: Springer

Sonett et al. (1996) studied sediments of different ages, concluding that the length of the terrestrial day 900 Ma ago was similar to 18 h. Further studies by Williams (2000) in Australian sediments give a mean rate of lunar recession of 1.24 ± 0.71 cm per year during most of the Proterozoic (2,450–2,620 Ma), suggesting that a close approach of the Moon did not occur during an earlier time. Figure 2.14 shows the variation of the distance to the Moon and the length of the Earth's day based on Touma and Wisdom (1998).

The Earth–Moon tidal effects are mutual. The Moon is the smaller object, and the effect of tidal friction has been to change the lunar rotation until its rotational period is equal to its orbital period about the Earth, the present situation. Ultimately, the Moon's action on Earth will produce a similar consequence. When Earth and Moon achieve full synchronism with each rotation being equal to their mutual orbital period, it is estimated that these periods will equal 55 present days and the Earth–Moon distance will be around 610,000 km. For monographs on this topic see Canup et al. (2000) and Lambeck (2005).

2.2.1.2 Late Heavy Bombardment

After planetary accretion, there was a period of quietness in the impact rate on the Earth. Valley et al. (2002) suggested that, during this period (4.4-4.0 Ga), the surface conditions led to liquid oceans and possibly an early emergence of life, which would have been truncated by a catastrophic event. At that time, bodies from the asteroid belt were ejected to the inner solar system by a gravitational perturbation of the outer planets (Strom et al. 2005) (see Chap. 8 for details on the stability of the Solar system).

Radiometric age dating of impact craters on the Moon indicated heavy bombardment with objects larger than 100 km around 3.85 Ga ago, and lasting from

2.2 The Precambrian Era (4,500–4,550 Ma BP)

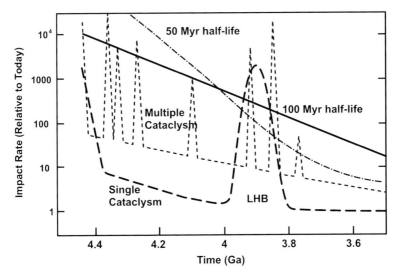

Fig. 2.15 Different concepts of the Late Heavy Bombardment based on Moon data. Continuous process: 50 Ma half-life (Wilhelms 1987); 100 Ma half-life (Neukum et al. 2001); Single cataclysm (Ryder 2002, 2003) and Multiple cataclysm (Tera et al. 1974). Source: Zahnle et al. (2007) Fig. 9. Copyright: Springer

20 to 150 Ma. Computer estimates suggest that, during this period, called the late heavy bombardment (LHB), Earth suffered impacts producing over 22,000 impact craters larger than 20 km, about 40 impact basins larger than 1,000 km, and several continent-sized, 5,000-km basins.[15]

Zahnle et al. (2007) summarizes the different histories proposed for the event. Wilhelms (1987) and Neukum et al. (2001) proposed a declining impact flux that extrapolates back in time. In contrast, Tera et al. (1974) and Ryder (2002, 2003) favoured different types of cataclysms (Fig. 2.15).

The cessation of the LHB coincides well with the first isotopic evidences for life on the Earth ∼3.8 Ga ago (Mojzsis et al. 1996). They found that elemental carbon trapped in old rocks of western Greenland have isotopic compositions that span much of the range found in living organisms.

2.2.1.3 The Early Crust and Mantle

Heat flow during the Archaean era was three times larger than today, leading to stronger convection in the mantle and associated higher rates of tectonism. Figure 2.16 shows the variation of the mean heat flow during the first billion years of Earth's history.

---

[15] Because Earth's effective cross section is 20 times bigger than the Moon, our planet must have suffered many more impacts than those recorded on the lunar surface.

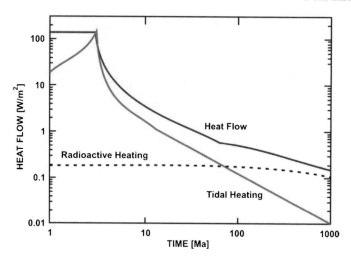

Fig. 2.16 Variation of the mean heat flow from internal energy sources during the first eon of the history of the Earth. Adapted from Zahnle et al. (2007) Fig. 8. Copyright: Springer

A hotter mantle is less viscous, convects faster and releases more heat. This implies that the oceanic crust would be thicker and more buoyant, and therefore difficult to be subducted. For this reason, the appearance of subduction marked the beginning of plate tectonics. There is a strong controversy on the timing, estimates going from the Hadean (Hopkins et al. 2008), late Archaean (Kusky et al. 2001, Sankaran 2006) to the Neoproterozoic (Stern 2005). In any case, the Archaean plates were most likely different from the present ones (De Wit and Hart 1993). An interesting view is that which considers the possibility of intermittency, with transitions from plate tectonics to stagnant-lid tectonics, a less efficient way to remove heat from the mantle. A first consequence would be the existence of jumps in the heat flow (see Sleep 2000; Silver and Behn, 2008).

The oldest rocks on Earth are located in the Isua belt in Greenland (>3.7 Ga), and in the belts of Barberton (South Africa) and Pilbara (Australia), dated in the range of 3.5–3.2 Ga. Zircon grains are the only remnant of the ancient crust.[16]

Cratons are old and stable parts of the continental crust, with deep roots that extend down into the mantle up to a depth of 200 km. The Cratonal lithosphere was formed by processes similar to modern tectonics including subduction (Sleep 2005). The Congo, Kaapvaal, Zimbabwe, Tanzania and West Africa cratons were built between 3.6 and 2.0 Ga ago, and make up for most of the current African continent.

---

[16] Zircon is a mineral belonging to the group of nesosilicates. Its chemical name is zirconium silicate and its corresponding chemical formula is $ZrSiO_4$.

2.2.1.4 The Young and Faint Sun

The Earth is not an isolated body and all the processes in its atmosphere are strongly dependent on the amount of solar energy received. Therefore, any change in the solar output may have deep consequences in our planet. The variability of the Sun is linked to the dissipation of the available energies (De Jager 1972). Different sources are characterized by distinct time-scales.

Solar energy is provided by thermonuclear reactions, mostly by the conversion of hydrogen into helium $(4^1H \rightarrow {}^4He)$. The mass of the resulting He nucleus is smaller than its constituents and 0.7% of the total mass is converted to energy (~ 26.5 MeV)[17].

The thermonuclear fusion process leads to a gradual increase in the molecular weight μ in the core. At present, about 50% of its central H content has been already transformed into He. The thermal pressure is given by $P \sim \rho RT/\mu$, where R is the gas constant. Thus, a lower mean molecular weight during the early phases of solar evolution implies a lower temperature (or density) in order to balance the gravitational force. Therefore, the theory of stellar evolution clearly predicts an increase in solar luminosity during its time on the main sequence (Gough 1981). As a consequence, the Sun was 30% dimmer 4.0 Ga ago than at present. This effect can be quantified, as seen below.

Stellar luminosity depends on mass, M, the mean molecular weight, μ, and the radius, R, according to the expression

$$L \propto M^{5.5} R^{-0.5} \mu^{7.5}. \tag{2.1}$$

The process of energy generation in the Sun during its stay on the main sequence is the nuclear transformation of hydrogen into helium. This produces an increase in the mean molecular weight and, according to (2.1), an increase in solar luminosity:

$$L(t) = [1 + 0.4(1 - t/t_0)]^{-1}.$$

The existence of an early phase of significant mass loss has also been suggested. Wood et al. (2005) obtained high-resolution Lyα spectra and found that the mass loss per unit surface is correlated with the level of magnetic activity. They derived a time dependence of the mass loss of the form

$$dM/dt = t^{-2.33 \pm 0.55},$$

which suggests that the wind of the active young Sun may have been around 1,000 times stronger than that at present. Sackmann and Boothroyd (2003) also proposed the existence of a more massive and larger Early Sun, with the planets being closer with respect to their present positions. In any case, the decay is not necessarily gradual, and a phase of rapid mass loss would lead to the present situation within a period of 1 Ga.

---

[17] The Sun transforms 4 million tons per second into energy and has lost about 1% of its original mass during its 4.5 billion years of evolution.

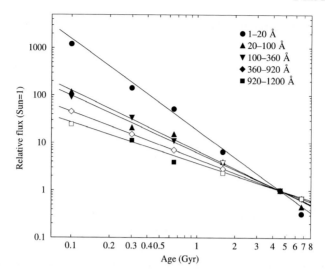

Fig. 2.17 Solar-normalized fluxes vs. age for different stages of the evolution of solar-type stars. Plotted here are the measurements for different wavelength intervals (*filled symbols*) and the corresponding fits using power-law relationships. Represented with empty symbols are the inferred fluxes for those intervals with no available observations. Adapted from Ribas et al. (2005). Reproduced by permission of the American Astronomical Society

The early Sun was not only fainter with respect to present times. Observations of younger Sun-like stars indicate clearly that our Sun was also rapidly rotating. This latter fact implies that the young Sun was a stronger emitter of high-energy radiation (Skumanich 1972; Messina and Guinan 2002). On the basis of the measurements of solar-like stars of different ages, Ribas et al. (2005) have reconstructed the high-energy irradiance in different spectral ranges along the Sun's evolution (Fig. 2.17).

We have already indicated that geological evidence indicates that oceans of liquid water have existed on the Earth at least since 4.4 Ga ago (Wilde et al. 2001), approximately coinciding with the end of the late bombardment period. Moreover, temperatures at the early times were probably much higher than that in our time.

Physical conditions of the atmosphere have changed along the history of the planet and have most likely been an important factor in the mass extinctions. At the geological scale, the climate is controlled by the balance between the variations in the solar output and the changes in the abundance of greenhouse gases in the atmosphere (Fig. 2.18).

Climate modelling indicates that the mean temperature of the Early Earth must have been below zero, a status called 'snowball Earth', from which it would have been impossible to escape (Newman and Rood 1977). This apparent contradiction, called *The faint Sun paradox*, has been explained classically in terms of an enhanced greenhouse effect produced by larger abundances of CO_2 in the primitive terrestrial atmosphere (Dauphas et al. 2007). However, the CO_2 levels were not likely to have been high enough to be the only factor involved in keeping the oceans from freezing (Kasting and Toon 1989; Rye et al. 1995).

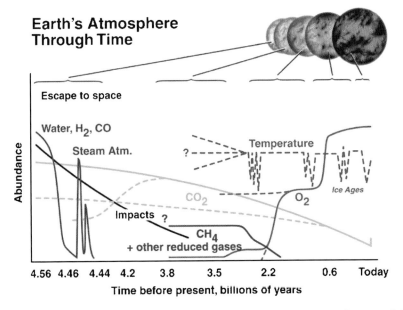

Fig. 2.18 Evolution of the chemical composition of the atmosphere. Adapted from an original drawing of D.D. Marais and K.J. Zahnle. Source: NASA Science News 2002

Güdel (2007) provides an excellent review on this topic. In Chap. 4 we study in more detail the consequences of the magnetic activity of the young Sun on the early atmosphere of our planet.

2.2.2 The Archaean and Proterozoic Times

Two main processes can be associated with the history of our planet during this period. One is related with the geological activity and the recycling of some relevant greenhouse gases, and the second with the effects produced by the appearance of life on its diverse forms. Let us start with the latter event, essential for describing our planet.

2.2.2.1 The Origin and Development of Life

It was first proposed in 1929 by A.I. Oparin (1894–1980) and J.B.S. Haldane (1892–1964) that the synthesis of organic compounds of biochemical significance took place in the primitive terrestrial environment. The prebiotic source of material was formed by contributions from endogenous synthesis in a reducing atmosphere (Miller 1953; Johnson et al. 2008) or neutral atmosphere (Cleaves et al. 2008), metal sulfide-mediated synthesis in deep-sea vents (Russell et al. 1988) and exogenous sources such as comets, meteorites and interplanetary dust (Oró et al. 2006). For further information about the processes leading to the origin of life, see Schopf (1983, 2002), Delaye and Lazcano (2005) and Chap. 5.

The main features of life's origins and development can be summarized in the following facts (see Schulze-Markuch and Irwin 2004; Ward 2005):

– Life arose relatively quickly
– Life tends to stay small and simple
– Evolution is accelerated by environmental changes
– Once life evolved on Earth, it proved to be extraordinarily resilient.
– Complexity inevitably increases but as the exception not the rule.
– Living things are placed in groups on the basis of similarities and differences at distinct levels.

The cell is the basic structure of life. According to this criterion, life forms were divided by Woese et al. (1990) into three main groups: Bacteria, Archaea and Eucaryotes (Fig. 2.19). All these life forms have a common ancestor, share the same genetic code and biochemistry and developed through the process of evolution. Archaea and bacteria differ mainly in aspects related with the biochemistry and the external parts of the cell. Many archaea are extremophiles. They can survive and thrive even at relatively high temperatures, often above 100°C, as found in geysers and black smokers. Others are found in very cold habitats or in highly saline, acidic or alkaline water. However, other archaea are mesophiles and have been found in environments like marshland, sewage, sea water and soil. Many methanogenic archaea are found in the digestive tracts of animals such as ruminants, termites and humans. Archaea are usually harmless to other organisms and none are known to cause disease. Archaea are usually placed into three groups based on their preferred habitat. These are the halophiles, methanogens and thermophiles.

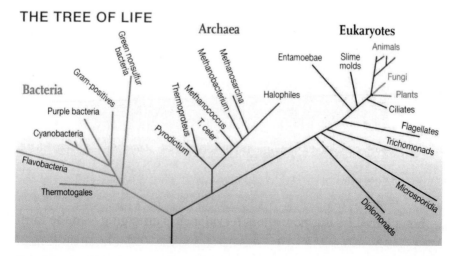

Fig. 2.19 A phylogenetic tree of terrestrial life in three domains proposed by Carl Woese in 1990

Unicellular life is composed of Bacteria and Archaea. About 3.5 Ga ago, Earth temperatures were probably in the range of 55–85°C (Knauth and Lowe 2003). This is the interval where the heat-loving organisms proliferate.

Eukaryotes are organisms in which the genetic material is organized into a membrane-bound nucleus. They appeared in the period 2.1–1.6 Ga ago (Knoll 1992) and were able to sexually reproduce. L. Margulis discovered the mechanism of endosymbiosis within eucaryote cells. Bacteria transformed in organelles[18] within the cell walls.

Unicellular life (microbes) has been the only life form around for close to seven eighths of Earth's history. They still constitute most of the global living mass on Earth today, exerting exclusive control over biomass turnover and biological activity in some Earth regions (deep biosphere, extreme deserts, polar areas etc.).

2.2.2.2 The Carbon Dioxide Cycle

The concentration of greenhouse gases in the atmosphere has been a dominant factor in the evolution of the Earth's climate. Apart from water vapour, carbon dioxide is the most abundant of theses gases in the atmosphere. Therefore, a large part of this chapter will be devoted to the study of its evolution over time.

At geological time scales, the concentration of CO_2 in the atmosphere is controlled by a cycle (Fig. 2.20), first studied by Ebelmen (1845) and Urey (1952). For updated descriptions see Walker et al. (1981), Berner et al. (1983) and Berner (2004). The cycle has two main components: weathering and metamorphism.

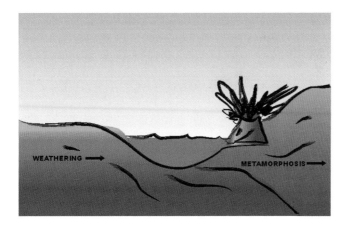

Fig. 2.20 The carbonate-silicate cycle. Adapted from an original drawing of Kasting (1993)

---

[18] The main organelles are Mitochondria, producing energy from oxygen and food, and Chloroplasts, converting sunlight into energy. Plastids provided eucaryotes with the ability to generate their own oxygen.

1. *Weathering and sedimentation*: Weathering refers to the transfer of carbon from the atmosphere to the rock record and the subsequent marine carbon sedimentation.

   First, microbial decomposition in the soils of the continents leads to a buildup of organic acids and CO_2. The carbonic acid weathers the rocks on the Earth's surface, releasing ions of calcium (Ca^{++}) and bicarbonate (HCO_3^-). The reaction is

   $$CO_2 \text{ (gas)} + 3H_2O + CaSiO_3 \longrightarrow Ca^{++} + 2HCO_3^- + H_4SiO_4.$$

   These products are carried out by groundwater to the rivers and finally to the sea. In the oceans they are precipitated, mostly biogenically, as calcium carbonate

   $$Ca^{++} + 2HCO_3^- + H_4SiO_4 \longrightarrow CaCO_3 + CO_2 + H_2O$$

   while the silicic acid precipitates as biogenic silica (quartz) following

   $$H_4SiO_4 \longrightarrow SiO_2 + H_2O.$$

   The whole process can be summarized by the reaction

   $$CO_2 + CaSiO_3 \longrightarrow CaCO_3 + SiO_2$$

   and similarly for magnesium silicates

   $$CO_2 + MgSiO_3 \longrightarrow MgCO_3 + SiO_2.$$

   Other subprocesses must also be considered. For example the weathering reaction of calcium carbonate, just the opposite of the precipitation of calcium carbonate into the oceans, which results in no net change of carbon dioxide

   $$CaCO_3 + CO_2 + H_2O \longrightarrow Ca^{++} + 2HCO_3^-$$

   Berner et al. (1983) adopt the following dependence of the weathering, W, on mean temperature

   $$f_W = \frac{W(T)}{W_0} = \left[1 + 0.087(T - T_0) + 1.86 \times 10^{-3}(T - T_0)^2\right],$$

   where the subindex 0 indicates present values: $W_0 = 3.3 \times 10^{14}$ g year^{-1} (Holland 1978) and $T_0 = 288$ K, the average temperature of the present Earth. An alternative expression is given by Walker et al. (1981) and Caldeira and Kasting (1992)

   $$f_W = \frac{W}{W_0} = \left[\frac{a_{H+}}{a_{H+,0}}\right]^{0.5} \exp\left(\frac{T_S - T_{S,0}}{13.7}\right),$$

where a_{H+} is the activity of the ion H+, which depends on the concentration of CO_2 in the soil and the temperature.

2. *Metamorphosis*: The calcium carbonate, $CaCO_3$, is eventually subducted down into the Earth (via plate tectonics), where high temperatures and pressures convert it back to carbon dioxide. The process is accomplished by metamorphic decarbonation reactions and by melting with the subsequent release of CO_2 by volcanic activity:

$$CaCO_3 + SiO_2 \longrightarrow CO_2 + CaSiO_3.$$

In the words of Stern (2002): *One can speculate that if subduction zones did not exist to produce continental crust, the large exposed surfaces of rock known as continents would not exist, the Earth's solid surface would be flooded, and terrestrial life, including humans, would not have evolved.*

The carbon cycle has been studied mainly in relation to its role as a thermostat of the global climate (Walker et al. 1981) and the habitability of planetary systems (Franck et al. 2000, 2002). The different phases are described in more detail below.

2.2.2.3 Sea-Floor Spreading and Continental Growth

The dynamic equilibrium of CO_2 fluxes is controlled by the balance between the spreading of the ocean floors, S, the area occupied by the continents, A_c, and the weathering process W (Kasting 1984)

$$\frac{W}{W_0} \frac{A_C}{A_{C,0}} = \frac{S}{S_0},$$

where S is given by

$$S(t) = \frac{q_m(t)^2 \pi \kappa A_{oc}(t)}{[2k(T_m(t) - T_{S,0}]^2}$$

q_m being the heat flux from the mantle, κ the thermal conductivity, k the thermal diffusivity, A_{oc} the area occupied by the ocean floors and T_m and T_s the temperatures of the mantle and the surface, respectively. Obviously at any time the planetary area is $A_E = A_{oc} + A_C$.

Franck et al. (1998) have modelled the evolution of q_m. Its value at the present time is 70 mW m^{-2}, assuming that 20% of the observed surface flow (87 mW m^{-2}) is related to radiogenic heat produced in the continental crust. In the Archaean era, the values of q_m, T_m and S were larger than today's values (by factors 2–3 and 4–9, respectively). Franck and Bounama (1997) discussed the thermal and outgassing history of the Earth for different theoretical models.

For continental growth, two broad types of models are generally assumed: (1) Constant growth with $A_c \propto t$ and (2) delayed growth models, with $A_c = 0$ for $t \leq t_{cr}$ and $A_c \propto t$ for $t > t_{cr}$. These models are summarized in Fig. 2.21. Note that according to Contie (1990), crustal growth had two major pulses in the Archaean and Proterozoic eras.

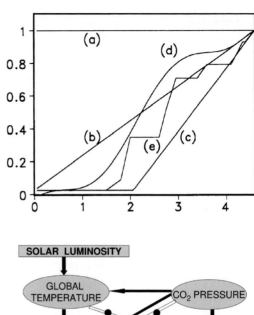

Fig. 2.21 Models of continental growth relative to the present: (**a**) Constant continental surface, (**b**) Linear growth, (**c**) delayed linear growth, (**d**) approximated growth function and (**e**) Episodic growth (Condie 1990). The time is given in Ga starting with at the Earth's origin. Adapted from Fig. 2.5 of Bounama (2007)

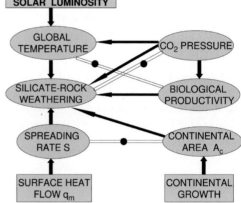

Fig. 2.22 Factors leading to the stability of conditions compatible with life and their mutual influences. *Arrows* indicate direct relationship and arrows with a bullet an inverse one

The balance between the two main processes, outgassing and weathering, has been expressed by Carver and Vardavas (1994) as

$$G(t) = P_{CO2}^n(t) A(t) W(t) \exp[(T - 288)/T_c],$$

where T_c is the current temperature (15°C) and n a fitting parameter with values between 0 and 1.

All the above mentioned factors influence the Earth's temperature. Figure 2.22 summarizes them with the main, positive and negative, links.

However, life was not a passive spectator of the CO_2 cycle. The presence of life has clearly influenced the weathering phase of the CO_2 cycle and therefore the environment. Schwartzman and Volk (1989) calculated that if today's weathering is 10, 100 or 1,000 times the abiotic weathering rate, then an abiotic Earth would be, respectively, approximately 15, 30 or 45°C warmer than today, suggesting that without biota the Earth today would be uninhabitable.

2.2 The Precambrian Era (4,500–4,550 Ma BP)

The biological productivity of photosynthetic organisms, Π, is dependent on the temperature and the partial pressure of CO_2 in the atmosphere. Therefore,

$$\frac{\Pi}{\Pi_{max}} = \Pi(T_S) \cdot \Pi(P_{CO2})$$

$$\Pi(T) = 1 - \left[\frac{T_S - 50}{50}\right]^2$$

$$\Pi(CO_2) = \frac{P_{CO2} - P_{min}}{P_{1/2} + (P_{CO2} - P_{min})},$$

where $P_{min} = 10^{-5}$ bar $= 10$ p.p.m is the minimum value of CO_2 pressure required for photosynthesis, and $P_{1/2}$ is the pressure at which Π reaches half the maximum productivity (see Volk 1987).

Therefore, we have

$$f_W(\text{biotic}) = \beta f_W(\text{abiotic}),$$

where

$$\beta = 1 - \sum_{i=1}^{n}\left(1 - \frac{1}{\beta_i}(1 - \Pi_i P_{i,0})\right),$$

where β is 1 for unicellular organisms and 3.6 for multicellular beings.

The three domains of life reached the maximum productivity at different temperature intervals (Table 2.3). Cooling of the Earth made the emergence of new types possible. Several crucial steps marked this evolution, namely: (1) the colonization of land surface by eucaryotes, (2) Diversification of large, hard-shelled animal life and (3) development of vascular land plants. All of them were associated with important changes in the environment.

Therefore, life probably evolved conditioned by the ambient climate (Schwartzman 1995; Schwartzman and Middendorf 2000). In other words, life evolved opportunistically on Earth in a simple interactive relationship with its environment, sustained by an external energy source, the Sun. The biosphere has been driven forward to greater complexity by the Earth's internal energy resources (plate tectonics). Without these resources and a hydrosphere, it would not be possible to sustain the necessary mass of carbon in the lower crust and mantle to drive the biosphere (Lindsay and Brasier 2002).

Table 2.3 Constants of biological productivity for the three main domains of life. From Franck et al. (2005)

	Procaryotes	Eucaryotes	Multicellular
T_{min} (C)	2	5	0
T_{max} (C)	100	45	30
Π_{max} (Gt/yr)	20	20	20
P_{min} (10^{-6} bar)	10	10	10
β	1	1	3.6

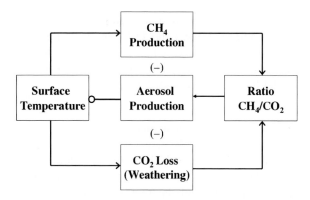

Fig. 2.24 Links between temperature and the abundances of methane and carbon dioxide. Adapted from J. Kasting

The loss of mass-independent fractination[20] in sulfur isotopes indicate a collapse of atmospheric methane, disappearing abruptly ca. 2.45 Ga ago (Zahnle et al. 2006). The methanogens and their role in the Earth's evolution were soon substituted by other organisms, the cyanobacteria, emitting another gas essential for understanding our planet: oxygen.

2.2.2.5 Oxygen, Ozone and Ultraviolet Radiation

In a primordial atmosphere dominated by carbon dioxide and water vapour, free oxygen atoms can be produced by the following reactions:

$$H_2O + \text{radiation } (\lambda < 240\text{nm}) \longrightarrow OH + H$$
$$OH + H \longrightarrow O + H_2O$$
$$CO_2 + \text{radiation } (\lambda < 230\text{nm}) \longrightarrow CO + O$$

The free oxygen will produce molecular oxygen and ozone through Chapman reactions (see Chap. 4). On the basis of UV measurements of T Tauri stars, Canuto et al. (1982) indicated that the O_2 surface mixing ratio was a factor 10,000–1,000,000 times greater than the standard value of 10^{-15}. Canuto et al. (1983) extended their calculations to other atmospheric components such as OH, H, HCO and formaldehyde (H_2CO).

At the beginning of the Proterozoic era, the cyanobacteria – photosynthetic procaryotes also known as blue-green algae – brought about one of the greatest changes

---

[20] It refers to any chemical or physical process that acts to separate isotopes, where the amount of separation does not scale in proportion to the difference in the masses of the isotopes.

2.2 The Precambrian Era (4,500–4,550 Ma BP) 73

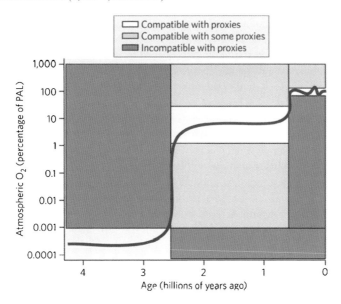

Fig. 2.25 Time evolution of atmospheric oxygen. The *red line* shows the inferred level of atmospheric oxygen bounded by the constraints imposed by the proxy record of atmospheric oxygen variation over Earth's history (Kupp 2008 Fig. 2). Reprinted by permission from Macmillan Publishers Ltd: Nature 451, p. 278, Copyright (2008)

this planet has ever known: a massive increase in the concentration of atmospheric oxygen through the reaction[21]:

$$CO_2 + H_2O \longrightarrow CH_2O + O_2.$$

The evolution of the concentration of oxygen in the atmosphere is not very well known, especially in early times (Kump 2008 and Fig. 2.25). Oxygen photosynthesis arose long before O_2 became abundant in the atmosphere, but oxygen levels in the atmosphere rose appreciably only around 2.5 Ga ago. The signature of mass-independent sulphur-isotope ($^{33}S/^{32}S$) behaviour[22] sets an upper limit for oxygen levels before 2.45 billion years ago (Pavlov and Kasting 2002) and a lower limit after that time. The record of oxidative weathering after 2.45 billion years ago sets a lower limit for oxygen levels at 1% of PAL (Present Atmopsheric Level), whereas an upper limit of 40% of PAL is inferred from the evidence for anoxic oceans during the Proterozoic. The tighter bounds on atmospheric oxygen from 420 million years ago to the present is set by the fairly continuous record of charcoal accumulation (Scott and Glasspool 2006).

---

[21] Here CH_2O is shorthand for more complex forms of organic matter.
[22] To preserve this signature three conditions are needed: very low atmospheric oxygen, sufficient sulphur gas in the atmosphere and substantial concentrations of reducing gases (methane).

The earliest evidence of anaoxygenic photosynthesis is dated around 3.4 Ga (Tice and Lowe 2004). However, prior to ∼2.5 Ga, oxygen did not leave oxidation signatures that are prevalent in the geological record to the present (cf. Canfield 2005), although cyanobacteria were already present since 2.7 Ga (Brocks et al. 2003). This transition is called the 'Great Oxidation Event' by Holland (2002), although it has also received the name of 'Oxygen Catastrophe'.

Kump and Barley (2007) proposed that this transition was favoured by the increase of subaerial volcanoes with respect to the submarine ones, reducing an important oxygen sink. The deep ocean may have become oxic at 1.8 Ga. Instead of a tectonic driver for this transition, Schwartzman et al. (2008) favoured a biological mechanism based on a methane atmosphere producing the emergence of oxygenic cyanobacteria at about 2.8–3.0 Ga. At that time the mean temperature was ∼60°C, the most adequate for cyanobacteria.

An apparent paradox arose because, though the terrestrial crust contains 1.1×10^{21} moles[23] of reduced carbon, the total amount of organic carbon to account for all the atmospheric oxygen is only 0.038×10^{21} moles. In other words, the atmosphere contains too little oxygen. Clearly, the missing oxygen is trapped in oxidized reservoirs such as sulphates and ferric iron.

The variations in oxygen content can be explained in terms of a balance between the following factors (Claire et al. 2006):

$$\frac{d}{dt}[O_2] = F_{Sources} - F_{Sinks} = (F_B + F_E) - (F_V + F_M + F_W).$$

$[O_2]$ is the total reservoir of molecular oxygen in units of teramoles, F_B the flux of oxygen due to organic carbon burial, F_E the flux of oxygen to the Earth due to hydrogen escape, F_V and F_M represent flux of reducing gases (i.e. H_2, H_2S, CO, CH_4) from volcanic/hydrothermal and metamorphic/geothermal processes, respectively, and F_W the oxygen sink due to oxidative weathering of continental rocks. Let us illustrate this balance with some examples.

The production of methane inevitably means a greater rate of hydrogen escape, which drives the oxidation of the lithosphere, a lowering of oxygen sinks, the rise of oxygen and, as a consequence, global cooling.

Given a certain amount of oxygen in the atmosphere, the formation of an ozone layer is an unavoidable process through the reactions schematized in Fig. 2.26.

The origin of the ozone layer is probably linked to the Great Oxidation (Goldblatt et al. 2006). The transition was caused by UV shielding, decreasing the rate of methane oxidation once oxygen levels were sufficient to form an ozone layer. As little as 0.01–0.1 present atmospheric levels (PAL, normalized to 1) of molecular oxygen may have been sufficient to produce an effective ozone shield (Kasting and Donahue 1980). The amount of ozone required to shield the Earth from biologically lethal UV radiation, at wavelengths from 200 to 300 nm, is believed to have been in existence ∼600 million years ago. At this time, the oxygen level was approximately

---

[23] One mole of an element is equal to 6.02×10^{23} atoms of that substance.

2.2 The Precambrian Era (4,500–4,550 Ma BP)

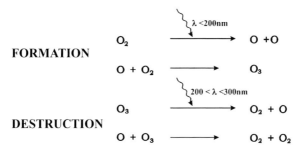

Fig. 2.26 Formation and destruction of ozone, according to the Chapman cycle

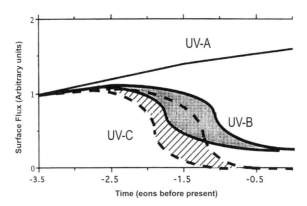

Fig. 2.27 Time variation of the UV radiation reaching the terrestrial surface. The scale is in relative arbitrary units of energy normalized to 1 at 3,500 Ma before present. Courtesy: F. García Pichel, Dept. of Microbiology, Arizona State University

10% of its present atmospheric concentration. Prior to that, life was restricted to the oceans. The presence of ozone enabled multicellular organisms to develop and live on land, playing a significant role in the evolution of life on Earth and allowing life as we presently know it to exist. Rye and Holland (2000) have raised the suggestion that land-dwelling microbes existed at least as early as 2.7 Ga ago. If this is true, our estimate of the abundance of ozone in the stratosphere at that times need to be revised, unless such microorganisms developed a strong defense mechanisms against UV action.

The increased concentration of oxygen has clearly decreased the amount of UV-B and UV-C fluxes arriving at the surface, while the UV-A range has experienced the commented increase of solar luminosity with time.[24] The implications of UV radiation for biological evolution have been studied by Cockell (2000).

García Pichel (1998) considers three main phases in the evolution of UV radiation at the Earth's surface (Fig. 2.27): (1) High environmental fluxes of UV-C

---

[24] The wavelength ranges of these three spectral bands are UV-C (100–280 nm), UV-B (280–315 nm) and UV-A (315–400 nm).

and UV-B restricting protocyanobacteria to refuges, coinciding with the period of heavy bombardment from interplanetary debris; (2) The appearance of true oxygen cyanobacteria, producing a compound called scytonemin, which screens out the UV radiation but allows through the visible radiation, essential for the photosynthesis; and (3) Gradual oxygenation and the formation of the ozone shield.

A detailed pattern and timing of the rise in complex multicellular life in response to the oxygen increases has not been established. Hedges et al. (2004) suggest that mitochondria and organisms with more than 2–3 cell types appeared soon after the initial increase in oxygen levels at 2,300 Ma. The addition of plastids[25] at 1,500 Ma, allowing eucaryotes to produce oxygen, preceded the major rise in complexity.

2.2.2.6 The Snowball Earth

It has been hypothesized that the Earth was subjected to two main glaciation periods at 2.7 and 0.7 Ga where the ice probably reached the equatorial regions (Harland, 1964). The global mean temperature would have been about −50°C because most of the Sun's radiation would have been reflected back to space by the icy surface. The average equatorial temperature would have been about −20°C, roughly similar to present Antarctica. Without the moderating effect of the oceans, temperature fluctuations associated with the day–night and seasonal cycles would have been greatly enhanced. The glacial deposits at low paleolatitudes have been explained in terms of a global ice cover, 'Snowball' (Hoffman et al. 1998), or ice-free tropical oceans, 'Slushball' (Hyde et al. 2000).

The existence of these events was mainly supported by the occurrence of negative carbon isotopic excursions in glacial marine deposits with thick carbonate (limestone) (see Shields and Veizer 2002 and Fig. 2.28). The carbon isotope ratio, $\delta^{13}C$, is expressed by the following ratio

$$\delta^{13}C = \left[\frac{^{13}C/^{12}C(\text{sample})}{^{13}C/^{12}C(\text{standard})} - 1 \right] \times 1000.$$

Life used the lighter ^{12}C isotopes more than the abiotic materials. Extinctions should be reflected in larger values of $\delta^{13}C$. The biological productivity of the oceans virtually ceased during the glacial periods. Together with anoxic conditions beneath the ice, this probably annihilated many kinds of eucaryotic life.

The main snowball event[26] was characterized by three broad intervals of widespread glaciation: the Sturtian glaciation, which occurred around 723 Ma, the

---

[25] Pastids are a group of organelles that play central roles in plant metabolism via photosynthesis, lipid and aminoacid synthesis (Wise and Hoober 2006).

[26] The first event probably took place around 2.5 Ga ago and it is based on glacial deposits in the Gowganda Formation in Canada and the Makganyene Formation in South Africa (Hilburn et al. 2005; Kopp et al. 2005).

2.2 The Precambrian Era (4,500–4,550 Ma BP)

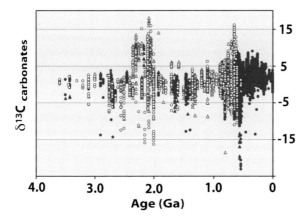

Fig. 2.28 Published $\delta^{13}C$ values for marine carbonates (Shields and Veizer 2002). Copyright: American Geophysical Union

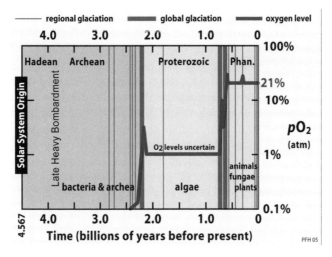

Fig. 2.29 Rise in atmospheric oxygen over geological time and its relationship to the snowball events (*blue bars*) and biological innovations. Courtesy: Paul F. Hoffman

Marinoan (Varanger) glaciation (659–637 Ma) and the Gaskiers glaciation occurring around 680 Ma ago. At that time, large portions of the continental land masses probably were within middle to low latitudes, a situation that has not been encountered at any subsequent time in the Earth's history (Kirschvink 1992). It was also a period of continental dispersal, involving the breakup of the supercontinent Rodinia and the aggregation of megacontinent Gondwana (Hoffman and Schrag, 2002).

The standard interpretation is based on reduced solar luminosity at that time (6% lower) combined with a burst in the oxygen production, leading to a decrease in the concentration of methane and carbon dioxide, two powerful greenhouse gases (Fig. 2.29) (Carver and Vardavas 1994).

Budyko (1969) shows the existence of the different types of stability of the climate system with respect to the ice coverage. When radiative forcing declined, the ice reaches $\sim 30°$ latitude, after which ice-albedo feedback is self-sustaining and ice lines move rapidly ($<10^3$ years) to the equator (Snowball Earth). After millions of years, dependent on the magnitude of the CO_2 hysteresis loop, normal volcanic outgassing combined with reduced silicate weathering caused CO_2 to reach the critical level for deglaciation. Meltdown occurs rapidly ($<10^4$), driven by reverse ice-albedo and other feedbacks, resulting in an ice-free state with greatly elevated CO_2, taking 10^5–10^7 years for this excess to be eliminated through silicate weathering of the glaciated landscape.

It is a matter of fact that photosynthetic life survived throughout the event, excluding models where thick ice could have prevented sunlight from reaching the underlying ocean. Sedimentary evidence provided by Allen and Etienne (2008) indicates that although ice was probably formed at low-latitudes, some of the Earth's oceans remained ice-free and permitted free exchange with the atmosphere. The planet escaped from a global glaciation probably through the link between the physical system and the carbon cycle. Peltier et al. (2007) suggested that when the atmospheric oxygen at surface temperature was drawn into the ocean, it could have remineralized a vast reservoir of dissolved organic carbon, which in turn caused atmospheric CO_2 levels to increase. As an additional mechanism, Kennedy et al. (2008) suggest that equatorial methane clathrates[27] were destabilized, providing additional warming through methane emission to the atmosphere.[28]

A completely different explanation of the event was proposed by Kirschvink et al. (1997) and Williams et al. (1998), suggesting that continental land masses moved far from the equator. The entire lithosphere reacted to bring them back to the equator much faster than the ordinary tectonic process, with the side effect that the continents have realigned with respect to the magnetic north pole. This hypothesis has been criticized by Torsvik and Rehnström (2001) and Levrard and Laskar (2003).

Stern (2005) proposed that the Neoproterozoic era marked the start of modern plate tectonics, resulting in a major increase in explosive volcanism and a cooling of the Earth's surface. The event was probably accompanied by a polar wander.

2.3 The Phanerozoic Era

The Phanerozoic era covers the Earth's history from 600 Ma to the present. Its start was characterized by a rapid flowering of multicellular life forms, the Cambrian Explosion. At that time the Earth's surface presented the distribution of continental masses shown in Fig. 2.30.

---

[27] Also called methane hydrate, this is a solid form of water that contains methane within its crystal structure. Its current locations are the continental margins and the permafrost of Siberia and Antarctica.

[28] Allen and Etienne (2008) quoted Louis Agassiz (1807–1873) saying: *the Earth may have avoided death enveloping all nature in a shroud.*

2.3 The Phanerozoic Era

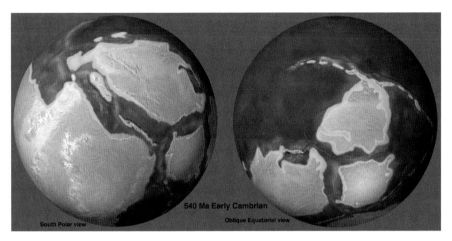

Fig. 2.30 Paleogeographic view of the Earth at the Early Cambrian era (540 Ma ago). Courtesy: Ron Blakey, Department of Geology, Northern Arizona University

The advent of sexual reproduction led to a rapid increase of eucaryotic microalgae between 1,100 and 900 Ma ago. From then to the close of the Precambrian era, microalgal activity declined as atmospheric CO_2 decreased and the climate became colder. Multicellular animals arose during this period, possibly 800–700 Ma ago, and the coelomic kinds that burrow through sediments were efficient producers of organic carbon. Burial of this carbon led to an increase in the oxygen content of the atmosphere (32 g of O_2 for every 12 g of carbon buried), and the subsequent increase of oxygen together with the decrease of carbon dioxide and temperature caused extinction of the microalgae.

The first evidence of complex multicellular animals is given by the so-called Ediacaran biota (McMenamin 1998), some 600 Ma ago, marking the start of the so-called Cambrian explosion, when all the major invertebrate phyla made their appearance. The first macroscopic land plants date back to the Devonian era (400 Ma ago).

Scott et al. (2008) have proposed that molybdenum depletion in the oceans, together with a similar oxygen deficit, may have produced the delay in the development of complex life. This element is used by some bacteria to convert the atmospheric nitrogen in a useful form for living things, a process known as nitrogen fixation.

Watson (2008) supported the hypothesis that complex life is separated from procaryotes by several unlikely steps. Probably this is not in contradiction with other theories that suggest that the Cambrian explosion was triggered by environmental changes (Marshall 2006). This leads us to consider this link in more detail.

The previous mentioned biotic influence of life on the cooling of the planet was amplified in this period by two processes: (1) the diversification of land plants led to increasing chemical weathering of rocks, and therefore an increasing flux of carbon from the atmosphere to rocks, and of nutrients from the continents to the oceans therefore to thus decreasing CO_2 levels, and (2) the presence of organisms, such as foraminifera, that increased the transportation of carbonates to the oceanic sediments.

2.3.1 The Drift, Breakup and Assembly of the Continents

The movement of plates has caused the formation and break-up of continents over time, including occasionally the formation of supercontinents (see Table 2.4). This process has a cycle of 250–500 Ma. See Rogers and Santosh (2004) and Nield (2007) for monographs on this topic.

In the mid 1960s, J. Tuzo Wilson (1908–1993) showed that continents show a cyclic history of rifting – drifting and collision, followed by rifting again, taking about 500 million years to complete a period. He described the process as a periodic opening and closing of oceanic basins. Plates divert apart and new ocean basins are born, followed by motion reversal, convergence back together and plate collision. Sea level is low when the continents are together and high when they are apart.

The last supercontinent, Pangaea, was composed by two subunits: Laurasia and Gondwana (Fig. 2.31). With the final breakup of Laurasia (60 Ma ago), the continents assumed their familiar configuration of our days.

Table 2.4 List of supercontinents

Name	Period of existence (Ma)
Ur	3,000
Kenorland	2,700
Columbia (Nuna)	1,800–1,500
Rodinia	1,100–750
Pannotia	650–540
Pangaea	300–180
Pangaea Ultima	250–400 Ma from now

The time in brackets is given in Ma

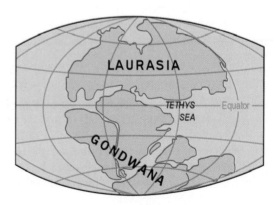

Fig. 2.31 The continents Laurasia-Gondwana at the Triassic, 200 million years ago

Vitousek et al. (1996) modelled the number of mammalian species that would be expected if all the continents were reunited in our days. They concluded that only half of the current species would be present. Plate tectonics clearly promotes biotic diversity.

Silver and Behn (2008) remark that at present most of the subduction zones are in the Pacific basin. At 350 Ma from now, this structure will disappear following the collision of Western America with Eurasia. As a consequence, plate tectonics may temporarily cease. As we have mentioned previously, such intermittency could also have happened in the past. Carbon burial and concomitant oxygen enrichment of the atmosphere could well have been associated with the supercontinent cycle (Lindsay and Brasier 2002).

2.3.2 Supereruptions and Hot Spots

Processes in the Earth's interior may release enough heat to generate plumes of material. As these plumes rise in the uppermost mantle, they spread beneath the lithosphere and begin to melt large plumes of basalt, which erupt in the surface. The mantle plume concept was first proposed by Morgan (1971) and was based on Wilson's (1963) ideas that stationary hot-spots in the shallow mantle underlay island/seamount chains in the deep ocean. Such events may pump large amounts of CO_2 into the atmosphere. According to Condie (1998, 2002), the most important of these events, the superplumes, took place at 2,700, 1,900 and 1,200 million years ago.

Increased production rates of juvenile crust correlate with formation of supercontinents and with superplume events. There may be two types of superplume events: catastrophic events, which are short-lived (<100 Ma), and shielding events, which are long-lived (200 Ma). Catastrophic events may be triggered by slab avalanches in the mantle and may be responsible for episodic crustal growth. Superplume events, caused by the shielding of the mantle from subduction by supercontinents, are responsible for relatively small additions of mafic components to the continents and may lead to supercontinent breakup (Condie 2004).

Inside tectonic plates, we also find transitory emergence of material, forming hot spots. Somewhere between 40 and 150 of these events have been described in both oceanic[29] and continental areas. For a list of hot spots during the last 250 million years, see Rampino and Stothers (1998) and Isley and Abbott (1999).

In summary, two convective processes drive heat exchange within the Earth: plate tectonics, which is driven by the sinking of cold plates of lithosphere back into the mantle asthenosphere, and mantle plumes, which carry heat upward in rising columns of hot material, driven by heat exchange across the core–mantle boundary. The sinking of vast sheets of oceanic lithosphere back into the mantle is the primary driving force of plate tectonics, where the sinking of these slabs is balanced

---

[29] The Hawaii and Canary Island archipelagos are probably a consequence of two such hotspots.

by the passive upwelling of asthenosphere along mid-oceanic ridges. In contrast, mantle plumes are narrow columns of material that rise more or less independently of plate motions.

2.3.3 The Connection Temperature-Greenhouse Gases

The climate of the late Precambrian was typically cold with glaciation spreading over much of the Earth. At this time the continents were assembled in a short-living supercontinent called Pannotia.

Figure 2.32 show the variation of climate along the Phanerozoic, with a main periodicity of 140 Ma. It oscillates between hot and cold phases, probably reflecting the supercontinent cycle. During the formation of supercontinents we have a lack of sea floor production and a cooler climate. On the other hand, during the break-up process we have a high level of sea floor spreading and high levels of greenhouse gases leading to a warm climate.

More in detail, potential causes for the climate changes at this time scale are the following: (1) Latitudinal position of continents. One key requirement for the development of large ice sheets is the existence of continental land masses at or near the poles. (2) Opening and closing of ocean basins, altering ocean and atmospheric circulation. When there were large amounts of continental crust near the poles, the records show unusually low sea levels during ice ages, because there were many polar land masses upon which snow and ice could accumulate. However, during times when the land masses clustered around the equator, ice ages had a smaller effect on sea level.

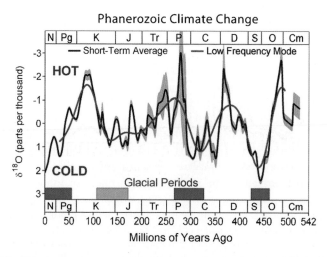

Fig. 2.32 500 million years of climate change based on the data of Veizer (2000). Courtesy: Robert A. Rohde, Global Warming Art

2.3 The Phanerozoic Era

However, the main drivers are the changes in CO_2 content of the atmosphere. Although this has been a matter of debate (Rothman 2002; Veizer et al. 2000; Ryer et al. 2004), recently Came et al. (2007) clearly established that times of minimal CO_2 coincide with the two main glaciations: Permo-Carboniferous 330–270 and the present starting 30 Ma (see also Berner 2004). Also periods of unusual warmth (Mesozoic 250–265 Ma) were accompanied by relatively large concentrations of CO_2 in the atmosphere (see Fig. 2.33). Royer et al. (2007) have estimated that a climate sensitivity greater than 1.5°C to a doubling in CO_2 concentrations has been a robust feature of the Earth's climate system over the past 420 million years. High-resolution CO_2 records of Fletcher et al. (2008) for the Mesozoic and early Cenozoic confirm the leading role of this greenhouse gas in controlling the global climate.

During the Phanerozoic era the oxygen content underwent an important increase during the Carboniferous period reaching values around 35%, accompanied by a decrease in CO_2 with its subsequent biological implications (Graham et al. 1995; Lane 2002).

Figure 2.34 shows the variation of the climate over the last 65 million years. The data are based on oxygen isotope measurements in benthic foraminifera (Zachos et al. 2001). Specially relevant is the Paleocene-Eocene Thermal Maximum, occurring at 55 Ma, a sudden warming of 6°C in only 20,000 years, associated with an increase in CO_2 atmospheric concentration and rise in sea level. The process was probably enhanced by the degassing of methane clathrates, which accentuated a pre-existing warming trend (Katz et al. 2001) . The increasing accumulation of or-

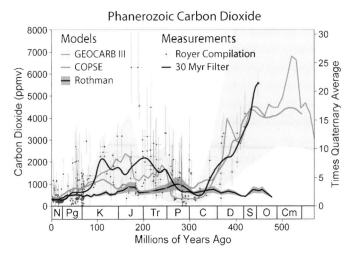

Fig. 2.33 Changes in carbon dioxide concentrations during the Phanerozoic era. Three estimates are based on geochemical modelling: GEOCARB III (Berner and Kothavala 2001), COPSE (Bergmann et al. 2004) and Rothman (2001). These are compared to the carbon dioxide measurement database of Royer et al. (2004) and a 30 Ma filtered average of those data. Error envelopes are shown when they were available. The right hand scale shows the ratio of these measurements to the estimated average for the last several million years (the Quaternary). Courtesy: Robert A. Rohde, Global Warming Art

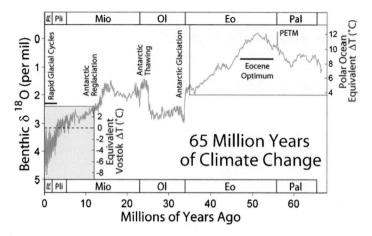

Fig. 2.34 The last 65 million years. Courtesy: Robert A. Rohde, Global Warming Art

ganic material in the deep sea may have cooled this hot climate, starting the current large-scale cooling period (Bains et al. 2000). This cooling is clearly associated with a large decrease in the CO_2 concentration, probably accelerated by the continuous buildup of the Tibetan plateau.

Some geological events during this last phase favoured the evolution towards the emergence of Homo Sapiens through their influence on climate. About 25 Ma ago, continental movements led to the collision of India against the Asian continent, giving rise to the formation of the Himalaya mountains. This originated dry winds blowing to Africa producing a dry season there. It is also worth mentioning that highly encephalized whales, dolphins and porpoises occurred with the drop of ocean temperatures 25–30 Ma ago (Schwartzman and Middendorf 2000).

2.3.4 Temporal Variations of the Magnetic Field

In molten igneous rocks, magnetic particles will align themselves with the Earth's magnetic field. Application of this technique, called paleomagnetism, allows variations in the field to be seen throughout history. The Earth's magnetic field reverses at intervals, ranging from tens of thousands to many millions of years, with an average interval of approximately 250,000 years. The last such event, called the Brunhes-Matuyama reversal, is theorized to have occurred some 780,000 years ago.

The frequency of reversals, the duration over which the reversals occur and the strength of the magnetic field, are well correlated (Glatzmaier et al. 1999). Over long time scales, a quiet magnetic period started 120 Ma ago, lasting 35 Ma. From when it ended (∼65 Ma) until the present time, the reversals have become more frequent. Figure 2.35 compares the reversal frequency with massive extinction events (Courtillot and Olson 2007). This correlation is explained in more detail in the following subsection.

2.3 The Phanerozoic Era

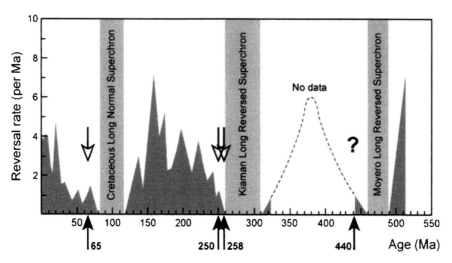

Fig. 2.35 Variation of geomagnetic polarity reversal frequency during Phanerozoic time compared with major faunal mass depletion events (*up arrows*) and associated trap eruptions (*down arrows*). Courtesy: V. Courtillot. Reprinted with permission from Elsevier

2.3.5 Mass Extinctions in the Fossil Record

2.3.5.1 Historical Introduction

The order that characterizes all the structures (including the living beings) existing in the Universe is subject to the conditions of an open thermodynamical system. This is also the case of our Earth. There is a continuous exchange with the media surrounding it, in the form of radiation in all the energy ranges (see Chap. 7 of Vázquez and Hanslmeier 2005) and bodies of different sizes (see Hanslmeier 2008). In this section we study how this interaction has affected, in a direct or indirect way, the evolution of life on our planet (Fig. 2.36). Two main theories have been put forward concerning the interaction between life and environment. The idea about an old but almost static and uniform Earth with only slight gradual changes was defended by James Hutton (1726–1797) in his 'Theory of the Earth' and Charles Lyell (1797–1875) with the reputed 'Principles of Geology'.[30] The principle that 'the present is the key to understanding the past' is a logical consequence of these views. In this context, Charles Darwin (1809–1882) proposed the evolution of living beings by natural selection. In any case, his theory was based on gradual changes, without abrupt jumps: 'nature non facit saltum'.

These principles were challenged by the catastrophist theory, mainly developed by G. Cuvier (1769–1832) in his *Discours sur les Révolutions de la surface du*

---

[30] Four types of uniformities were the base of the theory: laws, processes (actualism), rates (gradualism) and state.

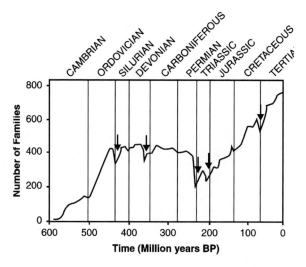

Fig. 2.36 Variation in the number of families along the Phanerozoic era. The major massive extinctions are marked with *arrows*. Reproduced from Steffen et al. (2005). Copyright: Springer Verlag

Globe. Based on geological observations of sediments, Cuvier assumed that biological evolution was driven by sudden events, producing the disappearance of some species and the emergence of new ones. Excluding astronomical causes, he favoured sudden changes in the positions of continents and oceans (see Rudwick 1997 for a translation of Cuvier's works with excellent commentaries on the source texts). According to this idea, processes operating in the past are not necessarily taking place today.

It soon becomes evident that the different biological species have rapidly evolved throughout time, some becoming extinct and others rapidly emerging. Alfred Wallace (1823–1913) expressed this with the words *We live in a zoologically impoverished world, from which all the hugest and fiercest and strangest forms have recently disappeared*. In short, the discussion was about the main driving agent in the biological evolution: chance (contingency) or necessity (natural selection), recalling Jacques Monod's (1917–1976) classic book 'Chance and Necessity'.[31]

The division of the history of the Earth in different periods (Table 2.1) has been marked, at least in the Phanerozoic era, by transitions coinciding with important changes in biological diversity. Living beings have a hierarchical classification: kingdoms, phyla, classes, orders, families, genera and species.

---

[31] In this context, the theory of 'punctuated equilibrium' should be mentioned, which supported the effect of chaotic events as a major force driving bursts in diversity. Evolutionary change occurs relatively rapidly in comparatively brief periods of environmental stress, separated by longer periods of evolutionary stability. This theory has been popularized in various books by S.J. Gould (1940–2002).

2.3.5.2 Biological Extinctions During the Phanerozoic Era

The record of the number of biological species reveals five important crises over the last 500 Ma, although many others are also evident (cf. Benton 1995; Hallan and Wignall 1997).

The extinction occurring at the end of the Ordovician period (443 Ma ago) was probably caused by a glaciation produced by the position of the supercontinent Gondwana, close to the South Pole (Sheehan 2001). Two pulses of extinction have been recorded, with the level of oceanic circulation playing a major role.

The end of the Devonian period was marked by an important decline in many biological species, the brachiopods and the foraminifera being specially affected (Wang et al. 1991 ; McGhee 1996). An impact at this time cannot be ruled out (Name 2003).

The end of the Permian period occurred 250 Ma before the present and was characterized by the extinction of 80% of all ocean-dwelling creatures and 70% of those on land (cf. Erwin 1994, 2006; Berner 2002; Benton 2003). Becker et al. (2001) discovery of fullerenes,[32] containing helium and argon with isotopic compositions similar to those in meteorites called carbonaceous chondrites, and the evidence that Becker et al. (2004) have recently found of an impact at the end of the Permian period in the crater Bedout (Australia) could point to the cause of this extinction.

At the boundary between the Triassic and Jurassic periods, about 200 million years ago, a mass extinction, occurring in a very short interval of time, destroyed at least half of the species on Earth (Ward et al. 2001). The thecodontians and many mammal-like reptiles became extinct, and this is the widely accepted view of how the dinosaurs attained dominance, as there were fewer predators to compete with them (Benton 1993). Olsen et al. (2002) claim to have found enhanced level of iridium at this stage. The Manicougan crater impact (210 ± 4 Ma old), located in Quebec (Canada), has been proposed as a possible scenario of the asteroidal impact.

However, the event which has been most studied is that which occurred at the end of the Cretaceous, no doubt because of its possible relation with the extinction of the dinosaurs.

2.3.5.3 The K/T Extinction

The Cretaceous extinction, hereafter called the K/T event,[33] is the one that has been most intensively studied. More than 75% of all the species present at that time became extinct, ranging across all families of organisms. De Laubenfels (1956) had already suggested that the extinction of the dinosaurs might have been caused by heat associated with the impact of a large meteorite. Urey (1973) and Hoyle and

---

[32] Large carbon compounds consisting of 60 or more carbon atoms, arranged as regular hexagons in a hollow shell. They are often called buckyballs, after Richard Buckminster Fuller (1895–1983), inventor of the geodesic dome, which their natural structure resembles.

[33] K stands for Cretaceous (from German) and T for the Tertiary Era – the Age of Mammals.

Wickramasinghe (1978) proposed, respectively, that ecodisasters such as the K/T extinction were probably caused by the collision or the close passage of a giant comet.

There was a surge of interest in the scientific and press media with the publication of the paper of Alvarez et al. (1980), which provided a degree of empirical association between massive biological extinction and the impact of a large (>10 km) extraterrestrial body, a comet or asteroid being the logical candidates (see Shoemaker 1983; Gehrels et al. 1994). Evidence of enhanced concentrations of iridium in the sediments corresponding to this period in the Italian site of Gubbio was verified by similar findings at other sites around the world (Orth et al. 1981). The publication of the Alvarez's paper immediately provoked different reactions supplying proof for and against this hypothesis.

After several failed attempts, the crater produced by the Cretaceous event was finally identified, the so-called 'smoking gun' being the Chicxulub crater on the Yucatan coast (Hillebrand et al. 1991). Kyte (1998) has tentatively identified materials that might have come from the impacting object. Figure 2.37 shows a view of the planet at this time.

These five large extinctions were not the only ones, and their magnitude is probably due to a combination of two or more external influences. The Earth mantle periodically undergoes instabilities, producing the eruption of the material to the surface in the form of giant volcanic events, called hotspots or superplumes

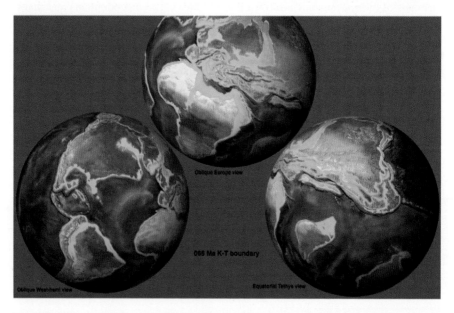

Fig. 2.37 Paleogeographic view of the Earth at the K/T transition. Courtesy: Ron Blakey, Department of Geology, Northern Arizona University

depending of their intensity and duration (Courtillot 1999; Condie 2001). The Permian (Siberian Traps) and Cretaceous (Deccan Traps) periods are associated with these manifestations, which lasted millions of years.

It has also been proposed (Abbott and Isley 2002) that large meteoric impacts could also trigger important volcanic eruptions. These authors proposed that the dating of the 38 known impact craters coincides with mantle eruptions. In principle, this could join the two main theories explaining the K/T mass extinction; however, the timing of mantle events, on the one hand, is not accurate enough and, on the other hand, they are characterized by quite different temporal extensions (see Palmer 1997). Purely biological explanations of mass extinctions should also be considered (see Raup 1992).

Physical conditions of the atmosphere have changed along the history of the planet and are likely to have been an important factor in the mass extinctions. At the geological scale, the climate is controlled by the balance between the variations in the solar output and the changes in the abundance of greenhouse gases in the atmosphere.

2.4 The Quaternary

The Quaternary Period is the geologic time period from the end of the Pliocene Epoch, roughly 1.8–1.6 million years ago to the present. The Quaternary includes two geologic subdivisions – the Pleistocene and the Holocene Epochs. This period is characterized by individual continents mainly accumulated in the Northern Hemisphere.

Significant growth of ice sheets did not begin in Greenland and North America until three million years ago, following the formation of the Isthmus of Panama by continental drift, thus preventing effective distribution of warm water in the North Atlantic Ocean. This triggered the start of a new era of rapidly cycling glacials and interglacials (Fig. 2.38 and Lisiecki and Raymo 2005).

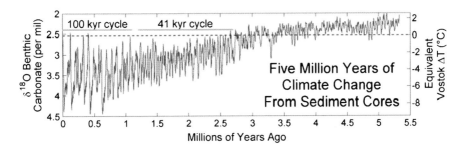

Fig. 2.38 Expanded view of climate change during the last five million years, showing the rapid oscillations in the glacial state. Present day is indicated by 0. Courtesy: Robert A. Rohde, Global Warming Art

2.4.1 The Ice Ages

The climate fluctuations in this period are mainly driven by variations in the orbital parameters of our planet, namely the inclination of the rotation axis, the eccentricity and the climate precession determining the time of the year where perihelion takes place. They are characterized by periods of 41,000, 100,000 and 19,000 years, respectively. In 1941, M. Milankovitch (1879–1958) calculated the insolation associated with these changes and proposed that the triggering of a glaciation was associated to the increase of ice-sheets in the northern hemisphere, produced at times of reduced temperature contrast between summer and winter.

Figure 2.39 shows the temperature record during the last half million years. The first two curves show local changes in temperature at two sites in Antarctica as derived from deuterium isotopic measurements on ice cores (EPICA Community Members 2004, Petit et al. 1999). The final plot shows a reconstruction of global ice volume based on measurements of oxygen isotopes on benthic foraminifera from a composite of globally distributed sediment cores and is scaled to match the scale of fluctuations in Antarctic temperatures (Lisiecki and Raymo 2005). The current decrease of the eccentricity of the Earth's orbit seems to delay a next glaciation, which may take place 50,000 years from now (Berger and Loutre 2002).

The temperatures oscillate in phase with CO_2 and CH_4 concentrations (Petit et al. 1999; Loulergue et al. 2008; Lüthi et al. 2008). Zeebe and Caldeira (2008) have shown that over the last 610,000 years the maximum imbalance between the

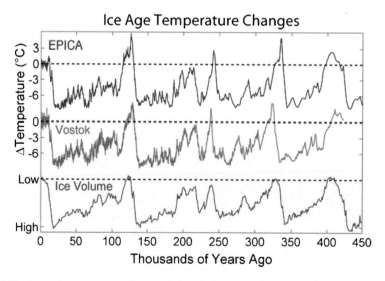

Fig. 2.39 Antarctic temperature changes during the last several glacial/interglacial cycles of the present ice age and a comparison to changes in global ice volume. The present day is on the *left*. Courtesy: Global Warming Project. Vostok data available at http://www.ngdc.noaa.gov/paleo/icecore/antarctica/vostok/vostok_data.html

supply and uptake of carbon dioxide (recall the phases of the CO_2 cycle) was 1–2% (~22 ppmv).[34] This long-term balance holds despite glacial–interglacial variations.

The 41,000 year periodicity, driven by obliquity changes, was dominant during the early Pleistocene. About 900,000 years ago this behaviour increased to a dominant 100,000 length of the glaciations linked to variation in the eccentricity. This is usually explained as an abrupt jump from one stable state of the climate system to another, typical of a non-linear response to a small external forcing (the insolation changes). Crowley and Hyde (2008) suggest that we are approaching to a new bifurcation, leading to a climate characterized by Antarctic-like 'permanent' ice sheets, which would shroud much of Canada, Europe and Asia. However, all these forecasts can be affected by the action of a new factor in the climate system: human intelligence.

2.4.2 The Present Warming: The Anthropocene

The last post-glacial era, known as the Holocene, was characterized by a stable climate (Fig. 2.40). During the last 7,000 years, the variations in the solar magnetic energy and the eruption of isolated volcanoes modulated the climate.

This period also signals the start of human civilization with the development of agriculture. Progressively, the activities of Homo sapiens became a significant force on the Earth (Burroughs 2005; Ruddiman 2005).

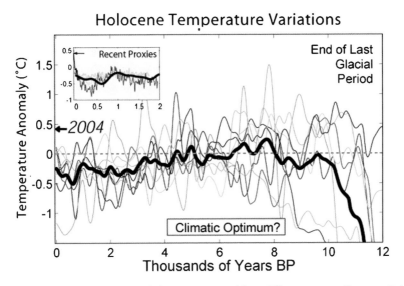

Fig. 2.40 Holocene temperature variations reconstructed from different sources. Courtesy: Robert Rohde

---

[34] ppmv: parts per million in volume.

We have previously seen how the living beings have clearly influenced the chemical composition of the atmosphere and the climate, through the emission of greenhouse gases resulting from their metabolism. At the time scale of a 1,000 years or less, the CO_2 cycle is dominated by two processes. One is when the gas is taken up via photosynthesis by green plants on the continents or phytoplankton in the oceans and later transferred to the soils or zooplankton, respectively; the plants, plankton and animals respire carbon dioxide and upon their death, they are decomposed by microorganisms with a subsequent production of CO_2. The other process is when carbon dioxide is exchanged between oceans and the atmosphere. As the cycle proceeds, concentrations of CO_2 can change as a result of perturbations in the cycle.

Probably the most characteristic event of the latter process started in the late 1700s, at which time the world entered the industrial era, which has continued into our days, an era in which many non-renewable resources are being used. The population of the planet exploded from 1 billion in 1850 to more than 6 billion by 2005, accompanied, for example by a growth in cattle population to 1,400 million (Mc Neill 2000). Crutzen and Stormer (2000) (see also Crutzen 2002a,b) named this period of the Earth history as the Anthropocene. Recently, Zalasiewicz et al. (2008) developed and updated this new concept.[35]

During this period the burning of fossil fuels produced large amounts of energy necessary for industry. As a consequence, residuals have been emitted to the atmosphere.

Georespiration is the principal process of atmospheric O_2 production at long time scales.

$$CO_2 + H_2O \longrightarrow CH_2O + O_2.$$

The opposite reaction is now occurring at a rate of about 100 times faster than that which would occur naturally.

$$CH_2O + O_2 \longrightarrow CO_2 + H_2O.$$

As a result, the long-term carbon cycle impinges on the short-time cycle, leading to an extremely fast rise in atmospheric CO_2 (Fig. 2.41). The CO_2 cycle cannot balance the disequilibrium produced by humans.

Because the carbon dioxide emitted in the combustion of fossil fuels is a greenhouse gas, we can expect a rise in mean temperatures. Figure 2.42 shows the increase of globally averaged temperatures during the instrumental record. The future warming will depend on future emissions of greenhouse gases. The recent IPCC report, 2007, established different scenarios with temperature increases in the range 2–5°C for 2100.

---

[35] Written by 21 members of a commission organized by the London Geological Society to elucidate if we have entered a new geological period. This commission unanimously answered 'yes' to the question: 'Are we now living in the Anthropocene?'.

2.4 The Quaternary

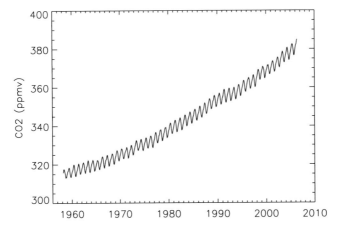

Fig. 2.41 Variation of the atmospheric concentration of carbon dioxide, as measured in Mauna Loa (Hawaii). Data: Earth System Research Laboratory (NOAA)

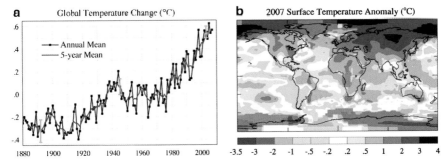

Fig. 2.42 (*Left*) Global annual surface temperatures relative to 1951–1980 (mean figures), based on surface air measurements at meteorological stations and ship and satellite measurements for sea surface temperature. (*Right*) Colour map of temperature anomalies in 2007 relative to the 1951–1980 mean. Areas that were warmest in 2007 are in *red*, and areas that have cooled are in *blue*. Courtesy: Goddard Institute for Space Studies

We may be at the beginning of a sixth massive extinction (Thomas 2004). During the mentioned Anthropocene period, only 40 animals and around 100 plants increased their numbers.[36]

The impacts of current human activities will continue over long periods and the climate may depart significantly from natural behaviour over the next thousands of years. We should remark that the Earth System has critical processes that are susceptible to abrupt changes triggered by human activities that will render the planet less hospitable for human life. The climate system, in particular the sea level, may be responding more quickly to climate change than the current generation of models indicates (Rahmstorf et al. 2007).

---

[36] They include humans, domesticated plants and animals and synanthropes (e.g. rats, rabbits etc.).

For further information on this topic, see the following monographs: Houghton 1997, 2005 and Stern 2006, and the successive IPCC reports[37] Houghton et al. 1996, 2001, IPCC 2007.

Referring to this topic, Carl Sagan commented in his book 'Pale Blue Dot': *Some planetary civilizations see their way through, place limits on what may and what must not be done, and safely pass through the time of perils. Others, not so lucky or so prudent, perish.*

2.5 The Future of Earth

We have already seen how the presence of life has produced a cooler Earth. A resulting feedback of this connection has been the extension of our estimates on the lifespan of the biosphere (Lenton and Van Bloh 2001). However, this process will also have a limit.

2.5.1 The End of Life

The long-term future of our planet will be driven by the continuous increase of solar luminosity (Fig. 2.43a). In response, the weathering rate will increase, lowering rapidly the concentration of CO_2 in the atmosphere and soil (Fig. 2.43b). At a given time both will reach minimum values for the survival of plants (10 ppm).

The disappearance of life will follow the reverse sequence to its appearance in our planet. Franck et al. (2005) have predicted that the extinction of multicellular life and eucaryotes will take place at 0.8 and 1.3 Ga from now, respectively. For earlier estimations of the biosphere life span, see Lovelock and Whitfield (1982) and Caldeira and Kasting (1992). Therefore, it seems that the time period in the Earth's history in which complex life exists is very short. The consequences of this process for the perspectives of finding alien complex life have recently been popularized by Brownlee and Ward (2000, 2004), with the 'Rare Earth' hypothesis. For critical overviews on this topic see Kasting (2000) and Darling (2002).

Two events will make the situation much more complicated. As a result of increasing temperatures, the Earth will enter a runaway greenhouse phase, similar to that suffered by Venus, probably 1.2 Ga from now; as a first consequence, the oceans will evaporate. Additionally, tectonic activity will be interrupted due to the progressive exhaustion of internal energy sources.

The end of life described is limited to photosynthetic organisms. Organisms that use chemical energy for their metabolism could probably survive, especially underground.

---

[37] Since its founding, in 1988, the Intergovernmental Panel on Climate Change (IPCC) has published four reports: 1990, 1995, 2001 and 2007.

2.5 The Future of Earth

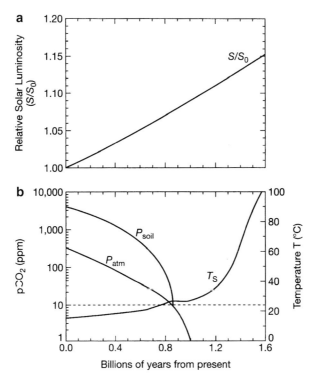

Fig. 2.43 Long term projections of (**a**) solar luminosity and (**b**) surface temperature and CO_2 concentration in the soil and in the atmosphere. Adapted from Caldeira and Kasting (1992) Fig. 2. Reprinted by permission from Macmillan Publishers Ltd. Nature Vol. 360 p. 721. Copyright (1992)

2.5.2 The End of the Earth

About 5.5 Ga from now, a significant amount of the hydrogen reserves of the Sun will have been spent, and our star will move to the Red Giant Branch (RGB) phase of its evolution. The solar diameter will expand to roughly 150 times its current diameter, reaching 0.77 AU, while the core will start to collapse under its own weight, getting hotter and denser. This phase will end with the ignition of Helium in the core, but this element is consumed rapidly and the star will contract again, leading to a second expansion of the outer layers (Asymptotic Giant Branch, AGB).

Mercury and Venus will clearly be affected during the RGB phase. The strong mass loss during this period will expand the Earth's orbit, probably to 1.6 AU, saving it temporarily from being destroyed. This will inevitably occur during the AGB phase, when the Sun undergoes a second expansion (see Rasio et al. 1996 for a consideration of the tidal decay in the final stage). The Earth will probably be vapourized, and the remains scattered through the interplanetary medium (Sackmann et al. 1993; Schröder et al. 2001 and Rybicki and Denis 2001). Recently, Schröder and Smith (2008) have presented a new RGB model, concluding that the

Earth's engulfment will happen during the late phase of the solar RGB evolution. The minimum orbital radius for a planet to be able to survive is found to be about 1.15 AU.

Villaver and Livio (2007) studied the survival of gas planets around stars with masses in the range 1–5 M_S, as these stars evolve off the main sequence. They show that planets with masses smaller than one Jupiter mass do not survive the planetary nebula phase if located initially at orbital distances smaller than 3–5 AU.

Retter and Marom (2003) interpreted an outburst, with at least three peaks, of the red giant star V838 Monocerotis to have been produced by the engulfment of three nearby planets. However, the death sentence of our planet is not yet final and two observations bring some hope. Silvotti et al. (2007) have detected at 1.7 AU a giant planet orbiting the red giant V391 Pegasi, which is already burning helium in its core, whereas Mullally et al. (2008) suggest the presence of a giant planet (2.1 Jupiter masses) orbiting a white dwarf at 2.36 AU. Moreover, we cannot forget that the first exoplanets were found orbiting the pulsar PSR1257+12, enduring a supernova explosion (Wolszczan and Frail 1992).

In any case, in a long-term perspective the human race must move into space to avoid extinction. Following the famous quote of K. Tsiolkovsky (1857–1935): *Earth is the cradle of humanity, but one cannot remain in the cradle forever.*

References

Abbott, D.H., Isley, A.E.: Extraterrestrial influences on mantle plume activity. Earth Planet. Sci. Lett. **205**, 53–62 (2002)

Allègre, C.J., Manhès, G., Göpel, C.: The age of the Earth. Geochimica et Cosmochimica Acta **59**, 1445–1456 (1995)

Allen, P.A., Etienne, J.L.: Sedimentary challenge to snowball Earth. Nat. Geosci. **1**, 817–825 (2008)

Alvarez, L.W., Alvarez, W., Asaro, F., Michel, H.V.: Extraterrestrial cause for the cretaceous tertiary extinction. Science **208**, 1095–1108 (1980)

Anguita, F.: Biografía de la Tierra: Historia de un planeta singular. Editorial Aguilar (2002)

Araki, T., et al.: Experimental investigation of geologically produced antineutrinos with KamLAND. Nature **436**, 499–468 (2005)

Arbab, A.I.: Evolution of angular momenta and energy of the Earth-Moon system. ArXiv Astrophysics e-prints (2003)

Bains, S., Norris, S.D., Corfield, R.M., Faul, K.L.: Termination of global warmth at the Palaeocene/Eocene boundary through productivity feedback. Nature **407**, 171–174 (2000)

Becker, L., Poreda, R.J., Basu, A.R., Pope, K.O., Harrison, T.M., Nicholson, C., Iasky, R.: Bedout: A possible end-permian impact crater offshore of Northwestern Australia. Science **304**, 1469–1476 (2004)

Becker, L., Poreda, R.J., Hunt, A.G., Bunch, T.E., Rampino, M.: Impact event at the permian-triassic boundary: Evidence from extraterrestrial noble gases in fullerenes. Science **291**, 1530–1534 (2001)

Benton, M.J.: Late triassic extinctions and the origin of dinosaurs. Science **260**, 769–770 (1993)

Benton, M.J.: Diversification and extinction in the history of life. Science **268**, 52–58 (1995)

Benton, M.J.: When life nearly died. Thames and Hudson, London (2003)

Berger, A., Loutre, M.F.: CLIMATE: An exceptionally long interglacial ahead? Sci. **297**, 1287–1288 (2002)

Berner, R.: The phanerozoic carbon cycle: CO2 and O2. Oxford University Press, Oxford (2004)

Berner, R.A.: Examination of hyphotheses for the permo-triassic boundary extinction by carbon cycle modeling. Proc. Natl. Acad. Sci. **99**, 4172–4173 (2002)

Berner, R.A., Kothavala, Z.: GEOCARBIII: A revised model of atmospheric CO2 over phanerozoic time. Am. J. Sci. **301**, 182–204 (2001)

Berner, R.A., Lasaga, A.C., Garrels, R.M.: The carbonate-silicate geochemical cycle and its effect on atmospheric carbon dioxide over the past 100 million years. Am. J. Sci. **283**, 641–683 (1983)

Betts, R.A.: Offset of the potential carbon sink from boreal forestation by decreases in surface albedo. Nature **408**, 187–190 (2000)

Bolfan-Casanova, N.: Water in the Earth's mantle. Mineral. Mag. **69**, 229–257 (2005)

Bounama, C.: Thermische Evolution und Habitabilität erdähnlicher Exoplaneten. Thesis, Potsdam University (2007)

Brocks, J.J., Buick, R., Logan, G.A., Summons, R.E.: Composition and syngeneity of molecular fossils from the 2.78 to 2.45 billion-year-old Mount Bruce Supergroup, Pilbara Craton, Western Australia. Geochim. Cosmochim. Acta **67**, 4289–4319 (2003)

Brownlee, D., Ward, P.D.: Rare Earth: Why complex life is uncommon in the Universe. Springer, Heidelberg (2000)

Brownlee, D., Ward, P.D.: The life and death of planet earth. Owl Books (2004)

Budyko, M.I.: Te effect of solar radiation variations on the climate of the Earth. Tellus **21**, 611–619 (1969)

Burchfield, J.D.: Lord Kelvin and the age of the Earth. University of Chicago Press, Chicago (1990)

Burroughs, W.J.: Climate change in prehistory. Cambridge University Press, London (2005)

Caldeira, K., Kasting, J.F.: The life span of the biosphere revisited. Nature **360**, 721–723 (1992)

Came, R.E., Eiler, J.M., Veizer, J., Azmy, K., Brand, U., Weidman, C.R.: Coupling of surface temperatures and atmospheric CO2 concentrations during the Palaeozoic era. Nature **449**, 198–201 (2007)

Canfield, D.E.: The early history of atmospheric oxygen: Homage to Robert M. Garrels. An. Rev. Earth Planet. Sci. **33**, 1–36 (2005)

Canup, R.M.: Dynamics of lunar formation. An. Rev. Astron. Astrophys. **42**, 441–475 (2004)

Canup, R.M., Asphaug, E.: Origin of the Moon in a giant impact near the end of the Earth's formation. Nature **412**, 708–712 (2001).

Canup, R.M., Righter, K., et al.: Origin of the earth and moon. University of Arizona Press, AZ (2000)

Canuto, V.M., Levine, J.S., Augustsson, T.R., Imhoff, C.L.: UV radiation from the young sun and oxygen and ozone levels in the prebiological palaeoatmosphere. Nature **296**, 816–820 (1982)

Canuto, V.M., Levine, J.S., Augustsson, T.R., Imhoff, C.L., Giampapa, M.S.: The young sun and the atmosphere and photochemistry of the early earth. Nature **305**, 281–286 (1983)

Carver, J.H., Vardavas, I.M.: Precambrian glaciations and the evolution of the atmosphere. Ann. Geophys. **12**, 674–682 (1994)

Cess, R.D., Zhang, M.H., Ingram, W.J., Potter, G.L., Alekseev, V., Barker, H.W., Cohen-Solal, E., Colman, R.A., Dazlich, D.A., Del Genio, A.D., Dix, M.R., Esch, M., Fowler, L.D., Fraser, J.R., Galin, V., Gates, W.L., Hack, J.J., Kiehl, J.T., Le Treut, H., Lo, K.K.W., McAvaney, B.J., Meleshko, V.P., Morcrette, J.J., Randall, D.A., Roeckner, E., Royer, J.F., Schlesinger, M.E., Sporyshev, P.V., Timbal, B., Volodin, E.M., Taylor, K.E., Wang, W., Wetherald, R.T.: Cloud feedback in atmospheric general circulation models: An update. J. Geophys. Res. **101**, 12,791–12,794 (1996)

Chamberlain, J.W.: Theory of planetary atmospheres. Academic, NY (1987)

Charlson, R.J., Schwartz, S.E., Hales, J.M., Cess, R.D., Coakley Jr., J.A., Hansen, J.E., Hofmann, D.J.: Climate forcing by anthropogenic aerosols. Science **255**, 423–430 (1992)

Claire, M.W., Catling, D.C., Zahnle, K.J.: Biochemical modelling of the rise in atmospheric oxygen. Geobiology **4**, 239–269 (2006)

Cleaves, H.J., Chalmers, J.H., Lazcano, A., Miller, S.L., Bada, J.L.: A reassessment of prebiotic organic synthesis in neutral planetary atmospheres. Origins of Life and Evolution of the Biosphere **38**, 105–115 (2008)

Cockell, C.S.: The ultraviolet history of the terrestrial planets – implications for biological evolution. Planet. Space Sci. **48**, 203–214 (2000)

Condie, K.C.: Episodic continental growth and supercontinents: a mantle avalanche connection? Earth Planet. Sci. Lett. **163**, 97–108 (1998)

Condie, K.C.: Mantle plumes and their record in earth history. Cambridge University Press, London (2001)

Condie, K.C.: Continental growth during a 1.9-Ga superplume event. J. Geodyn. **34**, 249–264 (2002)

Condie, K.C.: Earth as an evolving planetary system. Academic, NY (2004a)

Condie, K.C.: Supercontinents and superplume events: distinguishing signals in the geologic record. Phys. Earth Planet. In. **146**, 319–332 (2004b)

Courtillot, V.: Evolutionary catastrophes. Cambridge University Press, London (1999)

Courtillot, V., Olson, P.: Mantle plumes link magnetic superchrons to phanerozoic mass depletion events. Earth Planet. Sci. Lett. **260**, 495–504 (2007)

Crowley, T.J., Hyde, W.T.: Transient nature of late Pleistocene climate variability. Nature **456**, 226–230 (2008)

Crutzen, P.J.: The anthropocene. J. Phys. **12**, 1–5 (2002a)

Crutzen, P.J.: Geology of mankind. Nature **415**, 23 (2002b)

Crutzen, P.J., Stoermer, E.F.: The anthropocene. Global Change Newslett. **41**, 12–13 (2000)

Darling, D.: Life everywhere. Basic Books, London (2002)

Dauphas, N., Cates, N.L., Mojzsis, S.J., Busigny, V.: Identification of chemical sedimentary protoliths using iron isotopes in the > 3750 Ma Nuvvuagittuq supracrustal belt, Canada. Earth Planet. Sci. Lett. **254**, 358–376 (2007)

de Jager, C.: Solar Energy Sources. In: ASSL Vol. 29: The sun. Part 1 of solar-terrestrial physics/1970, pp. 1–8. D. Reidel, Dordrecht (1972)

De Laubenfels, M.W.: Dinosaur extinction: One more hypothesis. J. Paleontology **30**, 207–218 (1956)

de Pater, I., Lissauer, J.J.: Planetary sciences. Cambridge University Press, London (2001)

De Wit, M.J., Hart, R.A.: Earth's earliest continental litosphere, hydrothermal flux and crustal recycling. Lithos **30**, 309–335 (1993)

Delaye, L., Lazcano, A.: Prebiological evolution and the physics of the origin of life. Phys. Life Rev. **2**, 47–64 (2005)

Dickey, J.O., Bender, P.L., Faller, J.E., Newhall, X.X., Ricklefs, R.L., Ries, J.G., Shelus, P.J., Veillet, C., Whipple, A.L., Wiant, J.R., Williams, J.G., Yoder, C.F.: Lunar laser ranging – a continuing legacy of the apollo program. Science **265**, 482–490 (1994)

Diziewonski, A.M., Anderson, D.L.: Preliminary Earth reference model. Phys. Earth Plan. In. **25**, 297–356 (1981)

Ebelmen, J.J.: Sur les produits de la decomposition des especes minérales de la famile des silicates. Annu. Rev. Moines **12**, 627–654 (1845)

Ervin, D.H.: The permo-triassic extinction. Nature **367**, 231–236 (1994)

Erwin, D.H.: Extinction: How life on Earth nealy ended 251 million years ago. Princeton University Press, NJ (2006)

Fletcher, B.J., Brentnall, S.J., Anderson, C.W., Berner, R.A., Beerling, D.J.: Atmopsheric carbon dioxide linked with Mesozoic and early Cenozoic climate change. Nat. Geosci. **1**, 43–48 (2008)

Fourier, J.: Remarques Générales Sur les Temperatures Du Globe Terrestre et des Espaces Planétaires. Annales de Chemie et de Physique **27**, 136–167 (1824)

Franck, S.: Evolution of the global mean heat flow over 4.6 Gyr. Tectonophysics **291**, 9–18 (1998)

Franck, S., Block, A., von Bloh, W., Bounama, C., Schellnhuber, H.J., Svirezhev, Y.: Habitable zone for Earth-like planets in the solar system. Planet. Space Sci. **48**, 1099–1105 (2000)

Franck, S., Bounama, C.: Continental growth and volatile exchange during Earth's evolution. Phys. Earth Planet. In. **100**, 189–196 (1997)

Franck, S., Bounama, C., von Bloh, W.: Causes and timing of future biosphere extinction. Biogeosciences Discussions **2**, 1665–1679 (2005)

Franck, S., Kossacki, K.J., von Bloh, W., Bounama, C.: Long-term evolution of the global carbon cycle: historic minimum of global surface temperature at present. Tellus B Chem. Phys. Meteorol. **54**, 325 (2002)

Garcia-Pichel, F.: Solar ultraviolet and the evolutionary history of cyanobacteria. Origins of Life and Evolution of the Biosphere **28**, 321–347 (1998)

Gehrels, T., Matthews, M.S., Schumann, A.M. (eds.): Hazards due to comets and asteroids. University of Arizona Press, AZ (1994)

Glatzmaier, G.A., Coe, R.S., Hongre, L., Roberts, P.H.: The role of the Earth's mantle in controlling the frequency of geomagnetic reversals. Nature **401**, 885–890 (1999)

Glatzmaier, G.A., Roberts, P.H.: Rotation and magnetism of Earth's inner core. Science **274**, 1887–1891 (1996)

Goldblatt, C., Lenton, T.M., Watson, A.J.: Bistability of atmospheric oxygen and the great oxydation. Nature **443**, 683–686 (2006)

Gough, D.O.: Solar interior structure and luminosity variations. Sol. Phys. **74**, 21–34 (1981)

Graham, J.B., Dudley, R., Aguilar, N.M., Gaub, C.: Implications of the late Paleozoic oxygen pulse for physiology and evolution. Nature **375**, 117–120 (1995)

Güdel, M.: The Sun in time: Activity and environment. Living Rev. Sol. Phys. **4**, 3 (2007)

Hallan, A., Wignall, P.B.: Mass extinctions and their aftermath. Oxford University Press, Oxford (1997)

Hanslmeier, A.: Habitability and cosmic catastrophes. Springer, Heidelberg (2008)

Harries, J.E.: The greenhouse Earth: A view from space. Q. J. Meteorol. Soc. **122**, 799–818 (1996)

Hartmann, W.K., Davis, D.R.: Satellite-sized planetesimals and lunar origin. Icarus **24**, 504–514 (1975)

Hedges, S.B., Blair, J.M., Venturi, M., Shoe, J.L.: A molecular timescale of eukaryote evolution and the rise of complex multicellular life. BMC Evol. Biol. **4** (2004)

Herndon, J.M.: Sub-structure of the inner core of the Earth. Proc. Natl. Acad. Sci. **93**, 646–648 (1996)

Hess, H.: History of Ocean Basins. In: A.E. Engel, H.L. James, B.F. Leonard (eds.) Petrologic Studies, pp. 599–620. Geological Society of America, Co (1962)

Hilburn, I.A., Kirschvink, J.L., Tajika, E., Tada, R., Hamano, Y., Yamamoto, S.: A negative fold test on the Lorrain formation of the huronian supergroup: Uncertainty on the paleolatitude of the Paleoproterozoic Gowganda glaciation and implications for the great oxygenation event. Earth Planet. Sci. Lett. **232**, 315–332 (2005)

Hillebrand, A.R., Pemfield, G., Kring, D.A., Pilkington, M., Camargo, A., Jacobsen, S.B., Boynton, W.V.: Chicxulub crater: a possible cretaceous/tertiary boundary impact crater on the Yucatan peninsula, Mexico. Geology **19**, 867–871 (1991)

Hoffman, P.F., Kaufman, A.J., Halverson, G.P., Schrag, D.P.: A Neoproterozoic Snowball Earth. Science **281**, 1342–1346 (1998)

Hoffman, P.F., Schrag, D.P.: The snowball Earth hypothesis: Testing the limits of global change. Terra Nova **14**, 129–155 (2002)

Holland, H.D.: Volcanic gases, black smokers, and the great oxidation event. Geochim. Cosmochim. Acta **66**, 3811–3826 (2002)

Hopkins, M., Harrison, T.M., Manning, C.E.: Low heat flow inferred from > 4 Gyr zircons suggests Hadean plate boundary interactions. Nature **456**, 493–496 (2008)

Houghton, J.: The physics of atmospheres, 3rd edn. Cambridge University Press, London (2002)

Houghton, J.: Global warming. Rep. Progr. Phys. **68**, 1343–1403 (2005)

Houghton, J.T.: Global warming: The complete briefing. Cambridge University Press, London (1997)

Houghton, J.T., Meiro Filho, L.G., Callander, B.A., Harris, N., Kattenburg, A., Maskell, K.: Climate change 1995: The science of climate change. Cambridge University Press, Cambridge, UK (1996)

Hoyle, F., Wickramasinghe, C.: Comets, ice ages, and ecological catastrophes. Astrophys. Space Sci. **53**, 523–526 (1978)

Hyde, W.T., Crowley, T.J., Baum, S.K., Peltier, W.R.: Neoproterozoic 'snowball Earth' simulations with a coupled climate/ice-sheet model. Nature **405**, 425–429 (2000)

IPCC: Climate change 2007 – The physical science basis. Cambridge University Press, London (2007)
Isley, A.E., Abbott, D.H.: Plume-related mafic volcanism and the deposition of banded iron formation. J. Geophys. Res. **104**, 15,461–15,477 (1999)
Johnson, A.P., Cleaves, H.J., Dworkin, J.P., Glavin, D.P., Lazcano, A., Bada, J.L.: The miller volcanic spark discharge experiment. Science **322**, 404 (2008)
Kasting, J.: Comments on the BLAG model: The carbonate-silicate geochemical cycle and its effect on atmospheric carbon dioxide over the past 100 million years. Am. J. Sci. **284**, 1175–1182 (1984)
Kasting, J.F.: Peter Ward and Donald Brownlee's rare Earth. Perspect. Biol. Med. **44**, 117–131 (2000)
Kasting, J.F., Catling, D.: Evolution of a habitable planet. Ann. Rev. Astron. Astrophys. **41**, 429–463 (2003)
Kasting, J.F., Donahue, T.M.: The evolution of atmospheric ozone. J. Geophys. Res. **85**, 3255–3263 (1980)
Kasting, J.F., Ono, S.: Paleoclimates: The first two billion years. Phil. Trans. Roy. Soc. B **361**, 917–929 (2006)
Kasting, J.F., Siefert, J.L.: Life and the evolution of Earth's atmosphere. Science **296**, 1066–1068 (2002)
Kasting, J.F., Toon, O.B.: Climate evolution on the terrestrial planets, pp. 423–449. University of Arizona Press, AZ (1989)
Katz, M.E., Cramer, B.S., Mountain, G.S., Katz, S., Miller, K.G.: Uncorking the bottle: What trigered the Paleocene/Eocene thermal maximum methane release? Paleoceanography **16**, 549–562 (2001)
Kelvin, W.T.: On the secular cooling of the earth. Trans. Roy. Soc. Edinb. **23**, 157–170 (1863)
Kennedy, M., Mrofka, D., Von der Borch, C.: Snowball Earth termination by destabilization of equatorial permafrost methane clathrate. Nature **453**, 642–645 (2008)
Kirschvink, J.L.: Late Proterozoic low latitude glaciation: The snowball Earth. In: Schopf, J.W., Klein, C. (eds.) The Proterozoic Biosphere: A Multidisciplinary Study, pp. 51–52. Cambridge University Press, London (1992)
Kirschvink, J.L., Ripperdan, R., Evans, D.: Evidence for a large-scale reorganization of early cambrian continental masses by inertial interchange true polar wander. Science **277**, 541–545 (1997)
Kleine, T., Münker, C., Mezger, K., Palme, H.: Rapid accretion and early core formation on asteroids and the terrestrial planets from Hf-W chronometry. Nature **418**, 952–955 (2002)
Knauth, L.P., Lowe, D.R.: High Archean climatic temperature inferred from oxygen isotope geochemistry of cherts in the 3.5 Ga Swaziland Supergroup, South Africa. GSA Bull. **115**, 566–580 (2003)
Knoll, A.: The early evolution of eukaryotes: A geological perspective. Science **256**, 622–627 (1992)
Knoll, A.H.: A geological consequences of evolution. Geobiology **1**, 3–14 (2003)
Kopp, R.E., Kirschvink, J.L., Hilburn, I.A., Nash, C.: The Paleoproterozoic snowball Earth: A climate disaster triggered by the evolution of oxygenic photosynthesis. Proc. Natl. Acad. Sci. **102**, 11,131–11,136 (2005)
Korenaga, J.: Urey ratio and the structure and evolution of Earth's mantle. Rev. Geophys. **46**, G2007 (2008)
Kump, L.R.: The rise of atmospheric oxygen. Nature **451**, 277–278 (2008)
Kump, L.R., Barley, M.E.: Increased subaerial volcanisms and the rise of atmospheric oxygen 2.5 billion years ago . Nature **448**, 1033–1036 (2007)
Kump, L.R., Kasting, J.F., Crane, R.: The Earth system, 2nd edn. Pearson, Prentice-Hall (2004)
Kusky, T.M., Li, J.H., Tucker, R.D.: The Archean Dongwanzi Ophiolite Complex, North China Craton: 2.505-billion-year-old oceanic crust and mantle. Science **292**, 1142–1145 (2001)
Kyte, F.T.: A meteorite from the cretaceous/tertiary boundary. Nature **361**, 608–615 (1998)
Lambeck, K.: The Earth's variable rotation. Cambridge University Press, London (2005)

Lane, N.: Oxygen, the molecule that made the world. Oxford University Press, London (2002)

Lee, K.K.M., Jeanloz, R.: High-pressure alloying of potassium and iron: Radioactivity in the Earth's core? Geophys. Res. Lett. **30**(23), 230000–1 (2003)

Lenton, T.M., von Bloh, W.: Biotic feedback extends the life span of the biosphere. Geophys. Res. Lett. **28**, 1715–1718 (2001)

Levrard, B., Laskar, J.: Climate friction and the Earth's obliquity. Geophys. J. Int. **154**, 970–990 (2003)

Lindsay, J.F., Brasier, M.D.: Did global tectonics drive early biosphere evolution? Carbon isotope record from 2.6 to 1.9 Ga carbonates of Western Australian basins. Precambrian Res. **114**, 1–34 (2002)

Liou, K.N.: An introduction to atmospheric physics. In: International Geophysics Series, vol. 84, 2nd edn. Academic, NY (2002)

Lisiecki, L., Raymo, M.E.: A Pliocene-Pleiostocene stack of 57 globally distributed benthic 18O records. Paleoceanography **20** (2005)

Loulergue, L., Schilt, A., Spahni, R., Masson-Delmotte, V., Blunier, T., Lemieux, B., Barnola, J.M., Raynaud, D., Stocker, T.F., Chapellaz, J.: Orbital and millennial-scale features of atmospheric CH4 over the past 800,000 years. Nature **453**, 383–386 (2008)

Lovelock, J.: Gaia: A new look at life on Earth. Oxford University Press, London (1979)

Lovelock, J.E., Margulis, L.M.: Atmospheric homeostasis by and for the biosphere: The Gaia hypothesis. Tellus **26**, 2–10 (1974)

Lovelock, J.E., Whitfield, M.: Life span of the biosphere. Nature **296**, 561–563 (1982)

Lunine, J.I.: Earth: Evolution of a habitable world. Cambridge University Press, London (1999)

Lüthi, D., Le Floch, M., Bereiter, B., Blunier, T., Barnola, J., Raynaud, D., Jouzel, J., Fischer, H., Kawamura, K., Stocker, T.F.: High-resolution carbon dioxide concentration record 650,000-800,000 years before present. Nature **453**, 379–382 (2008)

Marshall, C.R.: Explaining the Cambrian explosion of animals. Ann. Rev. Earth Planet. Sci. **34**, 355–384 (2006)

Martin, P., van Hunen, J., Parman, S., Davidson, J.: Why does plate tectonics occur only on Earth? Phys. Educ. **43**, 144–150 (2008)

Mc Ghee, G.R.: The late devonian mass extinction. Columbia University Press, NY (1996)

Mc Neill, J.R.: Something new under the Sun. W.W. Norton Co., New York (2000)

McDonough, W.F.: Compositional model for the Earth's core. Treatise on Geochemistry **2**, 547–568 (2003)

McMenamin, M.: The Garden of Ediacara: Discovering the first complex life. Columbia University Press, New York (1998)

Messina, S., Guinan, E.F.: Magnetic activity of six young solar analogues I. Starspot cycles from long-term photometry. Astron. Astrophys. **393**, 225–237 (2002)

Miller, S.: A production of aminoacids under possible primitive Earth conditions. Science **117**, 528–529 (1953)

Mojzsis, S.J., Arrhenius, G., McKeegan, K.D., Harrison, T.M., Nutman, A.P., Friend, C.R.L.: Evidence for life on Earth before 3,800 million years ago. Nature **384**, 55–59 (1996)

Morbidelli, A., Chambers, J., Lunine, J.I., Petit, J.M., Robert, F., Valsecchi, G.B., Cyr, K.E.: Source regions and time scales for the delivery of water to Earth. Meteoritics Planet. Sci. **35**, 1309–1320 (2000)

Morgan, W.J.: Convection plumes in the lower mantle. Nature **230**, 42–43 (1971)

Mullally, F., Winget, D.E., Degennaro, S., Jeffery, E., Thompson, S.E., Chandler, D., Kepler, S.O.: Limits on planets around pulsating white dwarf stars. Astrophys. J. **676**, 573–583 (2008)

Name, A.N.: Impact eject layer from the mid-Devonian: possible connection to global mass extinctions. Science **300**, 1734–1737 (2003)

Neukum, G., Ivanov, B.A., Hartmann, W.K.: Cratering records in the inner solar system in relation to the lunar reference system. Space Sci. Rev. **96**, 55–86 (2001)

Newman, M.J., Rood, R.T.: Implications of solar evolution for the earth's early atmosphere. Science **198**, 1035–1037 (1977)

Nield, T.: Supercontinent: Ten billion years in the life of our planet. Harvard University Press, MA (2007)

North, G.R., Cahalan, R.F., Coakley Jr., J.A.: Energy balance climate models. Rev. Geophys. Space Phys. **19**, 91–121 (1981)

Oldham, R.D.: The constitution of the interior of the Earth as revealed by Earth quakes. Q. J. Geol. Soc. Lond. **62**, 456–472 (1906)

Olsen, P.E., et al.: Ascent of dinosaurs linked to iridium anomaly in the Triassic-Jurassic boundary. Science **296**, 1305–1307 (2002)

Oró, J., Lazcano, A., Ehrenfreund, P.: Comets and the origin and evolution of life. In: McKay, C.P. (ed.) Comets and the Origin and Evolution of Life, pp. 1–28 2nd edn. Advances in Astrobiology and Biogeophysics. Springer, Heidelberg (2006)

Orth, C.J., Gilmore, J.S., Knight, J.D., Pillmore, C., Tschudy, R., Fasset, J.E.: An iridium abundance anomaly at the palynological cretaceous-tertiary boundary in northern New Mexico. Science **214**, 1341–1343 (1981)

Palme, H., O'Neill, H.S.C.: Cosmochemical estimates of mantle composition. Treatise on Geochemistry **2**, 1–38 (2003)

Palmer, T.: Controversy: Catastrophism and evolution: The ongoing debate. Kluwer, Dordecht (1997)

Patterson, C.: Age of meteorites and the earth. Geochim. Cosmochim. Acta **10**, 230–237 (1956)

Pavlov, A.A., Brown, L.L., Kasting, J.F.: UV shielding of NH_3 and O_2 by organic hazes in the Archean atmosphere. J. Geophys. Res. **106**, 23,267–23,288 (2001)

Pavlov, A.A., Kasting, J., Brown, L., Rages, K.A., Freedmsan, R.: Greenhouse warming by CH4 in the atmosphere of early Earth. J. Geophys. Res. **105**, 11,981–11,990 (2000)

Pavlov, A.A., Kasting, J.F.: Mass-independent fractionation of sulfur isotopes in archean sediments: Strong evidence for an anoxic archean atmosphere. Astrobiology **2**, 27–41 (2002)

Peltier, W.R., Liu, Y., Crowley, J.W.: Snowball Earth prevention by dissolved organic carbon remineralization. Nature **450**, 813–818 (2007)

Petit, J.R., Jouzel, J., Raynaud, D., Barkov, N.I., Barnola, J., Basile, I., Bender, M., Chapellaz, J., Davis, M., Delaygue, G., Delmotte, M., Kotlyakov, V., Legrand, M., Lipenkov, V.Y., Lorius, C., Pepin, L., Ritz, C., Saltzman, E., Stievenard, M.: Climate and atmospheric history of the past 420,000 years from the Vostok ice core, Antarctica. Nature **399**, 429–436 (1999)

Pinti, D.L.: The Origin and evolution of the oceans. In: Lectures in Astrobiology, vol. 1, pp. 83–112. Springer, Heidelberg (2005)

Poirier, J.P.: Introduction to the physics of the Earth's interior. Cambridge University Press, London (1991)

Pollack, H.N., Hurter, S.J., Johnson, J.: Heat flow from the Earth's interior: analysis of the global data set. Rev. Geophys. **31**, 267–280 (1993)

Rahmstorf, S., Cazenave, A., Church, J.A., Hansen, J.E., Keeling, R.F., Parker, D.E., Somerville, R.C.J.: Recent climate observations compared to projections. Science **316**, 709 (2007)

Ramanathan, V., Cess, R.D., Harrison, E.F., Minnis, P., Barkstrom, B.R., Ahmad, E., Hartmann, D.: Cloud-radiative forcing and climate: Results from the Earth radiation budget experiment. Science **243**, 57–63 (1989)

Rampino, M.R., Stothers, R.B.: Flood basalt volcanism during the past 250 million years. Science **241**, 663–667 (1998)

Randall, D.A., Cess, R.D., Blanchet, J.P., Chalita, S., Colman, R., Dazlich, D.A., Del Genio, A.D., Keup, E., Lacis, A., Le Treut, H., Liang, X.Z., McAvaney, B.J., Mahfouf, J.F., Meleshko, V.P., Morcrette, J.J., Norris, P.M., Potter, G.L., Rikus, L., Roeckner, E., Royer, J.F., Schlese, U., Sheinin, D.A., Sokolov, A.P., Taylor, K.E., Wetherald, R.T., Yagai, I., Zhang, M.H.: Analysis of snow feedbacks in 14 general circulation models. J. Geophys. Res. **99**, 20,757–20,772 (1994)

Rasio, F.A., Tout, C.A., Lubow, S.H., Livio, M.: Tidal decay of close planetary orbits. Astrophys. J. **470**, 1187–1191 (1996)

Raup, D.M.: Extinctions: Bad genes or bad luck. W.W. Norton, NY (1992)

Raval, A., Ramanathan, V.: Observational determination of the greenhouse effect. Nature **342**, 758–761 (1989)

References

Raymond, S.N., Quinn, T., Lunine, J.I.: Making other earths: dynamical simulations of terrestrial planet formation and water delivery. Icarus **168**, 1–17 (2004)

Retter, A., Marom, A.: A model of an expanding giant that swallowed planets for the eruption of V838 Monocerotis. Mon. Not. Roy. Astron. Soc. **345**, L25–L28 (2003)

Ribas, I., Guinan, E.F., Güdel, M., Audard, M.: Evolution of the solar activity over time and effects on planetary atmospheres. I. High-energy irradiances (1-1700 Å). Astrophys. J. **622**, 680–694 (2005)

Richter, F.M.: Kelvin and the age of the Earth. J. Geology **94**, 395–401 (1986)

Rogers, J.W., Santosh, M.: Continents and supercontinents. Oxford University Press, London (2004)

Royer, D.L., Berner, R.A., Park, J.: Climate sensitivity constrained by CO_2 concentrations over the past 420 million years. Nature **446**, 530–532 (2007)

Ruddiman, W.F.: Plows, plagues and petroleum. Princeton University Press, NJ (2005)

Rudwick, M.: Georges cuvier, fossil bones, and geological catastrophes. The University of Chicago Press, Chicago (1997)

Russell, M.J., Hall, A.J., Cairns-Smith, A.G., Braterman, P.: Submarine hot springs and the origin of life. Nature **336**, 117 (1988)

Rybicki, K.R., Denis, C.: On the final destiny of the Earth and the solar system. Icarus **151**, 130–137 (2001)

Ryder, G.: Mass flux in the ancient Earth-Moon system and benign implications for the origin of life on Earth. J. Geophys. Res. (Planets) **107**, 5022 (2002)

Ryder, G.: Bombardment of the hadean Earth: Wholesome or deleterious? Astrobiology **3**, 3–6 (2003)

Rye, R., Holland, H.D.: Life associated with a 2.76 Ga ephemeral pond? Evidence from Mount Roe 2 paleosol. Geology **28**, 483–486 (2000)

Rye, R., Kuo, P.H., Holland, H.D.: Atmospheric carbon dioxide concentrations before 2.2 billion years ago. Nature **378**, 603–605 (1995)

Sackmann, I.J., Boothroyd, A.I.: Our Sun. V. A bright young sun consistent with helioseismology and warm temperatures on ancient Earth and Mars. Astrophys. J. **583**, 1024–1039 (2003)

Sackmann, I.J., Boothroyd, A.I., Kraemer, K.E.: Our Sun. III. Present and Future. Astrophys. J. **418**, 457–468 (1993)

Sankaran, A.V.: When did plate tectonics begin? Curr. Sci. **90**, 1596–1597 (2006)

Schellnhuber, H.J.: Earth system analysis and the second Copernican revolution. Nature **402**, C19–C23 (1999)

Schopf, J.: Earth's earliest biosphere, its origin and evolution. Princeton University Press, NJ (1983)

Schröder, K.P., Connon Smith, R.: Distant future of the Sun and Earth revisited. Mon. Not. Roy. Astron. Soc. **386**, 155–163 (2008)

Schroder, P., Smith, R., Apps, K.: Solar evolution and the distant future of Earth. Astron. Geophys. **42**, 26–29 (2001)

Schubert, G., Turcotte, D.L., Olson, P.: Mantle convection in the Earth and planets. Cambridge University Press, London (2001)

Schwartzman, D.: Temperature and the evolution of the Earth's biosphere. In: Shostak, G.S. (ed.) ASP Conf. Ser. 74: Progress in the Search for Extraterrestrial Life, pp. 153–164 (1995)

Schwartzman, D., Caldeira, K., Pavlov, A.: Cyanobacterial emergence at 2.8 Gya and greenhouse feedbacks. Astrobiology **8**, 187–203 (2008)

Schwartzman, D., Middendorf, G.: Biospheric cooling and the emergence of intelligence. In: Lemarchand, G., Meech, K. (eds.) ASP Conf. Ser. 213: Bioastronomy 99, pp. 425–430 (2000)

Schwartzman, D.W., Volk, T.: Biotic enhancement of weathering and the habitability of Earth. Nature **340**, 457–460 (1989)

Scott, A., Glasspool, I.J.: The diversification of Paleozoic fire systems and fluctuations in atmospheric oxygen concentration. Proc. Acad. Natl. Sci. USA **103**, 10,861–10,865 (2006)

Scott, C., Lyons, T.W., Bekker, A., Shen, Y., Poulton, S.W., Chu, X., Ambar, A.D.: Tracing the stepwise oxygenation of the Proterozoic ocean. Nature **452**, 456–459 (2008)

Sheehan, P.M.: The late ordovician mass extinction. Annu. Rev. Earth Planet. Sci. **29**, 331–364 (2001)

Shields, G.A., Veizer, J.: The Precambrian marine carbonate isotope database: Version 1.1. Geochem. Geophys. Geosyst. **3**, 1031 (2002)

Shoemaker, E.M.: Asteroid and comet bombardment of the earth. Annu. Rev. Earth Planet. Sci. **11**, 461–494 (1983)

Silver, P.G., Behn, M.D.: Intermittent plate tectonics? Science **319**, 85–88 (2008)

Silvotti, R., Schuh, S., Janulis, R., Solheim, J.E., Bernabei, S., Østensen, R., Oswalt, T.D., Bruni, I., Gualandi, R., Bonanno, A., Vauclair, G., Reed, M., Chen, C.W., Leibowitz, E., Paparo, M., Baran, A., Charpinet, S., Dolez, N., Kawaler, S., Kurtz, D., Moskalik, P., Riddle, R., Zola, S.: A giant planet orbiting the extreme horizontal branch star V391 Pegasi. Nature **449**, 189–191 (2007)

Skumanich, A.: Time scales for CA II emission decay, rotational braking, and lithium depletion. Astrophys. J. **171**, 565–567 (1972)

Sleep, N.H.: Evolution of the mode of convection within terrestrial planets. J. Geophys. Res. **105**, 17,563–17,578 (2000)

Sleep, N.H.: Evolution of the continental lithosphere. Ann. Rev. Earth Planet. Sci. **33**, 369–393 (2005)

Sonett, C.P., Kvale, E.P., Zakharian, A., Chan, M.A., Demko, T.M.: Late proterozoic and paleozoic tides, retreat of the moon, and rotation of the Earth. Science **273**, 100–104 (1996)

Stanley, S.M.: Exploring the Earth through time. W.H. Freeman, CA (1992)

Stanley, S.M.: Earth system history. W.H. Freeman, CA (1999)

Steffen, W., Sanderson, A., Tyson, P., Jäger, J., Matson, P., Moore, B., Oldfield, F., Richardson, K., Schellnhuber, H., Turner, B., Wasson, R.J.: Global change and the Earth system: A planet under pressure. Springer, Heidelberg (2005)

Stern, N.: The economics of climate change: The stern review. Cambridge University Press. (2006)

Stern, R.J.: Subduction zones. Rev. Geophys. **40**, 1012 (2002)

Stern, R.J.: Evidence from ophiolites, blueschists, and ultra-high pressure metamorphic terranes that the modern episode of subduction tectonics began in neoproterozoic time. Geology **33**, 557–560 (2005)

Stevenson, D.J.: Origin of the moon – The collision hypothesis. Ann. Rev. Earth Planet. Sci. **15**, 271–315 (1987)

Strom, R.G., Malhotra, R., Ito, T., Yoshida, F., Kring, D.A.: The origin of planetary impactors in the inner solar system. Science **309**, 1847–1850 (2005)

Taylor, F.W.: The Stratosphere. Phil. Trans. Roy. Soc. Lond. **361**, 11–22 (2003)

Tera, F., Papanastassiou, D.A., Wasserburg, G.J.: Isotopic evidence for a terminal lunar cataclysm. Earth Planet. Sci. Lett. **22**, 1–21 (1974)

Thomas, C.D.: Extinction risk from climate change. Nature **427**, 145–148 (2004)

Tice, M.M., Lowe, D.R.: Photosynthetic microbial mats in the 3,416-Myr-old oecan. Nature **431**, 549–552 (2004)

Torsvik, T.H., Rehnströhm, E.F.: Cambrian paleomagnetic data from Baltica:Implications for true polar wander and Cambrian paleogeography. J. Geol. Soc. Lond. **158**, 321–329 (2001)

Touboul, M., Kleine, T., Bourdon, B., Palme, H., Wieler, R.: Late formation and prolonged differentiation of the Moon inferred from W isotopes in lunar metals. Nature **450**, 1206–1209 (2007)

Touma, J., Wisdom, J.: Resonances in the early evolution of the Earth-Moon system. Astron. J. **115**, 1653–1663 (1998)

Trainer, M.G., Pavlov, A.A., Curtis, D.B., McKay, C.P., Worsnop, D.R., Delia, A.E., Toohey, D.W., Toon, O.B., Tolbert, M.A.: Haze aerosols in the atmosphere of Early Earth: Manna from heaven. Astrobiology **4**, 409–419 (2004)

Tyndall, J.: On the relation of radiant heat to aqueous vapor. Phil. Mag. 4 **26**, 30–54 (1863)

Urey, H.C.: The planets: Their origin and development. Yale University Press, New Haven (1952)

Urey, H.C.: Cometary collisions and geological periods. Nature **242**, 32–33 (1973)

Valley, J.W., Peck, W.H., King, E.M., Wilde, S.A.: A cool early Earth. Geology **30**, 351–354 (2002)

Veizer, J., Godderis, Y., François, L.M.: Evidence for decoupling of atmospheric CO2 and global climate during the Phanerozoic eon. Nature **408**, 698–701 (2000)

Villaver, E., Livio, M.: Can planets survive stellar evolution? Astrophys. J. **661**, 1192–1201 (2007)

Vitousek, P.M., D'Antonio, C., Loope, L., Westbrooks, R.: Biological invasions as global environmental change. Am. Sci. **84**, 468–478 (1996)

Volk, T.: Feedbacks between weathering and atmospheric CO2 over the last 100 million years. Am. J. Sci. **287**, 763–779 (1987)

Volk, T.: Gaia's body: Towards a physiology of Earth. MIT, MA (2003)

Walker, J.C.G., Hays, P.B., Kasting, J.F.: A negative feedback mechanism for the long-term stabilization of the earth's surface temperature. J. Geophys. Res. **86**, 9776–9782 (1981)

Wang, K., Orth, C.J., Attrep, M., Chatterton, B.D., Hou, H., Geldsetzer, H.H.: Geochemical evidence for a catastrophic biotic event at the Frasnian/Famennian boundary in South China. Geology **19**, 776–779 (1991)

Ward, P.D.: Life as we do not know it. Viking Penguin, NY (2005)

Ward, P.D., et al.: Sudden productivity collapse associated with the Triassic-Jurassic boundary mass extinction. Science **292**, 1148–1151 (2001)

Watson, A.: Implications of an anthropic model of evolution for the emergence of complex life and intelligence. Astrobiology **8**, 175–185 (2008)

Wilde, S., Valley, J., Peck, W., Graham, C.: Evidence from detrital zircons for the existence of continental crust and oceans on the Earth at 4.4 Gyr ago. Nature **409**, 175–178 (2001)

Wilhelms, D.E., McCauley, J.F., Trask, N.J.. The geologic history of the moon. USGS Professional Paper 1348 : For sale by the Books and Open-file Reports Section, US Geological Survey (1987)

Williams, D.M., Kasting, J.F., Frakes, L.A.: Low-latitude glaciation and rapid changes in the Earth's obliquity explained by obliquity-oblateness feedback. Nature **396**, 453–455 (1998)

Williams, G.D.: Tidal rhyhmites: Key to the history of the Earth's rotation and the lunar orbit. J. Phys. Earth **38**, 475–491 (1990)

Williams, G.E.: Precambrian length of day and the validity of tidal rhythmite paleotidal values. Geophys. Res. Lett. **24**, 421–424 (1997)

Williams, G.E.: Geological constraints on the Precambrian history of Earth's rotation and the Moon's orbit. Rev. Geophys. **38**, 37–60 (2000)

Wilson, J.T.: Did the Atlantic close and then re-open? Nature **211**, 676–681 (1966)

Wilson, T.J.: A possible origin of the Hawaiian islands. Can. J. Phys. **41**, 863–868 (1963)

Wise, R.R., Hoober, J.K. (eds.): The structure and function of plastids, advances in photosynthesis and respiration, vol. 23. Springer, Heidelberg (1994)

Woese, C.R., Kandler, O., Wheeler, M.L.: Towards a natural system of organisms: Proposal for the domains archaea, bacteria and eucarya. Proc. Natl. Acad. Sci. **87**, 4576 (1990)

Wolszczan, A., Frail, D.A.: A planetary system around the millisecond pulsar PSR1257 + 12. Nature **355**, 145–147 (1992)

Wood, B.E., Müller, H.R., Zank, G.P., Linsky, J.L., Redfield, S.: New mass-loss measurements from astrospheric Lyα absorption. Astrophys. J. **628**, L143–L146 (2005)

Wood, B.J., Walter, M.J., Wade, J.: Accretion of the Earth and segregation of its core. Nature **441**, 825–833 (2006)

Zachos, J., Pagani, M., Sloan, L., Thomas, E., Billups, K.: Trends, rhythms, and aberrations in global climate 65 Ma to present. Science **292**, 686–693 (2001)

Zahnle, K., Claire, M., Catling, D.: The loss of mass-independent fractionation in sulfur due to a Palaeoproterozoic collapse of atmospheric methane. Geobiology **4**, 271–283 (2006)

Zahnle, K., Arndt, N., Cockell, C., Halliday, A., Nisbet, E., Selsis, F., Sleep, N.H.: Emergence of a Habitable Planet. Space Sci. Rev. **129**, 35–78 (2007)

Zalasiewicz, J., et al.: Are now living in the Anthropocene? GSA Today **18**, 4–8 (2008)

Zeebe, R.E., Caldeira, K.: Close mass balance of long-term carbon fluxes from ice-core CO2 and ocean chemistry records. Nat. Geosci. **1**, 312–315 (2008)

Zhang, Y.: The age and accretion of the Earth. Earth Sci. Rev. **59**, 235–263 (2002)

Zharkov, V.: Interior structure of the Earth & planets. Harwood Academic Publishers, Switzerland (1986)

Chapter 3
The Pale Blue Dot

3.1 Globally Integrated Observations of the Earth

Over the past few years, advancements in astronomy have enabled us to discover planets orbiting around stars other than the Sun, and the number of detections is increasing exponentially. Even though we are not yet capable of detecting and exploring planets like Earth, ambitious missions are already being planned for the coming decades. In the near future, it is likely that Earth-size planets will be discovered, and efforts will then be directed toward obtaining images and spectra from them.

The first thing an extraterrestrial observer would find out about Earth would be that it is the third planet from the Sun and the largest of the rocky planets in the Solar System, in both diameter and mass. More detailed observations would identify its minimum (or absolute, depending on the observing techniques) mass and determine its orbital parameters. The determination of the Sun–Earth distance, the spectral type of the Sun and the mass of the planet would immediately point to the possibility of finding life, as Earth lies inside the habitable zone of the Sun (see Chap. 5). Thus, once curiosity is triggered, more detailed observations of the planet would be in order.

However, what these observations would reveal about our planet will be highly dependent on the techniques used, the temporal sampling, the spectral range and the timing. We have already seen in Chap. 2 how our planet underwent an evolution with time, and so an external observer would obtain a different view of the Earth if it were to be observed at different epochs. In reflected light, the observing geometry will determine whether what is being observed is either a full planet hemisphere or just a thin crescent. Finally, the spectral range is also important, as the light reflected or emitted by Earth is not uniformly distributed. In Fig. 3.1, the spatial distribution of the reflected and emitted light from Earth illustrates this point.

When observing the Earth from an astronomical distance, all the reflected starlight and radiated emission from its surface and atmosphere will be integrated into a single point. This also holds true for future observations of exoplanets. In that case, to interpret the resulting spatially integrated observations, we need to compare them with observations with similar resolution of the only planet that we know to be inhabited, namely Earth. However, despite the fact that Earth is our own planet

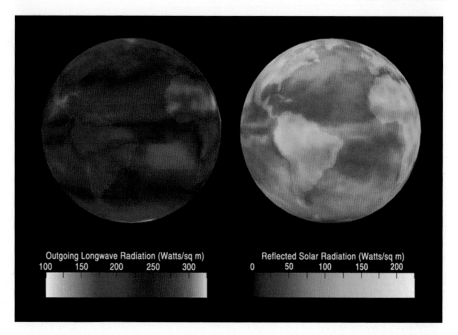

Fig. 3.1 Thermal radiation, or heat, emitted to space from Earth's surface and atmosphere (*left*) and sunlight reflected back to space by the ocean, land, aerosols and clouds (*right*). The lowest amount of sunlight reflected back to space, shown in *blue*, occurs over clear ocean areas. *Green* colours show gradually increasing amounts of reflected sunlight. The areas of greatest reflected solar energy, shown in *white*, occur both from the tops of thick clouds and from ice-covered regions on the Earth's surface during summer. Credit: NASA

and that we have a very detailed knowledge of its physical and chemical properties, Earth observations from a remote perspective are limited. They can be obtained only by integrating high-resolution Earth-orbiting satellite observations, from a very remote observing platform or from the ground by observing the earthshine reflected on the dark side of the Moon.

In Chap. 1, we revised the historical gathering of Earth observations, with ever deeper detail. In this chapter, we review the historical and modern observations of the Earth, but seen as a planet, that is integrating all its emitted and/or reflected light into a single point.

3.1.1 Earth Orbiting Satellites

The scientific and technological advances of the twentieth century have brought finer and finer detailed knowledge of our planet. Routinely, plankton blooms, city night lights and crop health indices are measured from space (Fig. 3.2). However, when observing an extrasolar planet, most of these features will disappear in the noise when global averages are taken.

3.1 Globally Integrated Observations of the Earth

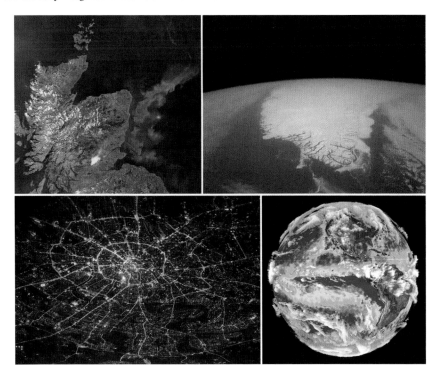

Fig. 3.2 A composite image illustrating a variety of low earth orbit (LEO) satellite observations. *Top left*: A green swirl of phytoplankton bloom in the North Sea off the coast of eastern Scotland captured by Envisat. *Top right*: A satellite view of Greenland. *Bottom left*: Moscow city lights. *Bottom right*: A 3D reconstruction of the cloud distribution and sea surface anomalies during the 1997–1998 El Niño phenomenon. Plankton blooms, ice extent, city lights, clouds and sea surface temperatures are only some of the hundreds of parameters regularly monitored from Earth-observing satellites. Images credit: ESA and NASA

In principle, one could combine measurements from Earth-observing satellites to derive a globally integrated perspective. However, the field of view of low Earth orbit satellites is quite small (typically of a few tens to a hundred km^2), and the geographical coverage is not always complete, and if it is, it takes time (sometimes days) to cover the whole of the Earth's surface. It is also necessary to change the geometry of the observations to that of a remote observer. Thus, there is a need to use models, complex inter-satellite calibrations and interpolations to recreate a global view.

Currently, more than 900 artificial satellites orbit around the Earth, a large number of them monitoring one or several properties of the Earth's surface or atmosphere. For the curious, NASA's J-Track 3D tool[1] allows you to find the position of each satellite in real time.

---

[1] http://science.nasa.gov/Realtime/JTrack/

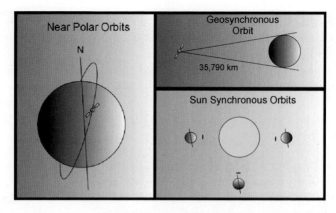

Fig. 3.3 An illustration of some of the most common orbits for Earth-observing satellites. Image credit: NASA

Earth-observation satellites move around the Earth in an orbit that depends on the mission the satellite was built for and determines the nature of the resulting dataset. The orbit is defined by three factors: the shape (circular or elliptical), the altitude (constant for a circular orbit and constantly changing for an elliptical orbit) and the angle with respect to the equator (large for polar orbiter satellites and small for satellites that stay close to the equator) (see Fig. 3.3). The main classes of orbits are the following:

Low Earth Orbit (LEO): Satellites in LEO orbit the Earth at altitudes of less than 2,000 km. Because they orbit so close to Earth, they must travel very fast so that gravity will not pull them back into the atmosphere. LEO satellites speed along at 27,359 km h^{-1} and can circle the Earth in about 90 min. Satellites in LEO can get very clear surveillance images of small regions and require little power to transmit their data to the Earth; however, they have a single observing geometry, in the sense that they measure every point of the Earth from a nadir point of view (retroflection). The majority of all satellites, as well as the Space Shuttle and International Space Station, operate from LEO.

Medium Earth Orbit (MEO): At an altitude of around 10,000 km a satellite is in MEO. MEOs are generally used for navigation and communication satellites.

Geosynchronous Orbit (GEO): Also known as geostationary orbits, satellites in these orbits circle the Earth at the same rate as the Earth spins. Geostationary orbits are useful because they cause a satellite to appear motionless in the sky. As a result, an antenna can point in a fixed direction and maintain a constant link with the satellite. The satellite orbits in the direction of the Earth's rotation at an altitude of approximately 35,786 km above ground. This altitude is significant because it produces an orbital period equal to the Earth's period of rotation, the sidereal day. The satellites are located near the equator, because at this latitude there is a constant force of gravity from all directions. At other latitudes, the bulge at the centre of the Earth would pull on the satellite.

Geosynchronous orbits allow the satellite to observe almost a full hemisphere of the Earth. These satellites are used to study large scale phenomena, such as hurricanes or cyclones, and have revolutionized global communications, television broadcasting and weather forecasting. The idea of a geosynchronous satellite for communication purposes was first published in 1928 by Herman Potocnik (1892–1929) in Potocnik (1928). The geostationary orbit was first popularized by Arthur C. Clarke (1917–2008), in Clarke (1945). As a result this is sometimes referred to as the Clarke orbit.

The main disadvantage of this type of orbit is that since these satellites are very far away, they have poor geographical resolution. The other disadvantage is that these satellites have trouble monitoring activities near the poles. Complete longitudinal coverage of the Earth could be achieved using five of the current GEO satellites, which would cover about 96% of the planet, leaving out the polar regions. However, because of the curvature of the Earth, data only up to about 50–60° is commonly used (Valero 2000). Currently, there are approximately 300 operational geosynchronous satellites.

Polar Orbit (PO): An orbit that goes over both the North and the South Pole is called a Polar Orbit, although the more correct term would be near polar orbits. These orbits have an inclination near 90°, which allows the satellite to see virtually every part of the Earth as the Earth rotates underneath it. It takes approximately 90 min for the satellite to complete one orbit. Most polar orbits are in LEO, but any altitude can be used for a polar orbit.

Sun Synchronous Orbit (SSO): These orbits allow a satellite to pass over a section of the Earth at the same time of day. As there are 365 days in a year and 360° in a circle, it means that the satellite has to shift its orbit by approximately 1° per day. These orbits are located at an altitude between 700 and 800 km and are used for satellites that need a constant amount of sunlight. Satellites that take pictures of the Earth will work best with bright sunlight, while satellites that measure long-wave radiation will work best in complete darkness.

Although none of these observing geometries offers us a view of the Earth as a planet, the high-resolution data provided by Earth-observing satellites are crucial to constructing Earth models. High resolution scene characterization data can be combined with measured angle-dependent reflection/emission functions to derive a global view, and to construct models, not only of the actual earth, but also of a broad range of possible planets.

3.1.2 Observations from Long-range Spacecrafts

Globally integrated observations of the Earth as a point, or as a planet, are sometimes taken by solar system probes, usually as an instrumentation test or as an effective public-relations event. Planetary flybys, while often required to eventually settle in a flying path, are also critical opportunities for the mission teams to test the spacecrafts and their scientific instruments. Over the years, a series of Earth

Table 3.1 A compilation of globally integrated observations of the Earth from remote satellite platforms

Mission	Date	Observations
Voyager 1[a]	1989	Pale blue dot image (6.4 billion km)
Galileo[b]	1990	Flyby Obs. Very low-resolution spectroscopy
Mars Global Surveyor, TES[c]	1996	Infrared spectroscopy
Mars Express, OMEGA[d]	2003	Low-res. visible and near-infrared spectra
ROSSETTA, Virtis[e]	2005	High-resolution visible and infrared imaging
MESSENGER[f]	2005	Imaging sequence of Earth during flyby
CASSINI[g]	2006	Image of Earth from Saturn's orbit
ROSSETTA, OSIRIS[h]	2007	Composite imaging, true colours
EPOXI, EPOCh[i]	2008	Composite imaging, true colours
Venus Express[j]	2007–2008	Repeated visible and infrared spectroscopy

More details can be found in the following:
[a] Sagan (1994)
[b] Sagan et al. (1993)
[c] Christensen and Pearl (1997)
[d] http://www.esa.int
[e] http://www.esa.int
[f] http://messenger.jhuapl.edu
[g] http://www.nasa.gov
[h] http://www.esa.int
[i] http://epoxi.umd.edu/4gallery/Earth-Moon.shtml
[j] Piccioni et al. (2008)

images and spectra, mostly as one-time observations, have been compiled. Some of the most significant are listed in Table 3.1.

These observations offer a unique perspective of our planet and will be discussed in more detail in the following sections. The first of such observations was taken by the Galileo spacecraft (Sagan 1994) and had a deep impact among both scientists and the general public. Recently, the EPOXI mission has released a set of images revealing changes in the visible and near-infrared reflectance as the Earth rotates, as well as the Moon's transit in front of the Earth (Fig. 3.4).

Although almost all the observations are taken only as one-time efforts or over short periods of time, it is worth noting the effort recently undertaken by the Venus Express Team. Venus Express started a program in May 2007 to continuously monitor the Earth spectra, with approximately monthly data (Piccioni et al. 2008). These data will be extremely useful in the near future for characterizing the most salient features of the Earth's spectra, and especially their variability with time and orbital configurations.

3.1.3 An Indirect View of the Earth: Earthshine

Ground-based measurements of the short-wavelength (visible and near infrared light) albedo of a planet in our solar system are relatively straightforward – except

3.1 Globally Integrated Observations of the Earth

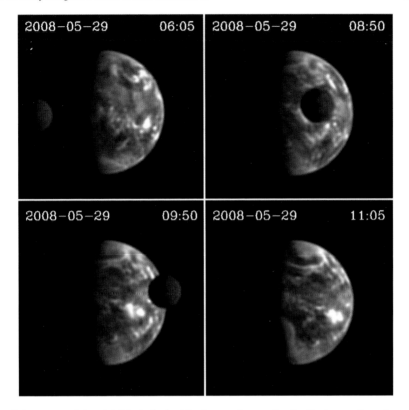

Fig. 3.4 As part of the EPOXI mission's objectives to characterize the Earth as a planet for comparison with planets around other stars, the spacecraft looked back at the Earth collecting a series of images. In the images, the Earth–Moon system is seen with the Moon transiting in front of the Earth. They were taken on 29 May 2008 through three filters: blue, green and orange, centred at 450, 550 and 650 nm respectively, while the spacecraft was 0.33 AU (49,367,340 km) away from Earth. Image Credit: NASA/JPL-Caltech/UMD/GSFC

for the Earth. However, we can determine the Earth's reflectance from the ground by measuring the earthshine, the ghostly glow visible on the dark side of the Moon.

Earthshine is sunlight that has reflected from the dayside Earth onto the dark side of the Moon and back again to Earth. The term 'dark side' refers to the portion of the lunar surface that, at any instant, faces the Earth but not the Sun. Both earthshine, from the dark side of the Moon, and moonlight, from the bright side of the Moon, are transmitted through the same airmass just prior to detection and thus suffer the same extinction and imposed absorption features. Their ratio is the averaged reflection coefficient (or albedo) of the global atmosphere; the global atmosphere being defined as the portion of the dayside Earth simultaneously visible from the Sun and the Moon (Qiu et al. 2003. Figure 3.5 illustrates the earthshine geometry.

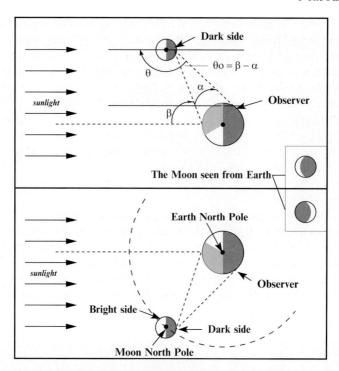

Fig. 3.5 A not-to-scale illustration of the Sun–Earth–Moon system viewed from the pole of Earth's orbit. In the *top* panel, the Earth's topocentric phase angle, α, with respect to the observer is defined. The plot also shows the Moon's selenographic phase angle, θ. β is the angle between the sunlight that is incident somewhere on the Earth and reflected, as earthshine. $\theta_0 (= \beta - \alpha)$ is the angle between the earthshine that is incident and reflected from the Moon. The path of the earthshine is indicated by the *broken lines*. θ_0 is of order 1° or less. In the *lower* panel the same diagram is drawn for a negative lunar phase angle, and extra features like the Moon's orbit around the Earth are indicated. On both panels the aspect of the Moon as would be seen from the nightside Earth is also indicated in a *box*. The *light-shaded* areas of the Earth indicate the approximate longitude range that contributes to the earthshine. Adapted from Qiu et al. (2003)

The Earth's differential cross-section is defined as the ratio between the scattered radiation per unit solid angle and the incident radiation of a given surface, and depends on its geometrical albedo and its phase function. The geometrical albedo (the ratio between the intensity of normally incident radiation reflected from a surface and the incident intensity) is independent of β, the Earth's phase angle; rather, it is proportional to the backscattered cross-section. At the top of the atmosphere, the differential cross-section of the reflected sunlight for scattering by an angle β is given by

$$\frac{d\sigma}{d\Omega} \equiv p_e f_e(\beta) R_e^2,$$

where R_e is the radius of the Earth, p_e is the geometrical albedo of the Earth and $f_e(\beta)$ is the Earth's phase function, defined such that $f_e(0) = 1$.

3.1 Globally Integrated Observations of the Earth

The bi-directional nature of the Earth's reflectance is included in $p_e f_e(\beta)$. This is a recognized central difficulty for satellite-based albedo estimates, where the solar zenith angle (function of latitude, date and time), the viewing zenith angle and the relative Sun–Earth scene-satellite azimuth must be taken into account with more or less reliable angular models (Loeb et al. 2003; Wielicki and Green 1989). The sampling is necessarily different, with earthlight on the Moon, but the problem does not go away because earthshine measurements are restricted to the light scattered to the orbital plane of the Moon around the Earth. To derive ideally perfect estimates of the Earth's reflectance, it would be necessary to observe reflected radiances from the Earth, from all points on the Earth and at all angles. Therefore, all measurements from which albedo can be inferred require assumptions and/or modelling to derive a good measurement.

Already in the twenty-sixth century, Leonardo da Vinci correctly deduced that Earthshine was due to sunlight reflected from the Earth toward the dark side of the Moon (Fig. 3.6). In Leonardo's Codex Leicester, circa 1510, there is a page entitled *Of the Moon: No solid body is lighter than air*. In there, he states his belief that the Moon possessed an atmosphere and oceans, and that it was a fine reflector of light, because it was covered with so much water. In fact, we now know that ocean areas as

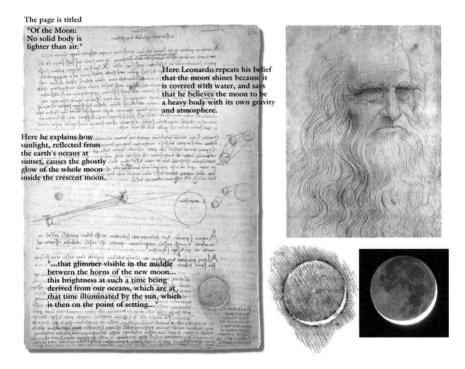

Fig. 3.6 Reproduction of a page of Leonardo Da Vinci's Codex Leicester, circa 1510. Also shown is a image of Da Vinci in his *Autoportrait* and a comparison of a real picture of the earthshine with one of his drawings. Credit: American Museum of Natural History Library

among the darkest (less reflective) surfaces on Earth. He also speculated about how storms on Earth could cause the Earthshine to become brighter or dimmer, which is indeed observable with modern instrumentation.

After Leonardo's ideas, others continued to observe the earthshine, in historical times. Among them, Galileo Galilei (1564–1642), based on previous studies of the reflectance properties of different materials made by his friend Paolo Sarpi (1552–1623), notes in his book *Dialogue Concerning the Two Chief World Systems* that: *When it is dawn in Italy the Moon faces a terrestrial hemisphere with fewer seas and more land, containing all Asia. When, by contrast, it is evening in Italy at the beginning of the month, the ashen light is much weaker, for the new Moon faces only the westernmost portions of Europe and Africa and then the immense stretch of the Atlantic ocean* (Byard 1999).

In the past few years, with the blooming of exoplanet detections and the plans for future missions, there has been renewed interest in the earthshine measurements. This has led several research groups to undertake observation campaigns of the earthshine in the visible and infrared ranges. Observations of the earthshine allow us to explore and characterize its photometric, spectral and polarimetric features, and to extract precise information on what are the distinctive characteristics of Earth, and life in particular. They also allow us to quantify how these spectral features change with time and orbital configurations.

3.2 The Earth's Photometric Variability in Reflected Light

The light reflected by the Earth toward the direction of an hypothetical, far away, observer will change in time depending on the orbital phase, rotation, seasonality (tilt angle) and weather patterns. Each observing perspective of Earth will have its unique features and will see a different photometric variability. Figure 3.7 illustrates this point: several views of Earth are represented for the exact same date and time, but judging from the visible scenery features, the three images could well represent three different planets. Note however, that the figures are misleading in the sense that clouds are missing.

A long-time series of the globally integrated variability in the Earth's reflected light is not available. There have been, however, a number of earthshine measurements that have documented changes with phase and rotation, but they are necessarily limited in their range of observations. Sporadic data have also been taken from remote platforms, usually on their way toward other solar system bodies, but without any temporal continuity. Earth-observing satellite data is usually obtained once or twice a day over each particular point of the Earth and a global photometric light curve, from the perspective of a distant observer, cannot be constructed from these data. Thus, to fully characterize the reflected light variability of the Earth, we need to combine the observations with reflectance models.

3.2 The Earth's Photometric Variability in Reflected Light 117

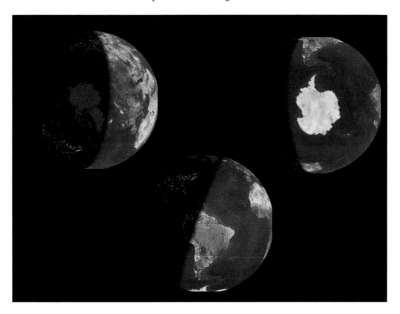

Fig. 3.7 The Earth from different viewpoints. The three images show the Earth for the exact same day and time (19/11/2003 at 10:00 UT) but from three different perspectives: from 90° above the ecliptic (north polar view) (*right*), from 90° below the ecliptic (south polar view) (*left*) and from within the ecliptic plane (*centre*)

3.2.1 Observational Data

The most important historical program of photometric earthshine measurements was carried out by André-Louis Danjon (1890–1967) from a number of sites in France. Previously, only Very (1915) reported on photographic spectra of the earthshine. From 1926 to 1935, Danjon made hundreds of earthshine measurements, and J. Dubois (1947) continued the program through 1960 from the observatory at Bordeaux. Using a 'cat's-eye' photometer,[2] Danjon stopped-down the light from the sunlit portion of the Moon to match the brightness of the ashen portion. This differential measurement removed many of the uncertainties associated with varying atmospheric absorption and the solar constant, allowing Danjon to achieve an estimated uncertainty of roughly 5% (Danjon 1954). He recorded the measurements using the so-called *Danjon scale*, in which zero equals to a barely visible Moon.

Danjon used his observations to estimate the mean global albedo. As the observations are only at visible wavelengths, they must be corrected for the balance of the short-wavelength radiation, most of which is in the near IR. Estimates of this

---

[2] A cat's-eye photometer produces a double image of the Moon, allowing the visual comparison of the intensities of two well-defined patches of the lunar surface – one in sunlight and the other in the earthshine – at various lunar phases.

correction were made by Fritz (1949) after taking into account the decrease of the Earth's albedo with increasing wavelength (our 'blue planet'). Fritz also attempted to correct for the geographical bias in Danjon's observations. The Earth eastern hemisphere (Asia, Russia), which was most frequently observed by Danjon, has a greater fraction of land than does the globe as a whole, implying that Danjon's value would be high because the sea is dark compared to land. Combining the decreases from the absence of the IR and geographical bias, Fritz found that Danjon's visual albedo of 0.40 corresponds to a Bond albedo (considering all the wavelengths and directions) of 0.36 (see Chap. 2 for a discussion about albedos).

Flatte et al. (1992) noted that a correction must be made for the 'opposition effect' present in lunar reflectance properties. Observations of the Moon show that the Moon's reflectivity has a strong angular dependence, which was unknown in Danjon's time (Fig. 3.8). This enhancement was once thought to be due to the porous nature of the lunar surface (Hapke 1971). More modern work has shown it to be caused by both coherent backscatter of the lunar soil (a non-linear increase in reflectance at small reflection angles) and shadow hiding (the disappearance of shadowed regions at small reflection angles, effectively increasing the reflecting

Fig. 3.8 The Moon showing the bright side and the earthshine. Half of the image (*the bright side*) was taken with a blocking filter to make the two portions of the Moon directly comparable. In the image the lunar phase is 115.9°, near a declining quarter Moon. Unlike the moonshine, the earthshine is flat across the disk. The flatness is due to the uniform, incoherent back scattering (non-Lambertian), in contrast to the forward scattering of sunlight occurring in the sunlit lunar crescent surface

area) in roughly equal amounts (Hapke et al. 1993, 1998; Helfenstein et al. 1997). In fact, an incorrect lunar phase function was the primary source of Danjon's overly large visual albedo (Pallé et al. 2003).

Danjon's and Dubois' results show a number of interesting features. The daily mean values of the observations vary more widely than would be expected on the basis of the variation of measurements on a single night. This can plausibly be attributed to daily changes in cloud cover or changes in the reflection because of changes in the Sun–Earth–Moon alignment during a lunar month. The typical lifetime of large-scale cloud systems (thousands of kilometre) is 3 days (Ridley 2001), but from one night to the next the Earth's area contributing to the earthshine changes (see Fig. 3.5). Unfortunately, extensive cloud-cover data were not available at the time of Danjon's and Dubois' observations.

Danjon (1928, 1954) also examined his observations to determine whether there was a long-term trend in albedo, but found none. Dubois' observations for some 20 years, ending in 1960, showed considerable annual variability, which he speculated was due to solar activity. His published monthly variations from 1940 to 1944, for example, showed a strong correlation with the 1941–1942 El Niño (Dubois 1958). In the past 40 years, there have been continuous observations of earthshine by Huffman et al. (1989) and sporadic observations by Franklin (1967) and Kennedy (1969). Hilbrecht and Kuveler (1985) analyzed about a year and a half of continuous observations of the earthshine by amateur astronomers. Dollfus (1957) also observed the earthshine and took some measurements of its linear polarization, which we discuss in Sect. 3.5.

At present, continuous photometric observations of the earthshine are carried out simultaneously from several observatories by the Earthshine Project, covering the spectral range from 400 to 700 nm (Qiu et al. 2003; Pallé et al. 2003). With the improvement in available technologies, such as CCD detectors, current measurements are about an order of magnitude more precise to those of Danjon's and Dubois's. This is illustrated in Fig. 3.9, where observations of the Earth's reflectance during two nights in 2004 are plotted. The changes in reflectance are due to the evolution of the cloud and geography pattern, over the sunlit Earth, during the observations. Figure 3.9 shows the cloud patterns (top) and the measured albedo for the same nights (bottom). A significant decrease or increase in the detected albedo is observed when large cloud-free ocean areas (with very low albedo) or large optically thick cloudy areas (with high albedo) appear, respectively, in the earthshine-contributing area of Earth. From these measurements the photometric variability of the Earth is found to be of the order of 10–15% from night to night and during one Earth rotation (Qiu et al. 2003). Goode et al. (2001) have shown that over a year and a half, the Earth's seasonal variation is larger than 10%.

The Earth's albedo also undergoes a seasonal cycle (Loeb et al. 2007), which is plotted in Fig. 3.10. The Earth appears brighter during the months from November to January, and has a secondary brightness peak in May–June. These global albedo variations occur because of the north–south asymmetry of the continental land distribution, the illumination changes in the polar regions, the seasonal changes in the extent of snow and ice cover and the seasonal changes in meteorological parameters,

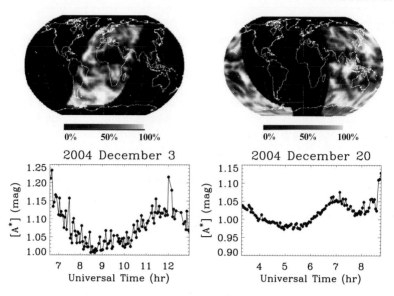

Fig. 3.9 *Top*: Earthshine-contributing areas of the Earth for two nights, 3 December 2004 (*left*), with phase angle −71°, and 20 December 2004 (*right*), with phase angle +60°. *Bottom*: Temporal anomalies in the apparent terrestrial albedo for these two nights. Adapted from Montañés-Rodríguez et al. (2007). Reproduced by permission of the American Astronomical Society

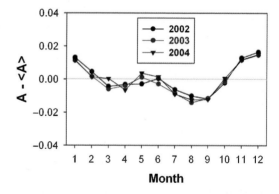

Fig. 3.10 The seasonal cycle of the Earth's global albedo. The variations are caused by changes in the Earth's obliquity, the extent of snow and ice and in forest leaf coverage. The data were collected by the CERES instruments aboard the Terra and Aqua satellites. Courtesy: N. Loeb

more specifically the clouds. Of all these factors, the polar caps' contribution is by far the largest. These changes are indicative, as all seasons are, of a non-zero planet obliquity, that is solar heating angle is what determines the cloud and weather patterns.

Nowadays, there are several satellite instruments that are able to monitor the Earth's reflectance. CERES[3] instruments, one of the more precise datasets, are flying onboard polar orbiter satellites, and map each point of the Earth twice a day, at

---

[3] http://science.larc.nasa.gov/ceres

midnight and noon, from a nadir perspective (looking straight down). Thus, the seasonal changes they observe are introduced only by changes in solar illumination and surface albedo (mostly snow/ice extent). For a different observer, with a fixed observational perspective, the seasonal cycle might differ, as we see in the next section.

3.2.2 Reflectance Models

For an observer outside the solar system, Earth observations will differ from the mostly equatorial perspective offered by the earthshine measurements or the global mean from polar orbiter satellites. The Earth's reflectance in the direction of β, where β is defined as the angle between the Sun–Earth and Earth–Observer vectors, can be expressed as

$$p_e f_e(\beta) = \frac{1}{\pi R_e^2} \int_{(\hat{R}\cdot\hat{S}, \hat{R}\cdot\hat{M}) \geq 0} d^2 R (\hat{R}\cdot\hat{S}) a (\hat{R}\cdot\hat{M}) L, \quad (3.1)$$

where $\hat{R}$ is the unit vector pointing from the centre of the Earth to a patch of Earth's surface, $\hat{S}$ is the unit vector pointing from the Earth to the Sun and $\hat{M}$ is the unit vector pointing from the Earth toward the observer. The integral is over all of the Earth's surface for which the sun is above the horizon (i.e. $\hat{R}\cdot\hat{S}$) and visible from the observer's perspective (i.e. $\hat{R}\cdot\hat{M} \geq 0$). Here a is the albedo of a given surface and L is the anisotropy function, dependent on surface type, cloud cover and geometric angles.

In the context of observing extrasolar planets, choosing the observer's position (β) is similar to fixing the inclination of the Earth's orbit with respect to the observer. Thus, the total reflected flux in a given direction, β, can be calculate using

$$F_e(\beta) = S\pi R_e^2 p_e f_e(\beta),$$

where S is the solar flux at the top of the Earth's atmosphere (1,370 W m^{-2}). There is a systematic variation of $p_e f_e(\beta)$ throughout the Earth's orbital period (sidereal year), and fluctuations of $p_e f_e(\beta)$ about its systematic behaviour are caused by varying terrestrial conditions, including weather and seasons (Pallé et al. 2004).

By taking into account the cloud distribution measured from satellite weather observations, Pallé et al. (2008) simulated the photometric variability of Earth from several viewpoints (Fig. 3.11). The Earth is much brighter when half of its surface is illuminated by the Sun, and it becomes much dimmer, from a remote observer's perspective, as it becomes a thinner and thinner crescent at small β angle. The small amplitude variability in the figure is due to the Earth rotation introducing changes in scenery. In the figure, the variation is normalized by the glare of the Sun. The Earth's reflected flux, F_e, that a distant observer would see is about 10^{-10} times smaller than the flux he would receive from the Sun.

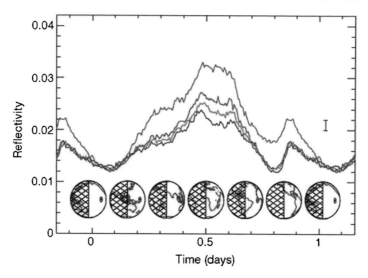

Fig. 3.13 The model light curve of the cloud-free Earth. The *pink, red, green* and *blue curves* correspond to wavelengths of 750, 650, 550 and 450 nm, and their differences reflect the wavelength-dependent albedo of different surface components. The images below the light curve show the viewing geometry (the cross-hatched region is not illuminated), a map of the Earth (*red*) and the region of specular reflection from the ocean (*blue*). At t = 0.5 day, the Sahara desert is in view, causing a large peak in the light curve owing to the reflectivity of sand, which is especially high in the near-infrared (*pink curve*). Adapted from Ford et al. (2001). Reprinted by permission from Macmillan Publishers Ltd: [Nature] Vol. 412, p. 885, copyright (2001)

the real cloudy Earth (right panels), as the data are subdivided into smaller integration periods, the albedo error at each rotational phase does not decrease, because of the random influence of clouds at short time scales (Pallé et al. 2008). It is also noticeable how the change in the shape of the light curve is smooth (ordered in time) from one series to the next in the case of a cloudless Earth, but it is random for the real Earth.

3.2.4 The Rotational Period

The diurnal light curve of the Earth (or any other astrophysical object) can be constructed only if the rotational period is known. To determine the rotational period of the Earth, one needs to analyze a time series of the reflected light and search for periodicities. In principle, weather patterns and/or the orbital motion of the Earth could pose a fundamental limitation that would prevent a remote observer from being able to make an accurate determination of the Earth's rotation period from the scattered light. Since the scattered light is dominated by clouds, it might be impossible to determine the rotation period if the weather patterns were completely random. Alternatively, even if the atmospheric patterns were stable over many rotation

3.2 The Earth's Photometric Variability in Reflected Light

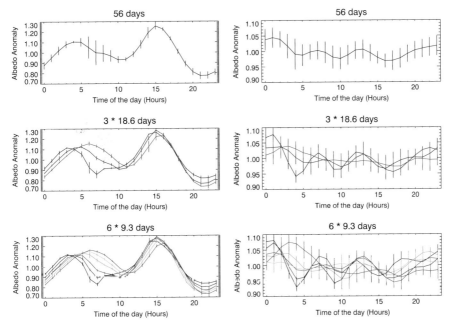

Fig. 3.14 Light curves of the Earth as observed from the ecliptic plane, at orbital phase 90° (a quarter of the Earth's surface is visible). The albedo anomaly scale is defined as the deviation from the mean value over the entire 2 month dataset. *Left* column are the light curves of a cloud-free Earth and *right* column are the light curves for the real Earth. In the *lower panels*, the simulations are divided in six sub-series, for analysis. Note the contrast between the uniformity of the light curves of an ideal (cloudless) Earth and the real light curves. Adapted from Pallé et al. (2008). Reproduced by permission of the American Astronomical Society

periods, observational determinations of the rotation period might not correspond to the rotation period of the planet's surface, if the atmosphere were rotating at a very different rate (e.g. Venus).

Clouds are common on the solar system planets, and even on satellites with dense atmospheres. In fact, clouds are also inferred from observations of free-floating substellar mass objects (Ackerman and Marley 2001). Hence, cloudiness appears to be a universal phenomenon. On Earth (Fig. 3.15), clouds are continuously forming and disappearing, covering an average of about 60% of the Earth's surface (Rossow 1996). This feature is unique in the solar system to Earth: some solar system planets are completely covered by clouds, while others have very few. Only the Earth has large-scale cloud patterns that partially cover the planet and change on timescales comparable to the rotational period. This is because the temperature and pressure on the Earth's surface allow for water to change phase with relative ease from solid to liquid to gas.

Pallé et al. (2008) found that scattered light observations of the Earth could accurately identify the rotation period of the Earth's surface. This is because large-scale time averaged cloud patterns are tied to the surface features of Earth, such as

Fig. 3.15 The Earth cloud system, a unique feature in the solar system. Image credit: NASA

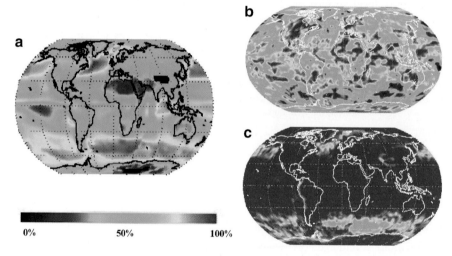

Fig. 3.16 Large-scale cloud variability during the year 2004 from ISCCP data. In panel (**a**) the 2004 yearly mean cloud amount, expressed in percentage coverage, is shown. In panels (**b**) and (**c**) cloud coverage variability (ranging also from 0 to 100%) is illustrated over a period of 2 weeks and 1 year, respectively. Note how the cloud variability is larger at weekly time scales in the tropical and mid-latitude regions than at high latitudes. Over the course of a whole year the variability is closer to 100% over the whole planet (i.e. at each point of the Earth there is at least a completely clear and a completely overcast day per year). One exception to that occurs at the latitude band near $-60°$, an area with heavy cloud cover, where the variability is smaller, i.e. the stability of clouds is larger. Adapted from Pallé et al. (2008). Reproduced by permission of the American Astronomical Society

continents and ocean currents. This relatively fixed nature of clouds (illustrated in Fig. 3.16) is the key point that would allow Earth's rotation period to be determined from afar. Figure 3.16 shows the averaged distribution of clouds over the Earth's surface for the year 2004. The figure also shows the variability in the cloud cover during a period of 2 weeks and over the whole year 2004. The lifetime of large-scale cloud systems on Earth is typically about a week (roughly 10 times the rotational period). In the latitude band around 60° south, there is a large stability produced by the vast, uninterrupted oceanic areas, which will perhaps be the characteristic of possible 'ocean planets' (see Chap. 7).

Among other important physical properties, the identification of the rotation rate of an exoplanet with relative accuracy will be important for several reasons: to understand the formation mechanisms and dynamical evolution of extrasolar planetary systems; to improve our analysis of future direct detections of exoplanets, including photometric, spectroscopic and potentially polarimetric observations; to recognize exoplanets that have active weather systems; and to suggest the presence of a significant magnetic field.

If the rotation period of an Earth-like planet can be determined accurately, one can then fold the photometric light curves at the rotation period to study regional properties of the planet's surface and/or atmosphere. With phased light curves we could study local surface or atmospheric properties with follow-up photometry, spectroscopy and polarimetry to detect surface and atmospheric inhomogeneities and to improve the sensitivity to localized biomarkers. Exoplanets, however, are expected to deviate widely in their physical characteristics and not all exoplanets will have photometric periodicities. If they have no strong surface features, as is the case for Mercury or Mars, or they are completely covered by clouds, as is the case for Venus and the giant planets, determining the rotational period may be an impossible task. In fact, Earth may well be the only one of the major planets for which a rotational period can be easily established from a distance of several Angstrom unit.

3.2.5 *Cloudiness and Apparent Rotation*

A long time series of photometric Earth observations can be subdivided into several subsets and analyzed for significant periodicities. In Fig. 3.17, modelled photometric observations of Earth (Pallé et al. 2008) spanning 2 months are subdivided in several equal-length subperiods (e.g. six periods each of about 9 days) and analyzed independently, so that the changes in viewing geometry are minimized. In this case, several peaks appear in the Fourier periodogram near 12 and 24 h. For Earth models with clouds, the best-fit rotation period shifts slightly to shorter periods. The shifts in the best-fit periodicity from the true periodicity are completely absent when considering an Earth model free of clouds for the same dates and times, even when including added noise. Therefore, Pallé et al. (2008) concluded that they are produced by variable cloud cover.

The shifts are introduced by the large-scale wind and cloud patterns (Houghton 2002). Since clouds are displaced toward the west (in the same direction of the Earth's rotation) by the equatorial trade winds (and to a minor extent by the polar easterlies), the apparent rotational period should be shorter than the rotation period

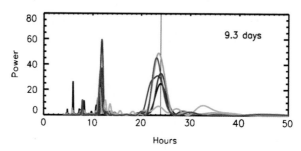

Fig. 3.17 Periodogram analysis of the Earth's scattered light, $p_e f_e(\beta)$, from an equatorial viewing angle. Two months of reflected light simulations (56 days), subdivided in six equally long time series, are used to calculate the periodograms. In the figure, different colours indicate different data subperiods. Note the appreciable decrease in the retrieved rotation rate for some of the time series, detectable with both autocorrelation and Fourier analysis. Adapted from Pallé et al. (2008). Reproduced by permission of the American Astronomical Society

of the surface. On the other hand, when clouds are moved toward the east (in the opposite direction of the Earth's rotation) by the westerly winds at mid-latitudes, the apparent rotational period should be longer than the rotation period of the surface.

In principle, both longer and shorter periodicities could be present in the periodograms, depending on the particular weather patterns. The models, however, often find shorter apparent rotation rates, but not longer (Pallé et al. 2008). The explanation probably lies in the different mechanisms of cloud formation on Earth. In the tropical regions most of the clouds develop through deep convection. This deep convective clouds have a very active cycle and a short lifetime, in other words, these cloud systems do not travel far. At mid-latitudes, however, deep convection does not occur and large weather and cloud systems remain stable (and moving) for weeks (Xie 2004).

For an extrasolar Earth-like planet, photometric observations could be used to infer the presence of a 'variable' surface (i.e. clouds), even in the absence of spectroscopic data. This would strongly suggest the presence of liquid water on the planet's surface and/or in the planet's atmosphere, especially if the effective temperature of the planet was also determined by other means.

3.2.6 Glint Scattering

Most planetary surfaces in the Solar System are diffuse scatters and are often approximated by an isotropic (Lambertian) scattering law, the notable exceptions being liquids such as water on Earth or ethane on Titan, smooth ices, and optically thin clouds (Campbell et al. 2003). Although the reflectivity of water is very low at high and medium angles of incident light, it increases tremendously at small angles. This is observable from space on the sun-illuminated side of the Earth near the terminator and it is known as the *glint*.

3.2 The Earth's Photometric Variability in Reflected Light

Consider a planet of radius R_p at a distance d_p from a star of luminosity L_* that reflects an amount of light F_p in the direction of Earth equal to

$$F_p = \frac{L_*}{4\pi d_p^2} \frac{f_A \pi R_p^2}{2\pi} \left[f_c a_c + f_l a_l + f_s a_s + \frac{f_w p_w a_w}{f_\Omega} \right],$$

where the parameter $f_A = (1 - \cos(\theta) \sin(i))/2\pi$ is the fraction of the visible planetary disk illuminated by the star, i is the orbital inclination and θ is the orbital phase. The parameters f_c, f_l, f_s, and f_w are the disk area fractions covered by clouds, land, snow/ice and water, respectively, and a_c, a_l, and a_s are the mean diffusive albedos of the first three surfaces. The albedo of clean seawater is $a_w = R_\perp^2 + R_\parallel^2$, where $R_\perp$ and $R_\parallel$ are the Fresnel reflection coefficients for two polarization directions (Griffiths 1998) and depends strongly on illumination angle. Water is extremely dark ($a \sim 0.04$) for zenith angles <45°, but is mirror-like at angles approaching glancing incidence when its albedo climbs steeply toward 100%.

Specularly reflected light from seawater is scattered into a solid angle equal to the area angle of the star on the planet, $\sim 6.3 \times 10^{-5}$ sr for the Sun on the Earth. Glint from a wavy ocean surface arrives from an area considerably wider than this, approximately 30° wide covering 0.214 sr (Williams and Gaidos 2008). This is comparable to what is observed in satellite images of the Earth, like the one illustrated in Fig. 3.18.

Fig. 3.18 Sunglint in the Sargasso Sea, obtained by GOES-West, on 22 June 2000. The *top corner* illustrates the glint phenomenon at a smaller, more familiar, scale. Credit: NASA

Williams and Gaidos (2004) and Gaidos et al. (2006) examined the unpolarized variability of the sea–surface glint from an Earth-like extrasolar planet. When starlight is incident on a planet surface from overhead, only a very small fraction of the ocean area reflects the starlight to a distant observer; the rest contributes nothing to the specular reflection of direct starlight. However, for planets in crescent phase with orbital inclinations near 90°, the ocean is obliquely illuminated and specular scattering is due to waves of small slope, which occur with greater probability than large ones. In addition, the albedo of ocean water a_w at glancing incidence grows to >0.9 or more than 20 times the albedo at normal incidence. Thus, the specular term in the last equation becomes much larger than the diffuse signal at this phase angle.

To assess the detectability of specular reflection from an ocean in the (disk-averaged) reflected light from a distant planet, Williams and Gaidos (2008) used a model to simulate the orbital brightness variation, or 'light curve', of Earth seen at an optimum ($i = 90°$) edge-on viewing geometry. Phase-weighted light curves are shown in Fig. 3.19 for a model Earth with no clouds and with 50% clouds. An important difference between the Lambertian planet and the cloudy Earth is the anomalous brightening of the model Earth relative to the Lambertian planet near inferior conjunction ($\theta = 0°$). At these phase angles, the planet is a thin crescent and demonstrates efficient specular reflection of starlight that is incident on the ocean surface at an oblique angle. Thus, ocean-bearing planets with edge-on orbital geometries could exhibit large changes in their apparent reflectivity due to specular reflection (Fig. 3.20). The statistical probability of a random planet having an orbital inclination less than i_0 is $(1 - \cos(i_0))/2$. Thus, 86.7% of extrasolar planets will have inclinations between 30 and 150°, and half of all planets will have inclinations in the range 60–120° where specular reflection is most pronounced. The effect can also be seen on a modelled planet like Earth with 50% cloud cover.

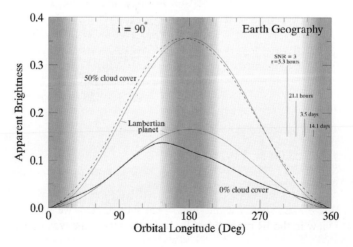

Fig. 3.19 Albedo variation of the model Earth seen at different orbital inclinations. The real Earth geography is used. The case of Earth with 50% cloud cover and 90° inclination is indicated with a *dashed line*. Earthshine data are from Pallé et al. (2003). Reprinted from Williams and Gaidos (2008) with permission from Elsevier

3.3 Earth's Infrared Photometry

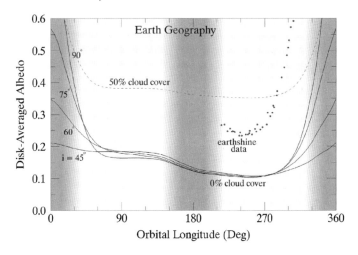

Fig. 3.20 Light curve of a model Earth (*solid black line*, 0% clouds; *dashed line*, 50% clouds) compared to an idealized Lambertian planet (*gray line*) having a uniform albedo equal to 0.35. Earth is brightest near $\theta = 150°$ when the illuminated area of northern-hemisphere snow and ice is maximum. Strong specular reflection relative to a Lambertian planet of albedo unity is apparent in the pair of bumps near $\theta = 20°$ and $340°$. Vertical gray zones denote regions where a planet would be too close to its star to resolve (assuming a working angle $4\lambda/D$) with a coronagraph telescope from a distance of 10 parsecs. Reprinted from Williams and Gaidos (2008) with permission from Elsevier

Similar phase brightening of the crescent Earth has been empirically identified in earthshine data collected from the gibbous Moon (Pallé et al. 2003). The earthshine data in Fig. 3.20 shows that the real Earth is fainter and a better specular reflector than the model Earth under fractional cloud cover, implying that the illuminated Earth beneath the Moon had fewer diffuse-scattering clouds at the time of the observations. Earthshine has also been found to be strongly polarized (see Sect. 3.5 for a detailed discussion on polarization), which is further indication of specular reflection of sunlight off the oceans (Stam et al. 2006). The Earth's glint is also captured in the spectacular flyby data[4] collected by the MESSENGER spacecraft in 2005.

3.3 Earth's Infrared Photometry

In Chap. 1, several pictures of the Earth at infrared wavelengths, taken by the Mars Odyssey and the MESSENGER missions, were shown. Here we show in Fig. 3.21 a full Earth view in the IR range taken by Rossetta, where the weather patterns are more prominent.

---

[4] http://messenger.jhuapl.edu

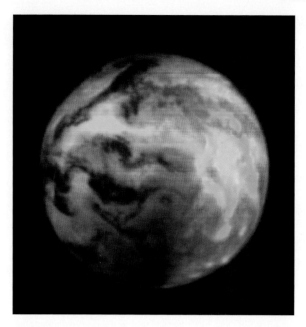

Fig. 3.21 Composite image of Earth from three different infrared wavelengths. This image was taken on 5 March 2005 after Rosetta's closest approach to Earth by VIRTIS from a distance of 250,000 km and with a resolution of 62 km per pixel. It is worth noticing that by selecting the three different wavelengths in the continuum and avoiding the water spectral bands (see Fig. 1.17), the difference between the illuminated and dark side of the Earth is still apparent. Image credit: ESA

For extrasolar planets, IR photometry has not been so far considered an useful tool in characterizing their atmospheric/suface signature, because the variability is smaller than in the visible range. The infrared emission of a planet as a whole is very dependent on its surface and atmospheric properties. A planet with an atmosphere and oceans, such as the Earth, exhibits a small range of emission temperatures between day and night, due to the very large thermal inertia of the ocean and air (Gaidos et al. 2006). On the other hand, the diurnal variability that a planet without atmosphere will present will be very large, mainly dependent on the thermal inertia of its surface.

It is, however, impossible to determine planetary properties unambiguously from IR variability alone. A planet with marked IR variability may point to a lack of atmosphere, a geologically old surface or high orbital obliquity. On the other hand, lack of variability may point to a planet like Earth, a barren planet with low obliquity or a thick atmosphere Venus-like world (Gaidos et al. 2006).

Gómez et al. (2009) have simulated the infrared photometric variability of the Earth observed from an astronomical distance, based on measured hourly data of the broadband outgoing longwave radiation ($\lambda > 4\,\mu m$) at the top of the atmosphere and a geometrical averaging model. The radiation data were obtained from

3.3 Earth's Infrared Photometry

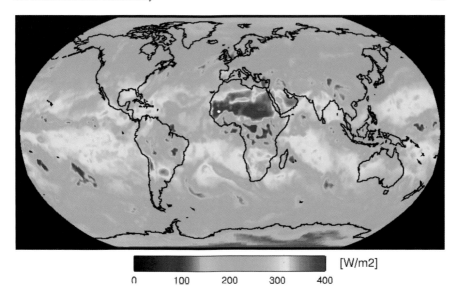

Fig. 3.22 Outgoing longwave radiation map for 1st April 1999 between 15:00 and 18:00 UT

the Global Energy and Water-cycle Experiment (GEWEX). In Fig. 3.22, a snap-shot of the outgoing longwave radiation for 1st April 1999 between 15:00 and 18:00 UT is shown.

On average, the daily maximum integrated IR flux, in the direction of a distant observer, is found when the Sahara region is within the observer's field of view. The minimum emitted flux is found when the Pacific and Asia/Oceania regions are in the field of view. Thus, IR emission from hot desert surfaces is the dominant factor, although the infrared flux is altered by the presence of clouds. Gómez et al. (2009) found that the determination of the rotational period of the Earth from IR photometric observations is feasible in general, but it becomes impossible at short time scales if cloud and weather patterns dominate the emission.

Infrared radiation is routinely monitored from satellites, and in Fig. 3.23, the seasonal cycle of the Earth's globally averaged infrared emission is given. Data were collected by the CERES instruments from polar orbiter satellites. There is a peak in thermal emission from Earth in July and August when the overall emitted flux is upwards by about $8\,\mathrm{W\,m^{-2}}$ with respect to December and January. This difference is mainly accounted for due to the different northern and southern hemisphere continental distribution.

This seasonal cycle would also be appreciable for a remote observer. Along the year, a 4–5% change in the IR flux emitted in the direction of a remote observer in the plane of the ecliptic is observed, with a peak during July and August (Fig. 3.24). There is, however, a large variability from the day to day and from 1 year to the next due to the presence of clouds (Gómez et al. 2009).

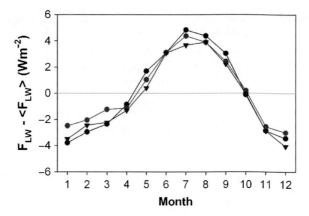

Fig. 3.23 The seasonal cycle of the Earth's global thermal infrared emission. The variations are caused by changes in the Earth's obliquity, but more importantly by the different continental distribution between the two hemispheres. The data were collected by the CERES instruments aboard the Terra and Aqua satellites. Image courtesy: Norman Loeb

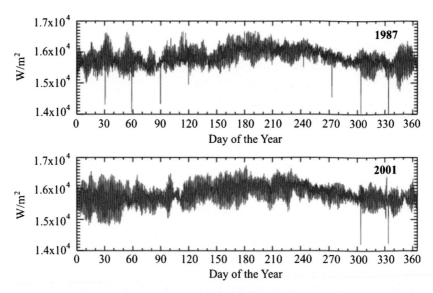

Fig. 3.24 Earth outgoing longwave radiation time series for two separate years 1987 and 2001, observed as a single point

3.4 Spectroscopy of Planet Earth

Globally integrated spectroscopic measurements are of great importance because they can reveal information about the composition and chemistry of the Earth's atmosphere and surface. Some of these spectral features are common to the rest of the solar system planets, and some are characteristic only to Earth and to the life it harbours.

In the past, there have been numerous attempts to observe the spectrum of the pale blue dot from the distance in different spectral regions. Among other efforts, observations of the Earth as a planet from afar have been made in the visible range from the Galileo spacecraft (Sagan et al. 1993), and from the Mars Global Surveyor spacecraft, at a distance of 4.7 million km in the thermal infrared (Christensen and Pearl 1997). More recently, the TES spectrograph onboard the Mars Global Surveyor took a spectrum of planet Earth on 24 November 1996.

With the aim of improving the spectral resolution and the sampling of seasonality and phase changes, spectroscopic measurements of the earthshine have also been taken from ground-based observatories and analyzed at visible wavelengths (Woolf et al. 2002; Montañés-Rodríguez et al. 2005, 2006) and in the near infrared wavelengths (Turnbull et al. 2006).

3.4.1 The Visible Spectrum

The overall shape of the Earth's spectrum in the visible region shows interesting and peculiar signatures. The most prominent is the Rayleigh scattering, the main source of opacity in the Earth's atmosphere, which shows an enhancement towards the blue part of the spectra (Fig. 3.25).

The Belorussian astronomer Gavriil A. Tikhov (1875–1960) discovered the blue colour of the Earthshine and correctly interpreted it as being due to the Rayleigh scattering in the atmosphere (Tikhov 1914). The Rayleigh scattering, named after Lord Rayleigh (1842–1919), is produced by particles much smaller than the wavelength of the light they scatter. It occurs when light travels in transparent solids and liquids, but is most prominently seen in gases. The amount of Rayleigh scattering that occurs to a beam of light depends upon the size of the particles and the wavelength of the light. In particular, the scattering coefficient varies inversely with the fourth power of the wavelength. Thus, the Rayleigh scattering cross section, σ_s, is given by

$$\sigma_s = \frac{2\pi^5}{3} \frac{d^6}{\lambda^4} \left(\frac{n^2 - 1}{n^2 + 2} \right)^2,$$

where n is the refractive index of the particle, d is the diameter of the particle, and λ is the light's wavelength. The strong wavelength dependence of the scattering (λ^{-4}) means that blue light is scattered much more than red light. The Rayleigh scattering of sunlight by the Earth's atmosphere is the main reason why the sky is blue, and our planet is known as the blue planet.

Except Neptune and Uranus, no other solar system body shows this strong Rayleigh feature in the blue. However, Neptune's and Uranus' predominant blue colour does not come from Rayleigh scattered light, but from the absorption of red and infrared light by methane gas in their atmospheres. For extrasolar planets, it would be relatively easy to distinguish between a Neptune-size and an Earth-size planet.

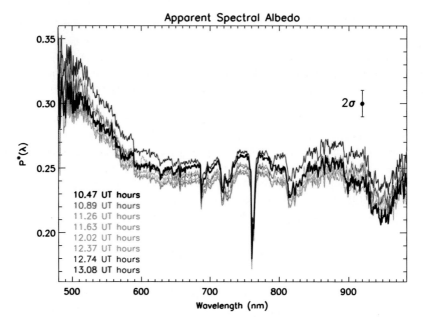

Fig. 3.25 Several Earthshine spectra, taken with the 60″ Telescope at Palomar Observatory on November 2003. Different colours indicate different observing times along the night. The main features of the Earths reflectance in this region include an enhancement due to the Rayleigh scattering in the blue, part of the Chappuis O_3 band, which contributes to the drop above 500 nm. Atmospheric absorption bands due to oxygen – the sharpest, A-O_2 at 760 nm – and water vapour are clearly detected. The surface vegetation edge, which is expected to show an apparent bump in the visible albedo above 700 nm, is not strong, and neither does it seem to vary appreciably or systematically throughout the night. Adapted from Montañés-Rodríguez et al. (2005). Reproduced by permission from the American Astronomical Society

At short wavelengths (<310 nm) ozone absorption dominates over the Rayleigh scattering, causing a strong decrease in reflectance, which has also been measured through earthshine observations (Hamdani et al. 2006).

Hitchcock and Lovelock (1967) pointed out that the remote detection of life forms might be possible by studying the atmospheric composition of a planet. In this spectral range, the presence of H_2O, O_2 and O_3 in the Earth's atmosphere is clearly marked. This particular combination of atmospheric constituents are in disequilibrium and strongly signals to the presence of life on Earth, the driver of the disequilibrium.

Indications of a sharp increase in reflectivity near $0.720\,\mu$m, and longward, have also been identified as due to chlorophyll in vegetation (Montañés-Rodríguez et al. 2006; Arnold et al. 2002). However, vegetation is not the only possible source of the bump in the 680–740 nm region; cloud effects may also cause it, because varying the quantity or type of clouds in the scenery will produce differential changes in albedo along the spectra (Tinetti et al. 2006). A detailed discussion about the vegetation signatures and other biomarkers on Earth is given in Chap. 5.

3.4.2 The Infrared Spectrum

In the very near-infrared (1–2 μm), earthshine observations are still possible, and the reflected Earth spectrums is dominated by water absorption bands (Turnbull et al. 2006). However, at longer wavelengths, earthshine observations cannot be obtained due to the strong absorption of the Earth's atmosphere and the emission from the Moon. Thus, observations of the infrared spectrum of Earth are available only from occasional remote spacecraft observations.

On the night of 3 July 2003, the Mars Express spacecraft was pointed backwards toward the Earth to obtain a view of the Earth–Moon system from a distance of eight million kilometre while on its way to Mars. During a series of instrument tests, the OMEGA spectrometer on board Mars Express acquired a low-resolution spectra of the Earth and the Moon in visible and near-infrared light. The spectrum shown in Fig. 3.26 corresponds to the entire Earth's illuminated crescent, dominated by the Pacific Ocean, and indicates the molecular composition of the atmosphere, the ocean and some continents.

Earth observations in the thermal infrared are also available. The thermal emission spectrometer (TES) on the Mars Global Surveyor spacecraft acquired observations of the Earth from a distance of 4.7 million kilometre for instrument performance characterization on 24 November 1996 (Christensen and Pearl 1997). The data provided the first known whole-disk thermal infrared spectral observations of the Earth (see Fig. 3.27). An infrared spectrum of the Earth was also taken by VIRTIS on 5 March 2005 after Rosetta's closest approach to Earth.

The thermal emission of the Earth dominates the IR spectrum, corresponding to its effective temperature of 288 K. The IR brightness peaks around 10 μm and then

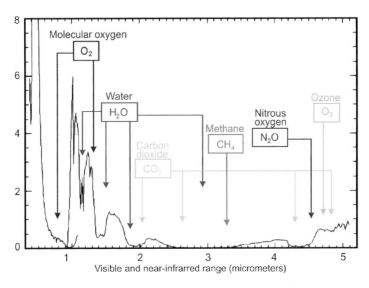

Fig. 3.26 Composition of the Earth as seen by the Mars Express OMEGA spectrometer in July 2003. Credit: ESA

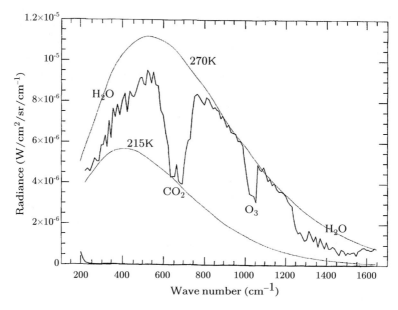

Fig. 3.27 Normalized calibrated spectral radiance of the Earth (*central heavy curve*). The data have been normalized to the highest signal. The noise equivalent spectral radiance is the *lower heavy curve* near the x axis. *Dotted curves* show blackbodies at 215 and 270 K. Note the presence of CO_2, O_3 and H_2O absorption features. The structure in the water vapour bands is real. To convert from wavenumber (cm^{-1}) to nanometres, we use the following equation: $1\,cm^{-1} \times 10^7 = 1\,nm$. Thus the data in the figure covers the spectral range from 6,250 to 50,000 nm. Adapted from Christensen and Pearl (1997)

decays slowly. It is superposed by different molecular bands corresponding to some of the most important atmospheric components. Spectral features of the gases carbon dioxide (CO_2), water vapour (H_2O) and ozone (O_3) dominate the Earth's spectra in this range, as well as methane (CH_4) and several other minor constituents. Radiation at the centre of the CO_2 band arises mainly from the lower stratosphere, near 650 and 700 cm^{-1} from near the tropopause and further into the band wings from the troposphere and surface. Thus, in the disk-averaged sense, the spectrum indicates a warm stratosphere above a tropopause somewhat colder than 215 K. The atmospheric window between approximately 800 and 1,200 cm^{-1} is relatively featureless, as expected, given the observing geometry centred over the Pacific Ocean and the partial obscuration by clouds.

Because of the lower contrast between a star and a planet in the infrared range, compared to the visible, the infrared is a good spectral range to study the atmospheric composition of the atmosphere of exoplanets. In Fig. 3.28 the infrared spectra of the different bodies of the Solar System are shown for comparison. These measurements allow us to distinguish the main characteristics and composition of their atmospheres and to derive estimates of the amounts of each atmospheric component for the averaged atmospheric depth (Traub et al. 2003). For well mixed gases,

3.4 Spectroscopy of Planet Earth

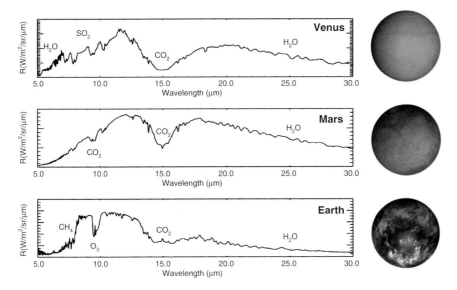

Fig. 3.28 Infrared spectral albedo for different planets of the Solar System

we can get their mixing ratios independently of the fact that we are not able to sample down to the planet's surface. For non-well-mixed gases we get a relative idea of their abundances, which can be improved with models if other variables such as mass, radius and temperature profile are known.

Note that the visible and infrared spectra shown in the previous sections is not a permanent property of our planet, rather it has changed as the Earth evolved. Some of the features identified in the modern-Earth observations may not have been detectable at all time scales, while other biosignatures might have been more dominant in the past. More detailed discussion on the evolution of the Earth spectra is given in Chap. 5.

3.4.3 The Earth's Transmission Spectrum

When a planet passes in front of its parent star, part of the starlight passes through the planet's atmosphere and contains information about the atmospheric species it encounters. This is called the transmission spectra of a planet. Transmission spectra are very interesting because they provide only the current methodology to characterize exoplanetary atmospheres (see Chap. 6 for details on methods and accomplishments).

The characterization of spectral features in our planet's transmission spectra can be achieved through observations of the light reflected from the Moon during a lunar eclipse, which resemble the observing geometry during a planetary transit.

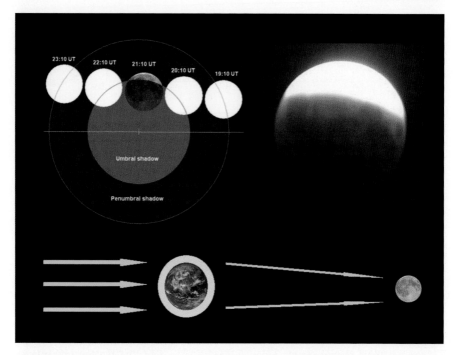

Fig. 3.29 *Top left*: The evolution of the partial lunar eclipse on 16th August 2008 (from NASA eclipse page at http://eclipse.gsfc.nasa.gov) *Top right*: An image of the Moon taken at 21:10 UT, the time of Greatest eclipse. *Bottom*: A not-to-scale illustration showing the path of the sunlight through the Earth's atmosphere before reaching the Moon during the eclipse. Considering the Earth's radius and the mean Sun–Earth distance, in order to reach the umbral shadow centre, the sunlight transmitted through the Earth's atmosphere must be refracted by an angle smaller than 2°

At that time, the reflected sunlight from the lunar surface within the Earth's umbra will be entirely dominated by the fraction of sunlight that is transmitted through an atmospheric ring located along the Earth's day–night terminator (see Fig. 3.29).

Observations of the lunar illumination during an eclipse have a long history (Slipher 1914; Moore and Brigham 1927). However, spectroscopic lunar eclipse observations, in the visible and near-infrared ranges, with the aim of retrieving the transmission spectra, only have been taken recently. Pallé et al. (2009) made observations of the lunar eclipse on 16 August 2008, which allowed them to characterize the Earth's spectrum as if it had been observed from an astronomical distance during a transit in front of the Sun. The Earth transmission spectrum was obtained from the brightness ratio of the light reflected by the lunar surface when in the umbra, penumbra and out of the eclipse.

In Fig. 3.30, the Earth's transmission spectrum from visible to near-infrared wavelengths is shown. It is common knowledge that the Earth transmission spectra is red, as can be inferred from simple naked-eye observations of a lunar eclipse or of a sunset/sunrise. Rayleigh scattering makes the reflected spectra of the Earth blue due to the larger scattering at shorter wavelengths but it also makes the transmission

3.4 Spectroscopy of Planet Earth

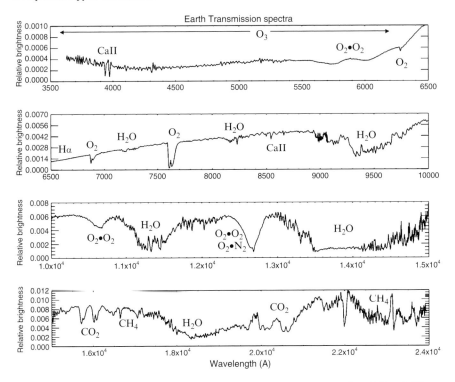

Fig. 3.30 A visible to near-infrared atlas of the Earth's transmission spectra from 360 to 2,400 nm, where the major atmospheric features of the spectra have been marked

spectra red by scavenging short wavelength radiation through a long atmospheric path. This redness is easily seen in Fig. 3.31, where the very deep absorption of the blue light marks the more spectacular feature of the transmission spectra. The atmospheric spectral bands of O_3, O_2, H_2O, CO_2 and CH_4 are also readily distinguishable.

An interesting feature of the Earth's transmission spectra is that it presents a marked N_2 signature feature at 1.26 μm, a band produced by oxygen collision complexes. Oxygen collision complexes are van der Waals molecules, also know as dimers (Calo and Narcisi 1980). These loosely bound species, held together by inter-molecular attractions rather than by chemical bonds, are present in all gases and have been observed experimentally in a variety of systems. In the natural atmosphere, certain dimer species have mixing ratios of the same order of magnitude as minor atmospheric constituents such as carbon dioxide, ozone, water vapour or methane (Calo and Narcisi 1980).

In the earth's transmission spectra, the oxygen collision complex band at 1.26 nm is produced by $O_2 \bullet O_2$ and $O_2 \bullet N_2$ collision complexes, and can be used in combination with other oxygen bands at 0.69, 0.76 and 1.06 nm to derive an averaged atmospheric column density of N_2. The strength of these bands in the transmission

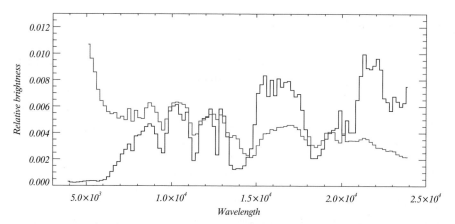

Fig. 3.31 A comparison between the Earth's transmission (*black*) and reflected (*blue*) spectra. The Earth's reflectance spectra is a proxy for the observations of Earth as a planet by direct observation, while the transmission spectra of the Earth is a simile for Earth observations during a transit. Both spectra have been degraded to a spectral resolution of 200 nm. Note the opposite wavelength dependence of the continuum and the different strength of the absorption bands. Adapted from Pallé et al. (2009) Fig. 1. Reprinted by permission from Macmillan Publishers; Nature Vol. 459, p. 814. Copyright (2009)

spectra implies that atmospheric dimers may become a major subject of study for the interpretation of exoplanets transmission spectra and their atmospheric characterization, although there is a clear need for further experimental developments of collision complexes molecular databases (Solomon et al. 1998).

Finally, the presence of the Earth's ionosphere is also revealed in the transmission spectrum, through the atomic transition lines corresponding to Ca^{+2}, at 395 and 850 nm, which originate in the upper layers of the Earth's atmosphere (Kopp 1997).

It is illustrating to compare the transmission and reflected spectra of Earth, two very different spectra for a single planet. While the transmission spectra is a simile for Earth characterization during a transit, the reflected spectra (obtained from earthshine observations) simulate that which one would obtain if one was able to isolate the light of the Earth from its parent stars, either using coronographic or interferometric techniques.

In Fig. 3.31, both the transmitted and reflected spectra have been degraded to spectral resolution of 200 nm and scaled for easy comparison. It is readily seen from the figure that the main difference between the reflected and transmitted spectra is the shape of the continuum. Thus, in transmission, the *pale blue dot* becomes the *pale red dot*. The atmospheric spectral bands of O_3, O_2, H_2O, CO_2 and CH_4 are readily distinguishable, even if the spectra is degraded. It is also noticeable how most of the molecular spectral bands are weaker, and some are missing, in the reflected spectra.

3.5 Polarimetry of Planet Earth

An electromagnetic (light) wave is a transverse wave, which has both an electric and a magnetic component. If the light wave is vibrating in more than one plane, it is referred to as unpolarized light. Such light waves are created by electric charges which vibrate in a variety of directions, thus creating an electromagnetic wave which also vibrates in a variety of directions. Light emitted by the Sun, for example, is unpolarized. However, when the sunlight interacts with the Earth atmosphere, it becomes polarized by either transmission, reflection, refraction or scattering (Hecht and Zajac 1997). Light polarization can be either linear, elliptical or circular. Light is linearly polarized when its electric field component is in a plane, and circular and elliptical polarization occurs when two or more linearly polarized waves add up in such a way that the electric field of the net wave rotates. For circularly polarized light, the direction of the electric field rotates but its magnitude stays the same, while for elliptically polarized light both the magnitude and the direction of the electric field vary.

The degree of polarization of the globally integrated light from Earth depends on the cloud coverage, together with the degree of polarization of each of the exposed areas of its surface. The contribution of each surface type is a combination of Rayleigh scattering above the surface and the reflection at the surface and on the cloud tops. Based on polarization observations of Earth made from the Polder satellite (Fig. 3.32), Wolstencroft and Breon (2005) determined that the degree of linear polarization for Earth, at 443 nm and 90° scattering angle, is 23% for averaged cloud cover conditions (55%).

In the Earth, water surfaces provide the only significantly polarized natural sources of thermal infrared radiation, while emission from the atmosphere and ground is almost always unpolarized to any practical degree (Shaw 2002). The actual degree of polarization from water depends upon a balance of orthogonally polarized emission and background reflection terms, weighted by atmospheric transmittance. For practical purposes, the degree of polarization of the infrared flux emitted from Earth, observed as a planet, is negligible.

3.5.1 Linear Polarization

Lyot and Dollfus (1949) studied the polarization of the earthlight by means of a coronagraph and a polarizer at the Pic du Midi Observatory. They concluded that there is a strong polarization, 1.2 times larger over the dark seas as over the brighter regions, it reaches a maximum when the Moon's phase angle is near 80° and then it exceeds 10% over the seas. Danjon (1954) explained these results as follows: the light scattered by the Earth will be strongly polarized; scattering by the Moon's surface will reduce this polarization, since it will be multiplied by a coefficient of depolarization which varies inversely as the albedo. Dollfus (1957) found such depolarization on volcanic ashes (the material that closest resembled the lunar regolith to him) to be about 0.3. Danjon concluded then that, 'the polarization curve of

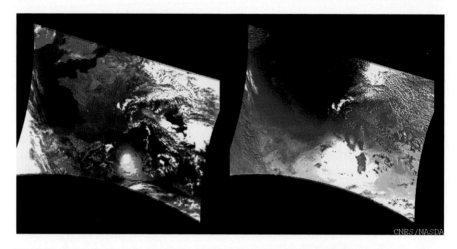

Fig. 3.32 This pair of images shows data from 16 September 1996 acquired over France in natural light (*left image*) and polarized light (*right image*). Each image is a blue, green and red colour composite of POLDER measurements at 443, 670 and 865 nm. On the conventional image, marked differences appear between the clouds (in *white*) and different types of surfaces: vegetation in *red*, soil in *brown* and *yellow*, sea in *dark blue*. The bright spot in the Mediterranean Sea, west of the island of Sardinia, is the sunglint pattern. On the polarized image the colour *blue* prevails and the geographic contours can hardly be recognized. This is because the polarized light mainly results from scattering in the atmosphere, which increases at shorter wavelengths. The clouds still appear in *grey* or *white* and the sunglint spot in the Mediterranean Sea corresponds to a strong polarized signal. Image credit: Japan Aerospace Exploration Agency

the Earth therefore will have a high maximum for the phase angle 100°, amounting to perhaps 30%, which is interpreted as due principally to scattering by the atmosphere'.

Dollfus (1957) continued with observations of the earthshine polarization, and obtained direct measurements of polarization from the ground from balloon observations. From his measurements, Dollfus concluded that the atmospheric polarization is by far larger than that introduced from the ground and noted that, 'for an extra-terrestrial observer it would be diluted by the intense, but little polarized, light from the background formed by the ground, the sea and the clouds. The resulting polarization for the combined terrestrial light may be roughly calculated; for phase angle 90° it is as high as 36%. The amount varies with phase similar to the polarization of the earthshine' (Fig. 3.33).

The rotation of a planet with surface features such as continents and oceans should modulate the polarized reflectances in a simple and predictable manner. The models (McCullough 2006) predict the shape of the phase function of the earthshine's linear polarization observed by Dollfus (1957); the maximum polarization agrees with Dollfus' observations but is approximately twice as large as that predicted by Wolstencroft and Breon (2005). Thus, linear polarization could be a potentially useful signature of oceans and atmospheres of Earth-like extrasolar planets. As with other techniques, global polarimetric observations of our planet will be the guideline.

3.5 Polarimetry of Planet Earth

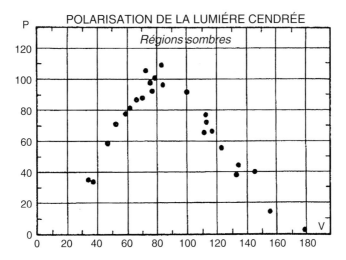

Fig. 3.33 The degree of polarization of the earthshine, plotted as a function of phase. These are the only existing globally integrated measurements of the Earth's polarization, and they are only indicative, because reflection from the Moon introduces a variable depolarization factor. Adapted from Dollfus (1957)

3.5.2 Circular Polarization

Biotic material, with its helical molecular structure, is known to produce circular polarization of reflected light in the visible range (Pospergelis 1969; Wolstencroft and Raven 2002). Homochirality, that is the exclusive use of L-amino acids and D-sugars in biological materials, causes a significant induction of circular polarization in the diffuse reflectance spectra of biotic material (Bustamante et al. 1985). Among many other bio-polymers, photosynthetic pigments (e.g. chlorophyll) induce between 0.1 and 1% circular dichroism in its absorption bands (Houssier and Sauer 1970). Thus, it is interesting to search for circular polarization in the spectrum of the Earth as induced by chiral molecules of living material on its surface.

Sparks et al. (2005) conducted a search for chiral signatures on Mars with negative results, which can most likely be explained by the absence of biotic material in large enough quantities (or at all) on the Martian surface. Organic material on Earth, however, is abundant, but its detectability, for example through the 'Vegetation Red Edge', is difficult (see Chap. 5).

Observed on the Moon, chiral signatures in the earthshine are expected of the order of $V/I \sim 10^{-4...-5}$, assuming a dilution of the circular polarimetric signal caused by vegetation due to partial cloud coverage (factor 10) and by depolarization on the lunar surface up to a factor 10 (DeBoo et al. 2005). Preliminary measurements of non-zero Stokes V/I during a specific phase of the earthshine were observed by Sterzik and Bagnulo (2009) but could not unambiguously be interpreted as the signature of biotic homochirality, and therefore life, on Earth.

The study of the several biosignatures present on the Earth's spectra will provide a benchmark for future attempts to detect biotic material on other astronomical objects, and will guide our search for life outside the solar system. Chapter 5 of this book is devoted entirely to this topic: the remote detection of biotic material (or biosignatures) on Earth and elsewhere.

References

Ackerman, A.S., Marley, M.S.: Precipitating Condensation Clouds in Substellar Atmospheres. Astrophys. J. **556**, 872–884 (2001)
Arnold, L., Gillet, S., Lardière, O., Riaud, P., Schneider, J.: A test for life on exoplanets: the terrestrial vegetation detection in the earthshine spectrum. In: B.H. Foing, B. Battrick (eds.) Earth-like Planets and Moons, ESA Special Publication, vol. 514, pp. 259–262 (2002)
Bustamante, E.G., Anabitarte, E., Calderon, M.A.G., Senties, J.M., Vegas, A.: Polarization Effects for E_{cr} in a Longitudinally Magnetized Plasma Waveguide. In: J.S. Bokos, Z. Sörlei (eds.) Phenomena in Ionized Gases, XVII International Conference, Vol. 1, p. 76 (1985)
Byard, M.M.: Painting the Heavens: Art and Science in the Age of Galileo. Renaiss. Q. **22**, 884 (1999)
Calo, J.M., Narcisi, R.S.: Van der Waals molecules - Possible roles in the atmosphere. Geophys. Res. Lett. **7**, 289–292 (1980)
Campbell, D.B., Black, G.J., Carter, L.M., Ostro, S.J.: Radar Evidence for Liquid Surfaces on Titan. Science **302**, 431–434 (2003)
Christensen, P.R., Pearl, J.C.: Initial data from the Mars Global Surveyor thermal emission spectrometer experiment: Observations of the Earth. J. Geophys. Res. **102**, 10,875–10,880 (1997)
Clarke, A.: Extraterrestrial Relays: Can Rocket Stations give world-wide radio coverage? Wireless World **51**, 305–308 (1945)
Danjon, A.: La Lumière cendrée et l'albedo de la terre. Annales de l'Observatoire de Strasbourg **2**, 165–184 (1928)
Danjon, A.: Albedo, color and polarization of the Earth. The Earth as a Planet, G.P. Kuiper (ed.) (1954)
Deboo, B.J., Sasian, J.M., Chipman, R.A.: Depolarization of diffusely reflecting man-made objects. Appl. Optic. **44**, 5434–5445 (2005)
Des Marais, D.J., Harwit, M.O., Jucks, K.W., Kasting, J.F., Lin, D.N.C., Lunine, J.I., Schneider, J., Seager, S., Traub, W.A., Woolf, N.J.: Remote Sensing of Planetary Properties and Biosignatures on Extrasolar Terrestrial Planets. Astrobiology **2**, 153–181 (2002)
Dollfus, A.: Étude des planètes par la polarisation de leur lumière. Supplements aux Annales d'Astrophysique **4**, 3–114 (1957)
Dubois, J.: Sur l'albedo de la terre. Bull. Astron. **13**, 193–196 (1947)
Dubois, J.: Au sujet de la luminescence lunaire. L'Astron. **72**, 267 (1958)
Flatte, S., Koonin, S., MacDonald, G.: Global change and the dark of the moon. In: S. Flatte, S. Koonin, G. MacDonald (eds.) Final Report Mitre Corp., McLean, VA (1992)
Ford, E.B., Seager, S., Turner, E.L.: Characterization of extrasolar terrestrial planets from diurnal photometric variability. Nature **412**, 885–887 (2001)
Franklin, F.A.: Two-color photometry of the earthshine. J. Geophys. Res. **72**, 2963–2967 (1967)
Fritz, S.: The Albedo of the Planet Earth and of Clouds. J. Atmos. Sci. **6**, 277–282 (1949)
Gaidos, E., Moskovitz, N., Williams, D.M.: Terrestrial Exoplanet Light Curves. In: C. Aime, F. Vakili (eds.) IAU Colloq. 200: Direct Imaging of Exoplanets Science Techniques, pp. 153–158 (2006)
Gómez-Leal, I., and Pallé, E.: Astrophysical Journal, submitted (2010)

References

Goode, P.R., Qiu, J., Yurchyshyn, V., Hickey, J., Chu, M.C., Kolbe, E., Brown, C.T., Koonin, S.E.: Earthshine observations of the earth's reflectance. Geophys. Res. Lett. **28**, 1671–1674 (2001)

Griffiths, D.J.: An Introduction to Electrodynamics. 3rd edn. 363-368 (Prentice-Hall, Englewood Cliffs, NJ) (1998)

Hamdani, S., Arnold, L., Foellmi, C., Berthier, J., Billeres, M., Briot, D., François, P., Riaud, P., Schneider, J.: Biomarkers in disk-averaged near-UV to near-IR Earth spectra using Earthshine observations. Astron. Astrophys. **460**, 617–624 (2006)

Hapke, B.: Optical properties of the lunar surface. Physics of Atoms and Molecules pp. 155–211 (1971)

Hapke, B., Nelson, R., Smythe, W.: The Opposition Effect of the Moon: Coherent Backscatter and Shadow Hiding. Icarus **133**, 89–97 (1998)

Hapke, B.W., Nelson, R.M., Smythe, W.D.: The opposition effect of the moon - The contribution of coherent backscatter. Science **260**, 509–511 (1993)

Helfenstein, P., Veverka, J., Hillier, J.: The Lunar Opposition Effect: A Test of Alternative Models. Icarus **128**, 2–14 (1997)

Hetch, E., Zajac, A.: Optics. Addison Wesley Publishing Company; 3rd edn (1997)

Hilbrecht, H., Kuveler, G.: Variations of the Ashen Light of Moon and the Earth's Albedo. Earth Moon Planets **33**, 229–230 (1985)

Hitchcock, D.R., Lovelock, J.E.: Life detection by atmospheric analysis. Icarus **7**, 149–159 (1967)

Houghton, J.T.: The Physics of Atmospheres. Cambridge University Press, London (2002)

Houssier, C., Sauer, K.: Circular dichroism and magnetic circular dichroism of the chlorophyll and protochlorophyll pigments. J. Am. Chem. Soc. **92**, 779–791 (1970)

Huffman, D., Weidman, C., Twomey, S.: Colloq. Andre Danjon. In: S. Debarbat (ed.) Journees 1990, Paris Observatory, p. 111 (1989)

Kennedy, J.R.: MS Thesis. Fresno state college, unpublished (1969)

Kopp, E.: On the abundance of metal ions in the lower ionosphere. J. Geophys. Res. **102**, 9667–9674 (1997)

Loeb, N.G., Loukachine, K., Manalo-Smith, N., Wielicki, B.A., Young, D.F.: Angular Distribution Models for Top-of-Atmosphere Radiative Flux Estimation from the Clouds and the Earth's Radiant Energy System Instrument on the Tropical Rainfall Measuring Mission Satellite. Part II: Validation. J. Appl. Meteorol. **42**, 1748–1769 (2003)

Loeb, N.G., Wielicki, B.A., Rose, F.G., Doelling, D.R.: Variability in global top-of-atmosphere shortwave radiation between 2000 and 2005. Geophys. Res. Lett. **34**, 3704 (2007)

Lyot, B., Dollfus, A.: Polarisation de la lumiere cendree de la Lune. Memoires et communications de l'academie des sciences **222**, 1773–1775 (1949)

McCullough, P.R.: Models of polarized light from oceans and atmospheres of earth-like planets. arXiv:astro-ph/0610518 (2006)

Montañés-Rodriguez, P., Pallé, E., Goode, P.R., Hickey, J., Koonin, S.E.: Globally Integrated Measurements of the Earth's Visible Spectral Albedo. Astrophys. J. **629**, 1175–1182 (2005)

Montañés-Rodríguez, P., Pallé, E., Goode, P.R., Martín-Torres, F.J.: Vegetation Signature in the Observed Globally Integrated Spectrum of Earth Considering Simultaneous Cloud Data: Applications for Extrasolar Planets. Astrophys. J. **651**, 544–552 (2006)

Moore, J.H., Brigham, L.A.: The Spectrum of the Eclipsed Moon. Publ. Astron. Soc. Pac. **39**, 223–226 (1927)

Pallé, E., Ford, E.B., Seager, S., Montañés-Rodríguez, P., Vazquez, M.: Identifying the Rotation Rate and the Presence of Dynamic Weather on Extrasolar Earth-like Planets from Photometric Observations. Astrophys. J. **676**, 1319–1329 (2008)

Pallé, E., Goode, P.R., Montañés-Rodríguez, P., Koonin, S.E.: Changes in Earth's Reflectance over the Past Two Decades. Science **304**, 1299–1301 (2004)

Pallé, E., Goode, P.R., Yurchyshyn, V., Qiu, J., Hickey, J., Montañés Rodriguez, P., Chu, M.C., Kolbe, E., Brown, C.T., Koonin, S.E.: Earthshine and the Earth's albedo: 2. Observations and simulations over 3 years. J. Geophys. Res. **108**, 4710 (2003)

Pallé, E., Zapatero Osorio, M.R., Barrena, R., Montañés Rodríguez, P., Martin, E.L.: Earth's transmission spectrum from lunar eclipse observations. Nature **459**, 814–816 (2009)

Piccioni, G., Drossart, P., Cardesin, A., VIRTIS-VenusX team: Is Earth an habitable planet? Earth as seen from VIRTIS-VenusX. In: Venus Express Workshop, La Thuile, Italy (2008)

Pospergelis, M.M.: Spectroscopic Measurements of the Four Stokes Parameters for Light Scattered by Natural Objects. Sov. Astron. **12**, 973 (1969)

Potocnik, H.: Das Problem der Befahrung des Weltraums – der Raketen-Motor. Richard Carl Schmidt (1928)

Qiu, J., Goode, P.R., Pallé, E., Yurchyshyn, V., Hickey, J., Montañés Rodriguez, P., Chu, M.C., Kolbe, E., Brown, C.T., Koonin, S.E.: Earthshine and the Earth's albedo: 1. Earthshine observations and measurements of the lunar phase function for accurate measurements of the Earth's Bond albedo. J. Geophys. Res. **108**, 4709 (2003)

Ridley, J.: Variations in Planetary Albedo. In: W.R.F. Dent (ed.) Techniques for the detection of planets and life beyond the solar system, 4th Annual ROE Workshop, held at Royal Observatory Edinburgh, Scotland, pp. 28 (2001)

Rossow, W.B., Walker, A.W., Beuschel, D.E., Roiter, M.D.: International Satellite Cloud Climatology Project (ISCCP): documentation of New Cloud Datasets. WMO/TD-No. 737, World Meteorological Organization, Geneva, pp. 115 (1996)

Sagan, C.: Pale blue dot : a vision of the human future in space. New York : Random House, 1st edn (1994)

Sagan, C., Thompson, W.R., Carlson, R., Gurnett, D., Hord, C.: A Search for Life on Earth from the Galileo Spacecraft. Nature **365**, 715 (1993)

Shaw, J.A.: Infrared polarization in the natural Earth environment. In: D.H. Goldstein, D.B. Chenault (eds.) Polarization Measurement, Analysis, and Applications V. Proceedings of the SPIE, Presented at the Society of Photo-Optical Instrumentation Engineers (SPIE) Conference, vol. 4819, pp. 129–138 (2002)

Slipher, V.M.: On the Spectrum of the Eclipsed Moon. Astron. Nachr. **199**, 103–104 (1914)

Solomon, S., Portmann, R.W., Sanders, R.W., Daniel, J.S.: Absorption of solar radiation by water vapor, oxygen, and related collision pairs in the Earth's atmosphere. J. Geophys. Res. **103**, 3847–3858 (1998)

Sparks, W.B., Hough, J.H., Bergeron, L.E.: A Search for Chiral Signatures on Mars. Astrobiology **5**, 737–748 (2005)

Stam, D.M., de Rooij, W.A., Cornet, G., Hovenier, J.W.: Integrating polarized light over a planetary disk applied to starlight reflected by extrasolar planets. Astron. Astrophys. **452**, 669–683 (2006)

Sterzik, M.F., Bagnulo, S.: Search for Chiral Signatures in the Earthshine. In: K. Meech, M. Mumma, J. Siefert (eds.) Biostronomy 2007: Molecules, Microbes and Extraterrestrial Life. ASP Conference Series (2009)

Tikhov, G.A.: Etude de la lumiere cendree de la lune AU moyen des filtres selecteurs. Izviestiia Nikolaevskoi glavnoi astronomicheskoi observatorii **6**, 15 (1914)

Tinetti, G., Meadows, V.S., Crisp, D., Kiang, N.Y., Kahn, B.H., Fishbein, E., Velusamy, T., Turnbull, M.: Detectability of Planetary Characteristics in Disk-Averaged Spectra II: Synthetic Spectra and Light-Curves of Earth. Astrobiology **6**, 881–900 (2006)

Traub, W.A.: The Colors of Extrasolar Planets. In: D. Deming, S. Seager (eds.) Scientific Frontiers in Research on Extrasolar Planets, Astronomical Society of the Pacific Conference Series, vol. 294, pp. 595–602 (2003)

Turnbull, M.C., Traub, W.A., Jucks, K.W., Woolf, N.J., Meyer, M.R., Gorlova, N., Skrutskie, M.F., Wilson, J.C.: Spectrum of a Habitable World: Earthshine in the Near-Infrared. Astrophys. J. **644**, 551–559 (2006)

Valero, F.P.J.: Triana – A deep space earth and solar observatory report. National Academy of Sciences reports (2000)

Vázquez, M., Montañes-Rodríguez, P., Pallé, E.: The Earth as an object of astrophysical interest in the search for extrasolar planets. Lecture Notes and Essays in Astrophysics, Springer **2**, 49–70 (2006)

Very, F.W.: The Photographic Spectrography of the earth-shine. Astron. Nachr. **201**, 353–398 (1915)

Wielicki, B.A., Green, R.N.: Cloud Identification for ERBE Radiative Flux Retrieval. J. Appl. Meteorol. **28**, 1133–1146 (1989)

Williams, D.M., Gaidos, E.: Specular Reflection of Starlight off Distant Planetary Oceans. Bull. Am. Astron. Soc. **36**, 1173 (2004)

Williams, D.M., Gaidos, E.: Detecting the glint of starlight on the oceans of distant planets. Icarus **195**, 927–937 (2008)

Wolstencroft, R.D., Breon, F.M.: Polarization of Planet Earth and Model Earth-like Planets. In: A. Adamson, C. Aspin, C. Davis, T. Fujiyoshi (eds.) Astronomical Polarimetry: Current Status and Future Directions, Astronomical Society of the Pacific Conference Series, vol. 343, pp. 211–212 (2005)

Wolstencroft, R.D., Raven, J.A.: Photosynthesis: Likelihood of Occurrence and Possibility of Detection on Earth-like Planets. Icarus **157**, 535–548 (2002)

Woolf, N.J., Smith, P.S., Traub, W.A., Jucks, K.W.: The Spectrum of Earthshine: A Pale Blue Dot Observed from the Ground. Astrophys. J. **574**, 430–433 (2002)

Xie, S.P.: Satellite Observations of Cool Ocean Atmosphere Interaction. Bull. Am. Meteorol. Soc. **85**, 195–208 (2004)

Chapter 4
The Outer Layers of the Earth

In the previous chapter, we studied the most important signatures of the Earth seen from space in visible and infrared light. All of them are related to processes occurring in the interior, the surface or the lower atmospheric layers. Now, it is time to discuss features in the electromagnetic spectrum concerning the upper atmospheric layers, which contain only 10% of the total mass of the atmosphere. Because of the relative faintness of these layers, even if present on exoplanets, they will not be observable until some time in the far future. In the meantime, we can apply the Earth–Exoplanet connection to this context. The names and characteristics of these atmospheric layers were briefly summarized in the Sect. 2.3.1.

In these outer layers, the gases are distributed in distinct strata by gravity according to their atomic weight. Thus higher mass constituents, such as oxygen and nitrogen, fall off more quickly than lighter constituents such as helium and hydrogen. Small components such as NO, CO, CO_2, N_2O, H_2O, O_3b and NO_2 are important for the photochemistry, energetics and emissions of the upper atmosphere. The books written by G. Herzberg (1904–1999) are the best known and most influential in the field of molecular spectroscopy (Herzberg 1939, 1945, 1966).

Figure 4.1 summarizes the main phenomena occurring in the Earth's upper layers, which are able to be observed from space. These phenomena result from the interaction of the main solar input with the Earth's atmosphere and magnetosphere.

Together with discrete molecular bands in other ranges, the extremes of the electromagnetic spectrum (X–UV rays and radiowaves) will be the best tools for observing these layers. For the sake of clarity, in Table 4.1 we give an overview of the different designations in the UV domain. Table 4.2 lists the main experiments observing the Earth in the XUV range.

For a comparative overview of the upper layers of the different planets of the Solar System, see Müller–Wodarg (2005).

4.1 Temperature Profile and the Energy Balance

Previously, in Chap. 2, we outlined the different layers that constitute the Earth's atmosphere. Here, we describe in more detail the physical processes happening in the outer layers, namely from the stratosphere to the thermosphere. This tenuous layer

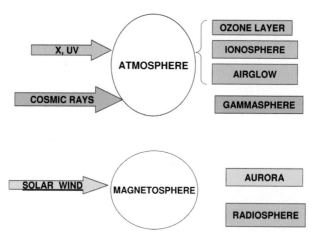

Fig. 4.1 Main phenomena observable in the upper layers of the Earth atmosphere

Table 4.1 UV radiation and its division into different ranges

Name	Range
UV	$100\,\text{nm} < \lambda < 400\,\text{nm}$
Extreme UV, EUV	$10\,\text{nm} < \lambda < 100\,\text{nm}$
Far UV, FUV	$100\,\text{nm} < \lambda < 200\,\text{nm}$
Middle UV, MUV	$200\,\text{nm} < \lambda < 300\,\text{nm}$
Near UV, NUV	$300\,\text{nm} < \lambda < 400\,\text{nm}$
UV–C	$100\,\text{nm} < \lambda < 280\,\text{nm}$
UV–B	$280\,\text{nm} < \lambda < 315\,\text{nm}$
UV–A	$315\,\text{nm} < \lambda < 400\,\text{nm}$

Table 4.2 List of satellites observing the Earth in the XUV range

Satellite	Operation	Experiment	Range	Scientific Target
TIMED	2001–	GUVI	115–180 nm	Thermosphere, Ionosphere
IMAGE	2000–2005	Extreme Ultraviolet Imager		
		Far Ultraviolet Imager		
POLAR	1996–	Ultraviolet Imager (UVI)	115–180 nm	
		PIXIE	2–60 keV	Aurora

Global Ultraviolet Imager (GUVI), Polar Ionospheric X-ray Imaging Experiment (PIXIE), Total Ozone Mapping Spectrometer (TOMS)

of neutral and charged particles shields the human habitat from high energy radiation and particles. Let us start by describing briefly how the radiation is absorbed in these layers.

Every component of the atmosphere is characterized by a specific spectrum of absorption bands. The strength and location of this spectrum depends on its relative abundance and also on the characteristics of the electronic, vibrational and rotational transitions. The latter two cases apply to molecules.

4.1 Temperature Profile and the Energy Balance

The temperature of a physical system is defined as the average energy of microscopic motions of its particles. By absorbing radiation from the exterior, the atmospheric material can be heated (Mlynczak et al. 2007). The radiation absorbed by the different atmospheric components configures what we call the neutral atmosphere (Fig. 4.2). Figure 4.3 shows the penetration depth of the different ranges of the electromagnetic radiation incident on the top of the atmosphere.

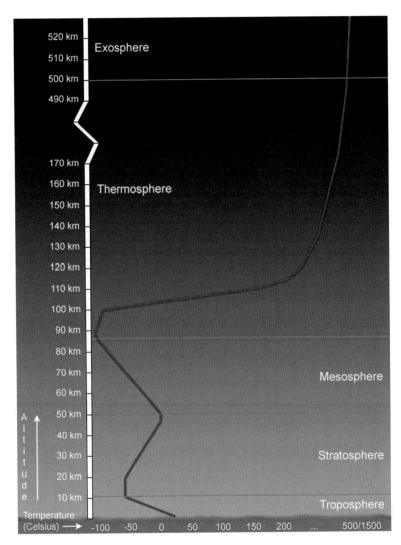

Fig. 4.2 Temperature profile of the Earth atmosphere. The source of this graph is 'Windows to the Universe', at http://www.windows.ucar.edu/ at the University Corporation for Atmospheric Research (UCAR). ©The Regents of the University of Michigan

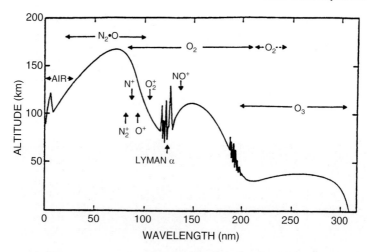

Fig. 4.3 Depth of penetration of solar radiation as a function of wavelength. Altitudes correspond to an attenuation of 1/e (63%). The principal absorbers and ionization limits are indicated: Schumann–Runge continuum and bands of O_2 and Hartley band of ozone. Adapted from Dickinson (1986)

Observed from space, the visible and infrared features are detected only over the background of the troposphere in some spectral lines. We can observe the uppermost layers in the XUV and the radio ranges.

Any change in the energy balance of these layers will give rise to a variation in the temperature.

$$\Delta T = \text{Energy Inputs} - \text{Energy Outputs} + \text{Energy Transport}$$

The main energy input is provided by solar heating. Other factors to be considered are Joule,[1] particle and chemical heating. Figure 4.4 shows the variation with the altitude of the mean heating rates calculated with the CMAT (coupled middle atmosphere–thermosphere general circulation model) for the solar minimum.[2]

The upper atmosphere shows an extreme spatial and temporal variability, driven by changes in the main energy inputs related with solar activity.

- Solar high energy radiation (γ and X-rays, UV radiation) enhanced by transitory events such as flares.
- Solar high energy particles, the solar wind, enhanced by transitory events such as the coronal mass ejections.
- Cosmic rays.

---

[1] Joule heating occurs when ions and electrons are accelerated by strong magnetic and electric fields.

[2] It covers a vertical range from 30 to approx. 300 km. See Harris (2001) and http://www.apl.ucl.ac.uk/cmat for a description.

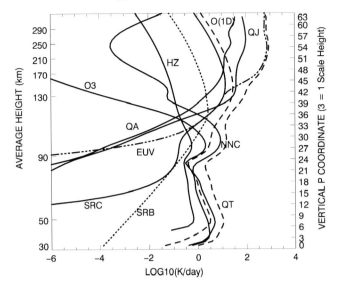

Fig. 4.4 Global mean heating rates. O_3 is the absorption of solar radiation by ozone; HZ is the absorption of solar radiation by O_2 in the Herzberg continuum; SRC is the absorption of solar radiation by O_2 in the Schumann–Runge continuum; QA is heating due to auroral electron precipitation; QJ is the Joule heating; NNC is heating due to exothermic neutral chemistry; and QT the total. Credit: Matthew Harris, Atmospheric Physics Laboratory, University College London

Radiation cooling is the main energy output, resulting in characteristic emission spectra as the airglow or infrared emissions of molecules such as CO_2, O_3, O_2, CO and NO (Rodgers and Walshaw 2006). These processes left signatures in the electromagnetic spectrum that we can observe from the ground and from space. Spectral features of the main polyatomic molecules configure the visible and IR spectrum of the Earth and were already studied in the previous chapter. Figure 4.5 shows the variation with the altitude of the mean cooling rates calculated with the CMAT for the solar minimum. The major cooling mechanism in the thermosphere is downward molecular heat conduction.

Finally, energy transport appears in different ways (Heat Advection, Molecular Heat Conduction, Chemical Heating etc.).

The flow of charged subatomic particles, cosmic rays and solar wind interact with the magnetosphere and the atmosphere of our planet. Together with the XUV radiation, this gives rise to the existence of an ionized atmosphere, the ionosphere, and to the emission of non-thermal radiation, located in the radio range. The atmosphere is heated by collisions with these particles and radiates also in the XUV range.

It is now time to describe in detail these upper layers and the processes that are involved in their thermal structure. We distinguish two main atmospheres, the neutral and the ionized.

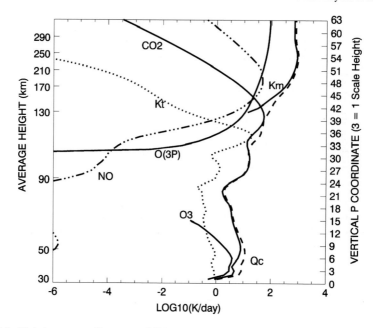

Fig. 4.5 Global mean cooling rates. QC is the total neutral gas cooling rate; Km the cooling rate due to downward molecular thermal conduction; KT is the cooling rate due to eddy thermal conduction; NO radiative cooling due to 5.3 μm emission; CO_2 radiative cooling due to 15 μm band emission; O(3p) radiative cooling due to 63 μm emission of atomic oxygen; and O_3 radiative cooling due to the 9.6 μm emission due to ozone. Credit: Matthew Harris, University College London

4.2 Stratosphere: The Ozone Layer

Using instruments aboard balloons, R. Assman (1845–1918) and Teisserenc de Bort (1855–1913) proved there was an increase in temperature from an altitude of 10 km, bringing on the discovery of a new atmospheric layer: the stratosphere, which extends from the top of the troposphere to about 52 km.

In 1921, F.A. Lindemann (1886–1957) suggested that it should be possible to obtain information about the vertical variation in temperature and density of the air from observations of meteor trails.[3] The records were interpreted by G. Dobson (1889–1976) as evidence of a warmer and denser stratosphere produced by the absorption of UV radiation by ozone (see Lindemann and Dobson 1922). This was soon confirmed by Whipple (1932) when studying the propagation of sound waves through the upper atmosphere, and marked the start of a growing pool of knowledge of the physical structure of our atmosphere.

---

[3] He was also one of the first to predict the existence of a solar wind composed of protons and electrons (Lindemann 1919).

4.2 Stratosphere: The Ozone Layer

In 1926 Dobson designed an instrument to measure the atmospheric vertical column of ozone from the ground (Dobson 1968). The Dobson spectrophotometer measures ultraviolet light from the Sun in 2–6 different wavelengths from 305 to 345 nm. The amount of ozone can be calculated by measuring UV light at two different wavelengths. One of the wavelengths used is absorbed strongly by ozone (305 nm), whereas the other wavelength (325 nm) is not. The ratio between the two light intensities is therefore a measure of the amount of ozone in the light path from the sun to the spectrophotometer.

The ozone content is measured in Dobson units (Fig. 4.6). The Dobson unit (DU) is defined to be a thickness of 0.01 mm at standard atmospheric pressure and temperature. 300 DU correspond to 8.07×10^{22} molecules m^{-2} or 6.42×10^{-3} kg m^{-2}. Because we use the Dobson Unit so commonly, we also express the density vs. altitude in units of DU per kilometre, where 10 DU km^{-1} is equal to 2.69×10^{18} molecules m^{-3} (number density) or 2.14×10^{-7} kg m^{-3} (density). For monographs on the ozone layer, see Dessler (2000), Christie (2000) and Vázquez and Hanslmeier (2005).

Table 4.3 lists the main experiments observing the ozone layer of the Earth.

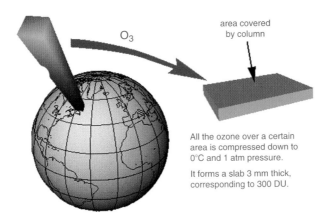

Fig. 4.6 The Dobson unit

Table 4.3 List of satellites observing the ozone layer and UV surface flux

Satellite	Operation	Experiment	Spectral range (nm)
NIMBUS 7	1978–	TOMS	
SME	1981–1988	UV Spectrometer	
EARTH PROBE	1996–	TOMS	
ER-2	1995	GOME	240–790
AURA	2004–	OMI	UV1 270–314
			UV2 306–380
ENVISAT	2002–	SCIAMACHY	240–2380

Total Ozone Mapping Spectrometer (TOMS); SME (Solar Mesosphere Explorer); GOME (Global Ozone Monitoring Experiment); ENVISAT (Environmental Satellite); SCIAMACHY (Scanning Imaging Absorption spectroMeter for Atmospheric Cartography).

4.2.1 Natural Processes of Ozone Formation and Destruction

In the second chapter we discussed the formation of the ozone layer and the subsequent influence on the amount of UV radiation at the Earth's surface. Now, we briefly describe how this layer is observed in our days.

Stratospheric ozone is created and destroyed primarily by ultraviolet radiation. When high energy photons strike molecules of oxygen (O_2), they split the molecule into two single oxygen atoms. The free oxygen atoms can then combine with oxygen molecules (O_2) to form ozone (O_3) molecules (Chapman 1930), as shown in Fig. 2.26. The main reactions are described in the following subsections.

4.2.1.1 Ozone Formation

Ozone production is driven by UV radiation. It occurs in the tropical stratosphere at heights between 20 and 50 km:

$$O_2 + \text{UV light} \, (\lambda < 242 \, \text{nm}) \longrightarrow O + O.$$

Absorption in the Schumann–Runge continuum ($135 < \lambda < 175$ nm) has a maximum cross-section at 142.5 nm and results in the production of two O atoms, one in the ground state and the other in the first excited state. Absorption in the weak Herzberg continuum (195 nm $< \lambda <$ 200 nm) produces two ground state atoms, giving rise to feeble absorption.

Subsequent recombination of the oxygen atoms may occur either directly,

$$O + O + M \longrightarrow O_2 + M,$$

or via the intermediate formation of ozone (the dominant process in the stratosphere),

$$O + O_2 + M \longrightarrow O_3 + M,$$

where M represents any other molecule (most probably N_2 or O_2) that may partially absorb the surplus of energy but is not changed in the process.

The lifetime[4] of oxygen atoms increases almost linearly with the altitude, being very brief in the stratosphere, typically less than 1 s. Hence, oxygen atoms almost immediately form ozone after they are dissociated.

The absorption of UV radiation heats the stratosphere and is therefore largely responsible for the formation of the stratosphere and the mesosphere.

4.2.1.2 Ozone Destruction

The same characteristic of ozone that makes it so valuable, its ability to absorb a range of ultraviolet radiation, also causes its destruction. When an ozone molecule

---

[4] Time required for the abundance of these atoms to decrease by about 63% (e-folding scale).

4.2 Stratosphere: The Ozone Layer

is exposed to ultraviolet or visible photons ($\lambda < 1200$ nm), it may break down into O_2 and O:

$$O_3 + \text{UV, visible NIR light} \longrightarrow O + O_2.$$

Absorption of solar UV radiation by ozone takes place in the following spectral bands: (a) Hartley, placed between 200 and 300 nm with a maximum at 255 nm, (b) Huggins, weak absorption band between 320 and 360 nm and (c) Chappuis, a weak diffuse system placed between 375 and 650 nm (Fowler and Strutt 1917).

The free oxygen atom may then combine with an oxygen molecule to create another ozone molecule, or it may take an oxygen atom from an existing ozone molecule to create two ordinary oxygen molecules:

$$O + O_3 \longrightarrow O_2 + O_2.$$

The cycle described above, proposed by Chapman (1930), gives levels of ozone much higher than those that are observed. Other reactions leading to ozone loss have been proposed involving chlorine, bromine, nitrogen etc. Moreover, there is an equator-to-pole stratosphere circulation, which we describe below.

4.2.1.3 Ozone Transport

Ozone is primarily produced by solar UV radiation at the tropics, but it is observed in larger quantities in polar regions (see Fig. 4.7). Also the ozone layer is higher in altitude in the tropics.

The main reason for these anomalies lies in the dynamics of the stratosphere, characterized by a meridional flow from the equator to the pole, the so-called

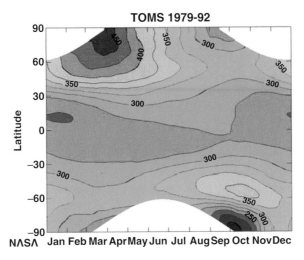

Fig. 4.7 Variation of global ozone content along the year for different latitudes, measured in Dobson units, by the TOMS (Average of 1978–1992). Credit: Stratospheric Ozone: An Electronic Textbook. Available at http://www.ccpo.odu.edu/SEES/ozone/oz_class.htm

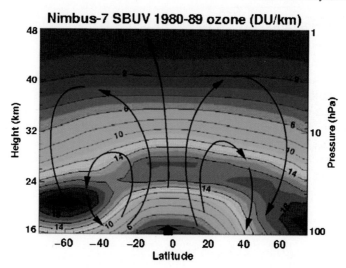

Fig. 4.8 Schematic flow pattern of the Brewer–Dobson circulation at the stratosphere (*black arrows*), superimposed by the ozone number density. Credit: Stratospheric Ozone: An Electronic Textbook

Brewer–Dobson circulation, named after its discoverers (see Fig. 4.8 for a diagram). As this slow circulation bends towards the mid-latitudes, it carries the ozone-rich air from the tropical middle stratosphere to the mid-and-high latitudes' lower stratosphere. The time needed to lift an air parcel from the tropical tropopause near 16 to 20 km is about 4–5 months. Even though ozone in the lower tropical stratosphere is produced at a very slow rate, the lifting circulation is so slow that ozone can build up to relatively high levels by the time it reaches 26 km.

This slow circulation has very important implications for the accumulation in polar regions of the stratosphere of human-produced atmospheric pollutants, such as chlorofluorocarbons, which have formed the ozone hole in the South Pole.

4.3 Mesosphere

The mesosphere is located from about 50 to 80–90 km altitude above the Earth's surface. Within this layer, the temperature decreases with increasing altitude due to decreasing solar heating and increasing cooling by CO_2 radiative emission. The minimum in temperature at the top of the mesosphere is called the mesopause, and is the coldest place in the atmosphere. Because it lies between the maximum altitude for aircraft and the minimum altitude for orbital spacecraft, this region of the atmosphere has been accessed only through the use of sounding rockets.

Polar mesospheric clouds, also called noctilucent clouds, are observed in these layers in summer at high latitudes, being both more frequent and brighter in the Northern Hemisphere than in the Southern Hemisphere. They seem to be connected

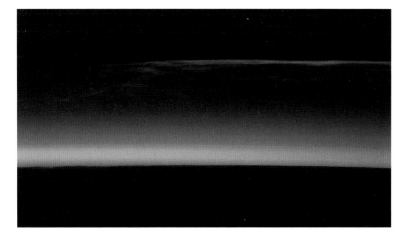

Fig. 4.9 This astronaut photograph of polar mesospheric clouds was acquired at an altitude of about 320 kilometres in the pre-dawn hours (18:24:01 Greenwich Mean Time) on July 22 2008, as the International Space Station (Expedition 17) was passing over western Mongolia in central Asia. Credit: NASA

with one of the associated processes connected with the current climate change, namely the cooling of the upper layers and the formation of ice aerosols, although the solar cycle appears to be the dominant natural influence (Thomas 2003; Thomas et al. 2003; Morris and Murphy 2008).

They are visible from the ground when illuminated by sunlight from below the horizon while lower layers of the atmosphere are in the Earth's shadow. Space-based observations to date have been made serendipitously by instruments designed for other purposes (Fig. 4.9). Now, satellites such as TIMED and AIM (Aeronomy of Ice in the Mesosphere) are providing information on this still enigmatic layer.

Millions of meteors burn up daily in the mesosphere as a consequence of collisions with the gas particles. This produces enough heat to vaporize almost all of the impacting objects, resulting in a high concentration of metallic dust here. Emissions from these metallic species, neutral and ionized, can be used to retrieve column densities in these layers (Scharringhausen et al. 2008). The mean total column density of the ionized metals is $4.4 \pm 1.2 \ 10^9 \ cm^{-2}$ in periods without special meteor shower activity, but increases by one order of magnitude during meteor showers (Kopp 1997).

4.4 The Thermosphere

In this layer, located at 90 km above the Earth's surface, the temperatures increase with altitude due to absorption of highly energetic solar radiation by the small amount of residual oxygen still present. Temperatures and densities are highly

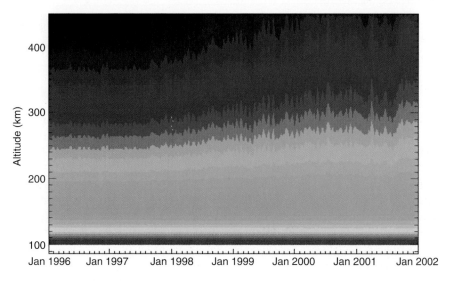

Fig. 4.10 Earth thermospheric densities from solar minimum to maximum produced with modelled SOLAR2000 spectral irradiances and the 1DTD physics-based model. Credit: Space Environment Technologies

dependent on solar activity (Fig. 4.10). In this layer, important processes of ionization occur, giving rise to the ionosphere and the appearance of auroras, phenomena that we will describe later in this chapter.

Weather balloons and research aircraft cannot reach the thermosphere. Although sounding rockets travel through the upper atmosphere, they can at best take short snapshots of a specific region, and do not provide global mapping. Furthermore, the Space Shuttle orbits in a region well above the lower thermosphere; upon reentry, the Space Shuttle passes through the lower thermosphere; however, it does so only briefly.

As the Earth rotates, absorption of solar energy in the thermosphere undergoes daily variation. Day-side heating causes the atmosphere to expand and at night to contract. This heating pattern drives a global circulation pattern: the atmospheric tides.

4.5 The Exosphere: Geocorona

The exosphere is the uppermost layer of the atmosphere, where collisions between particles are rare. Its lower boundary is estimated to be 500 to 1,000 km above the Earth's surface, and its upper boundary at about 10,000 km. It is only from the exosphere that atmospheric gases, atoms, and molecules can escape into space.

Exospheric temperatures of the planets do not decrease with distance from the Sun. It is evident that solar heating is not the only factor influencing the thermal

4.5 The Exosphere: Geocorona

structure of these layers (as discussed in Sect. 4.1). UV radiation therefore is the best way to observe these layers of the Earth's atmosphere from outer space.

Few observations of the Earth from the distance exist in ultraviolet light. Three days after the Mars Reconnaissance Orbiter launch (19 August 2005), the spacecraft was pointed toward Earth to acquire a set of ultraviolet images of the Earth and the Moon. From 1,170,000 km the apparent size of our planet was only 4.77 pixels across.[5] The Mars Color Imager instrument has two ultraviolet bands, centred at 320 and 260 nm, but our planet did not appear in the 260 nm band, most likely due to the large amount of ozone in our atmosphere.

However, monochromatic observations reveal a 'constant' glow produced at the uppermost layer of the atmosphere, the exosphere, as a consequence of the interaction between solar radiation and atmospheric components. The term 'geocorona' refers to the solar far-ultraviolet light (the Lyman α line 121.5 nm) that is scattered off the cloud of neutral hydrogen atoms that surrounds the Earth (see Fig. 4.11) and by neutral interstellar hydrogen entering the heliosphere. Solar far-ultraviolet

Fig. 4.11 The Earth's geocorona as viewed from the surface of the Moon. Image acquired during the Apollo 16 mission (April 1972) with the Naval Research Laboratory's far-ultraviolet camera/spectrograph. The part of the Earth facing the Sun reflects much UV light, but perhaps more interesting is the side facing away from the Sun. Here bands of UV emission are also apparent. These bands are the result of aurora and are caused by charged particles expelled by the Sun spiraling to Earth along magnetic field lines. Credit: NASA, Apollo 16, George Carruthers (Naval Research Laboratory) and the Far UV Camera Team

---

[5] The image is too small to be reproduced here. Those readers interested can consult it at the Planetary Photojournal (Image PIA04159).

photons scattered by exospheric hydrogen have been observed from a distance of approximately 100,000 km (∼15.5 Earth radii) from Earth (Carruthers et al. 1976).

During its Earth flyby, Galileo's UVS (Ultraviolet Spectrometer) conducted 11 scans of the space around the Earth and Moon. Galileo detected a huge hydrogen corona bulge surrounding the Earth to approximately 400,000 kms at the geotail, nearly to the Moon's orbit and four times the thickness of the traditional geocorona model. In fact, Galileo actually detected atomic hydrogen near the Moon at a level of approximately 1 atom cm^{-3}. Most of this hydrogen is probably associated with the extension of the Earth's geocorona, rather than an aspect of the Moon's tenuous atmosphere. Similar observations were carried out during the Cassini fly-by in 1999 (Werner et al. 2004). Recent observations from the Earth's orbit have been carried out by the IMAGE satellite (Burch 2000).

The present rate of escape of hydrogen atoms equals a layer of 1 mm depth every million years due to the evaporation of water from the oceans and the corresponding dissociation of water vapour molecules by UV radiation. This rate will grow in the future due to the increase in terrestrial temperatures following the increase of solar luminosity (see previous chapter). However, most of the glow spectrum is produced by the principal components of the atmosphere (nitrogen and oxygen).

We now describe in more detail the processes in the upper layers of the Earth's atmosphere leading to observable signatures.

4.6 Airglow

Non-thermal emission of radiation from excited states formed by processes resulting (directly or indirectly) from solar UV radiation produces *airglow*. It is one of the best remote observables for deriving the physical conditions of the upper layers of a planetary atmosphere. These emissions vary considerably with time because of solar activity. They also show a distinctive tropical brightness enhancement.

Most of this radiation emanates from the region about 70–300 km (mesosphere and thermosphere), and seen from space, it is a green bubble enclosing the planet. Airglow images have been taken by cameras on board different Space Shuttles looking back into the wake of the flight path. Similar observations can be obtained from aircraft platforms operating all-sky imagers. The International Space Station is also a good observatory of these phenomena. At the left in Fig. 4.12 we can see the faint glow encircling our planet.

Anders Angstrom (1814–1874) discovered green airglow light in 1868. Newcomb (1901) suggested that the emission of stars in the visible range makes only a part of the total night-sky emission. In 1909 at Groningen, L. Yntema presented his PhD thesis 'On the brightness of the sky and the total amount of starlight', where he drew attention to the long time variability of the phenomenon and showed that starlight scattered by atmospheric molecules was insufficient to explain the night-sky glow (Yntema, 1909).Robert John Strutt (1875–1947), son of the third

4.6 Airglow

Fig. 4.12 View of a full Moon photographed by Expedition 14 onboard the International Space Station. Earth horizon and airglow are visible at left. Courtesy of NASA

Lord Rayleigh, showed in 1922 that the geographical distribution of the strength of this line differed from that of aurorae. He gave the name of the units normally studied for these type of studies the Rayleighs.[6]

McLennan and Shrum (1925) identified the green line as due to atomic oxygen. It was in 1931 when Chapman (1931) suggested that airglow could result from chemical recombination. Chamberlain (1955) identified the UV airglow lines (340–380 nm) as due to oxygen Herzberg bands.

Chamberlain (1961, 1995), Mc Cormack (1971), Roach and Gordon (1973), Brasseur and Solomon (1994) and Slanger and Wolven (2002) provide a good background on this subject. Khomich et al. (2008) summarize the progress in this field and how the airglow can be used as an indicator of the structure and dynamics of the Earth's atmosphere.

Depending on the excitation mechanism, we can distinguish different components in the airglow, which we will describe in the following subsection. The main source is solar radiation, the immediate emission is called *dayglow* and the subsequent delayed emission *nightglow*.

---

[6] The Rayleigh is the unit commonly used to quantify the intensity of night-sky emission lines. It is defined as $1\,R = 1.583 \times 10^{-5}/(\lambda)\ \mathrm{erg\,s^{-1}\,cm^{-2}\,sr^{-1}}$, where λ is expressed in Å.

4.6.1 Nightglow

Nightglow is produced by chemiluminescence, the emission of radiation resulting from chemical reactions at a height of between 100 and 300 km. The main emphasis is on the molecular emissions of OH and O_2 from the mesosphere and lower thermosphere (Feldman and McNutt 1969; Llewellyn and Soldheim 1978). We have previously discussed the existence of different absorption features in the infrared range (CO_2 and O_3). The dynamics of the mesosphere[7] play an important role in structure the spatial distribution of this emission.

In the UV spectral region, the major contribution is from the O_2 Herzberg I band, with weaker contributions from the Herzberg II and Chamberlain bands (Chamberlain 1955):

$$O + O + M \longrightarrow O_2^* + M$$
$$O_2^* \longrightarrow O_2^{**} + \text{radiation (glow)}$$
$$O_2^* + M \longrightarrow O_2^{**} + M + \text{radiation (glow)}$$

where M is usually nitrogen. The speed of these reactions will depend on the identity of M.

Figure 4.13 shows a global view of UV nightglow. The NUV is brighter than the MUV, but fainter than the FUV. Rocket measurements by Greer et al. (1986) confirmed that the bulk of the UV nightglow comes from a region between 90 and 110 km.

Koomen et al. (1956) and Heppner and Meredith (1958) first measured altitude profiles of oxygen and sodium emission in Earth night airglow using data from sounding rockets. Bedinger et al. (1957) used sodium released by a rocket to quantify twilight and night airglow emission and measured mesospheric winds. López-Moreno et al. (1998), by observing the EUV night airglow, were able to identify the complete Lyman series up to Lyman ε. Table 4.4 gives the typical brightness of the most important UV spectral lines.

In the visible range we have the well-known green line of oxygen at 557.7 nm, covering a relatively narrow altitude region and strongly peaking around 100 km altitude. Some emission is also recorded at the red line at 630 nm.

Redward of 650 nm, we see the Meinel bands of OH produced by the reaction

$$H + O_3 \Longrightarrow OH^* + O_2^*,$$

where OH^* is a ro-vibrationally excited level of OH. It is limited at relatively high altitudes ($\sim$85 km) by the rapid fall in ozone concentration with height. Figures 4.14 and 4.15 show typical spectra of the nightglow in the visible and near infrared

---

[7] Atmospheric tides, internal gravity waves and planetary waves.

4.6 Airglow

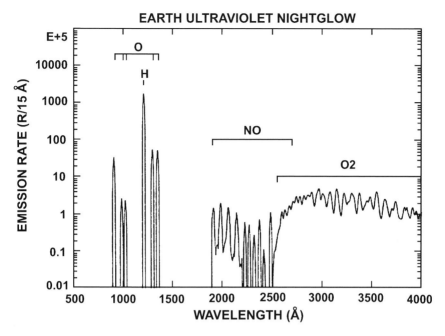

Fig. 4.13 The different regimes of UV nightglow. (source: R.R. Meier, 1991, Space Science Reviews 58, 19). Copyright: Springer

Table 4.4 UV nightglow spectral lines

Source	Wavelength (nm)	Height of emitting layer	Intensity
Lyman β	102.6	Geocorona	10 R
Ly α	121.6	Geocorona	3 kR (night)–34 kR (day)
O I	130.4	250–300 km	40 R (tropical airglow)
O I	135.6	250–300 km	30 R (tropical airglow)
O_2	300–400	90 km	0.8 R/Å

R stands for Rayleigh

ranges. Osterbrock et al. (1998) have recorded a high signal-to-noise spectrum in the range 572–881 nm using the Keck telescope at the Mauna Kea Observatory.

A natural global layer (usually about 5 km thick) of sodium atoms exists between about 80 and 105 km altitude. The sodium originates from the ablation of meteors. The atoms are naturally excited and emit a weak glow near a wavelength of 589 nm (yellow) known as 'the sodium D lines'. Above the layer, sodium exists in its ionized form (which does not emit yellow light), and below the layer, sodium exists as chemical compounds such as sodium oxide (which also do not emit yellow light). It was discovered by Slipher (1929) at the Lowell Observatory and later was explained by Chapman (1939).

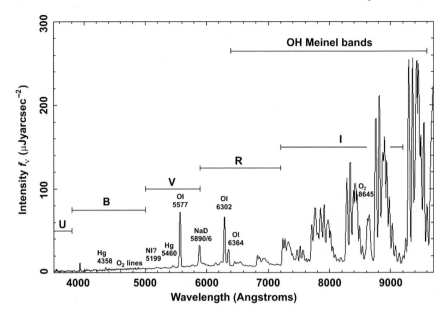

Fig. 4.14 Spectrum of the nightglow of La Palma sky on a moonless sky, taken with the Faint Object Spectrograph of the William Herschel Telescope in March 1991. Courtesy: C. Benn (ING, Roque de los Muchachos Observatory)

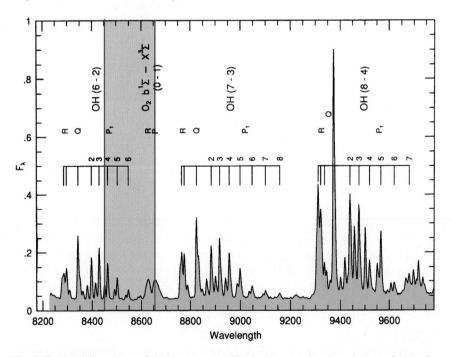

Fig. 4.15 Near infrared part of airglow spectrum. Hatched area marks spectral range of Argentine Airglow Spectrometer. Credit: Aeronomy Group, Instituto de Astronomía y Física del Espacio, Buenos Aires

4.6.2 Dayglow

Dayglow is produced when the atmosphere is illuminated by the Sun. The main processes involved are resonance scattering and fluorescence. Although it is intrinsically bright, dayglow is overwhelmed by direct and scattered sunlight. The UV spectrum is dominated by single ionized and neutral lines of oxygen and nitrogen, produced by UV photons and photoelectron impacts. Some lines are also present in the solar spectrum (He I 58.4 nm; O II triplet 83.4 nm; Lyman α 121.6 nm and β). Figure 4.16 shows a representative spectrum of the UV dayglow.

Various space experiments have measured the dayglow in the FUV range (Chakrabarti et al. 1983; Link et al. 1988 and Feldman et al. 2001). Craven et al. (1994), Meier et al. (2002) and Strickland et al. (2004) have, respectively, studied the dayglow response to an intense period of auroral activity, to a large solar flare and to short-term solar EUV variations.

MAHRSI is a middle ultraviolet (190–320 nm) spectrograph experiment that measures the OH density profile by observing the bright airglow produced by the resonance fluorescence of sunlight in the OH A-X (0,0) band around 310 nm. The difficulty with this measurement arises from sunlight Rayleigh-scattered by the ambient atmosphere, which generates an equally bright emission that contains the complex Fraunhofer spectrum formed when gases in the solar atmosphere absorb solar radiation. Separation of these two signals can best be achieved using very high spectral resolution. Observation of the Moon during the flight provided a

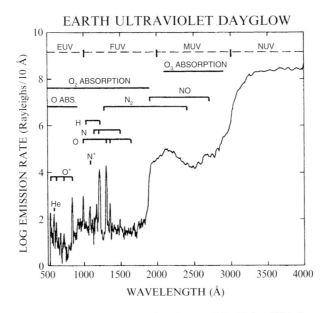

Fig. 4.16 The different regimes of UV dayglow. Source: R.R. Meier, 1991, Space Science Reviews Vol. 58. Copyright: Springer

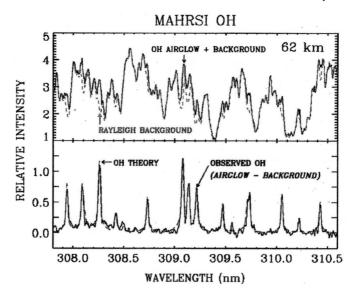

Fig. 4.17 The *upper panel* compares the observed spectrum at 62 km (*black curve*) with a background spectrum (*blue dashed curve*), scaled in magnitude to fit the wavelength regions of the atmospheric spectrum known to be free of OH lines. The *lower panel* shows the difference between the two spectra, that is, the observed intensity minus the background (*black curve*). Plotted over the remaining residual is the normalized theoretical prediction of the OH emission lines (*red dashed curve*), smoothed to the MAHRSI spectral resolution measured in the laboratory. Available from Offerman and Conway (1995) Fig. 2

convenient measure of the precise shape of the solar spectrum uncontaminated with atmospheric emissions or absorptions, including OH and O_3. The plots in Fig. 4.17 illustrate how the OH signal is retrieved (Offerman and Conway 1995).

4.6.3 Twilight Airglow

Twilight is the period between dawn and sunrise and between sunset and dusk. It is produced by sunlight scattered in the upper atmosphere illuminating the lower atmosphere. For astronomical purposes, it is defined as the period when the Sun reaches heights between $-12°$ to $-18°$ below the horizon.

A phenomenon known as twilight airglow may be observed at the transition between the dark and fully illuminated atmosphere as the shadow height moves vertically over the full range of emissive layers. Like airglow, it results from direct excitation by solar UV photons. Most of ultraviolet astronomical observations are taken from above the atmosphere by rockets or satellites. The viewing line of the spacecraft on the night side of the atmosphere may cross the terminator and continue through the sunlit parts of the atmosphere. Under these twilight conditions, dayglow features are easier to discern.

4.7 The Ionosphere

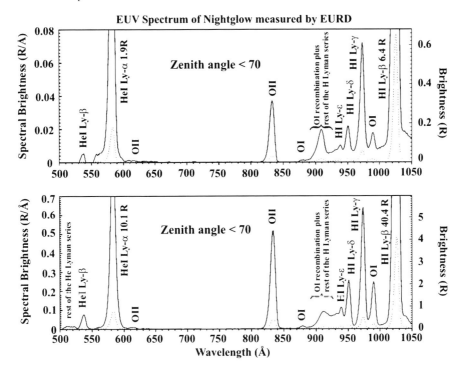

Fig. 4.18 Airglow spectrum in the Extreme Ultraviolet, taken by the EURD observing in the antisolar direction at two zenithal distances. Note the difference of scales. *Dotted line*: emission divided by 10. Courtesy of J.J. López Moreno (IAA, Granada)

During this time, yellow emissions from the sodium layer and red emissions from the 630 nm oxygen lines are dominant and contribute to the purplish colour sometimes seen during civil and nautical twilight. N_2 and NO molecular emissions from the mesosphere and thermosphere are observable above the Earth limb and at twilight since Rayleigh scattering is weaker there.

The EURD experiment on board Minisat (Giménez and Sabau-Graziatti 1996) has obtained high resolution spectra corresponding to nightglow and twilight conditions (see López-Moreno et al. (1998) and Fig. 4.18).

4.7 The Ionosphere

The troposphere and stratosphere are electrically neutral. However, in the mesosphere and thermosphere (approximately from 70 to 1000 km), the neutral components coexist with ionized particles produced by the Sun and cosmic rays. This layer, called the ionosphere, contains only a small fraction of the Earth's atmosphere (less than 1% of the mass above 100 km). Nagy and Cravens (2002) provide an overview of the Solar System ionospheres.

The first notions about an electric conducting layer in the terrestrial atmosphere were advanced by C.F. Gauss (1777–1855), in 1839, who proposed that small daily variations in the geomagnetic field could be explained by electric currents flowing in such a layer. Belfour Stewart (1828–1887) pictured such currents, around 1886, as arising from electromotive forces generated by periodic motions of the electric layer across the terrestrial magnetic field.

In 1901, G. Marconi (1874–1937) was able to establish a transatlantic radio-communication.[8] In 1902, Oliver Heaviside (1850–1925) and Arthur E. Kennelly (1861–1939), independently, explained this in terms of a reflection of the radiowaves by free charges in the high atmosphere. J.A. Fleming (1906)[9] proposed that solar UV radiation generates such free charges.

The first observational evidence of a new layer in the atmosphere, the ionosphere,[10] came from Edward Appleton (1892–1965) and M.A.F. Barnett in 1925. Later, this finding was verified by G. Breit (1899–1981) and M.A. Tuve (1901–1982) by using pulsed radio waves (Breit and Tuve 1926).

Hulburt (1928) proposed that the ultraviolet radiation shortwards of 123 nm might be the source of the ionosphere. Soundings with V-1 and V-2 rockets after World War II showed that this radiation shaped the bottom of the ionosphere.

4.7.1 General Structure

The components of the Earth's atmosphere may be ionized by capturing photons whose energy exceeds the corresponding ionization potential (see Table 4.5). Thus, radiation only with $\lambda < \lambda_{max}$ produces ionization; we therefore concentrate only on the X-ray and EUV regions of the spectrum.

At the highest levels of the Earth's atmosphere, the density is low and therefore the ionization rate is also low. As the altitude decreases, more gas atoms are present and so the ionization process increases. At the same time, however, an opposed process called recombination begins to take place in which a free electron is 'captured' by a positive ion if it moves close enough to it. As the gas density increases at lower altitudes, the recombination process accelerates as the gas molecules and ions are closer together. The point of balance between these two processes determines the degree of 'ionization' present at any given time and location.

Apart from high energy radiation, cosmic rays and solar particles also play an important role in the ionization balance. A number of monographs give deeper insight into the subject (Rishbeth and Garriot (1969); Bauer (1973); Kelley (1989); Hargreaves (1979); Hunsucker and Hargreaves (2002)).

---

[8] On 12 December 1901, he transmitted a Morse code signal from Cornwall (England) to Newfoundland (Canada), a distance of 2,900 km, a big surprise for the scientists that could not explain how the waves propagating as straight lines could curve over the 160 km bulge of the Earth.

[9] Fleming, J.A. 1906, On the electric radiation from bent antennae, Proceedings Physical Society London, 20, 409–426.

[10] This term was first used in 1926 by Robert Watson-Watt (1892–1973), the father of radar.

4.7 The Ionosphere

Table 4.5 Ionization potential of components of the terrestrial atmosphere

	I (ev)	λ_{max} (nm)
NO	9.25	134
O_2	12.08	102.7
H_2O	12.60	98.5
O_3	12.80	97.0
H	13.59	91.2
O	13.61	91.1
CO_2	13.79	89.9
N	14.54	85.3
H_2	15.41	80.4
N_2	15.58	79.6
Ne	21.56	57.5
He	24.58	50.4

From Hargreaves (1979) *The upper atmosphere and solar-terrestrial relations*, Van Nostrand Reinhold

The rate of change of electron density, N_e, is given by the continuity equation (see Hargreaves 1979)

$$\frac{\partial N_e}{\partial t} = q - L - \text{div}(N_e V),$$

where q is the rate of production, L the recombination rate and div $(N_e V)$ the change in electron density by movements with speed V.[11]

The rate of change in the ion density, q, can be expressed by

$$q = n F \sigma \eta,$$

where n is the number density of particles, F is the flux of the ionizing radiation ($J\,m^{-2}s^{-1}$), σ is the absorption cross section and η is the ionization efficiency (the number of electrons produced per absorbed photon). As F increases and n decreases with the altitude in the atmosphere, we expect there to be at a particular height a maximum q and hence a maximum in the electronic density.

Assuming a plane stratified atmosphere and monochromatic radiation, we can derive the altitude variation of this parameter:

$$q(z, \chi) = q_0 \exp\left(1 + \frac{h-z}{H} - \frac{1}{\cos \chi} e^{(h-z)}\right),$$

where z is the vertical altitude, h the height of maximum ionization, H the scale height[12] and χ the angle between the line of sight and the vertical. Figure 4.19 shows the distribution of the ionization normalized to the maximum density.

---

[11] It has a macroscopic component (advection) and a microscopic one (diffusion).

[12] An e-folding distance, commonly used to describe the fall off in atmospheric pressure or other related quantities.

Fig. 4.19 Various sources of ionization in the upper atmosphere. The *solid lines* show regular sources. During the day, EUV light is the main ionization source. At night, EUV scattered from the geocorona as well as from stars helps to maintain the E ionosphere. Source: Richmond (1987) Figure 5. Copyright: D. Reidel

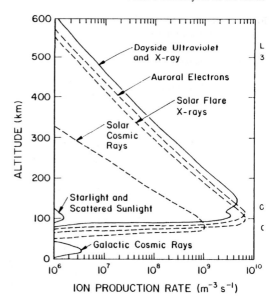

There are two main ways of producing loss of electrons. Occasionally, the electrons recombine directly with positive ions[13] following the relation $L = \alpha N_e^2$, where α is the recombination coefficient.[14] Another option is that electrons are removed by becoming attached to neutral molecules, Z ($e^- + Z \longrightarrow Z^-$), a process expressed by $L = \beta N_e$, β being an altitude-dependent attachment coefficient.

Figure 4.19 gives the variation with the altitude of the ion production rate from various sources (Richmond 1987).

The total electron content of the ionosphere is

$$N_T \equiv \int_0^\infty N_e \, dz.$$

Typically, N_T is about 10^{17} m^{-2}.

The major ionization sources are

$$O + \text{radiation} \longrightarrow O^+ + e^-$$
$$O_2 + \text{radiation} \longrightarrow O_2^+ + e^- \text{ or } O + O^+ + e^-$$
$$N_2 + \text{radiation} \longrightarrow N_2^+ + e^- \text{ or } N + N^+ + e^-$$

with the inverse process, recombination, returning the original constituents; for example, $O_2^+ + e^- \longrightarrow 2\,O$.

---

[13] We have two options: radiative recombination ($e^- + X^+ \longrightarrow X + \text{radiation}$) and dissociative recombination ($e^- + XY^+ \longrightarrow X + Y$).

[14] This law is valid only when the densities of ions and electrons are approximately the same.

4.7 The Ionosphere

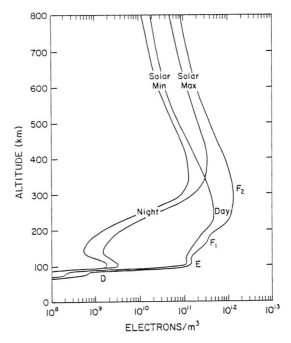

Fig. 4.20 Variation in the electronic density with the altitude during day and night for two different phases of solar activity Source: Richmond (1987) Fig. 1. Copyright: D. Reidel

The combination of the amount of incident radiation and the density of the atmospheric particles configures the extension of this layer (Fig. 4.20), which is subdivided into different regions, also called Chapman layers.

- *D layer*: The lowest, at altitudes between 50 and 80 km (mesosphere). It disappears at night. The main sources of ionization are Lyman-α (121.5 nm) on NO; EUV (102–111 nm) on oxygen and hard X-rays and cosmic rays on all atmospheric constituents.
- *E layer*: Found between 100 and 125 km. Also practically disappears at night. This layer is largely in photochemical equilibrium, a property shared with most of other planetary ionospheres. Ions in this region are mainly molecules such as O_2^+ and NO^+, produced through the following reactions:

$$O^+ + N_2 \longrightarrow NO^+ + N$$
$$N_2^+ + O \longrightarrow NO^+ + N$$
$$O^+ + O_2 \longrightarrow O_2^+ + O$$

- *F layer*: The ionization production is smaller than in lower layers, but the much longer electron lifetime permits large values of electron density to be reached. Its strength varies according to the time of the day, the season and the level of solar activity. In the daytime it is split into two sub-layers, with the F1 located around 180 km, where NO^+ and O_2^+ ions dominate, and the F2 at 400 km or more, with O^+ as the main contributor. It is controlled both by photochemistry and plasma transport and thus is not in photochemical equilibrium.

At low latitudes the largest N_e values are found in peaks on either side of the magnetic equator, called the equatorial anomaly. One would expect the largest concentration to occur at the equator because of the maximum of the solar ionizing radiation. This peculiarity can be explained by the special geometry of the magnetic field and the presence of electric currents.

The International Reference Ionosphere (IRI) is an empirical reference model of the physical parameters of this layer (Bilitza and Reinisch 2008). It is updated biannually and distributed by the National Space Science Data Center and World Data Center A for Rockets and Satellites. Figure 4.20 illustrates the change of N_e between day and night and two phases of solar activity.

The GUVI experiment onboard the TIMED satellite studied the spatial and temporal changes of the total electron content (TEC) of the atmosphere (Fig. 4.21). Variations in the solar output lead to changes in the total amount of ionization that occurs in the upper atmosphere, but the details of what happens after those electrons have been produced depend on the amount of heating at high latitudes, the Earth's seasonal wind pattern, the electric fields created in the upper atmosphere and the conditions that existed before a solar event occurred. These figures provide examples of how the ionosphere changes under the influence of the 11-year solar 'year', the Earth's season, the local times of the observation, the time that a storm started, the magnitude of the storm, whether the Earth experienced a coronal mass ejection or some other change in the solar wind and many other factors.

Finally, Fig. 4.22 compares the structure of the neutral atmosphere with that of the ionosphere.

4.7.2 Ionosphere Indicators

Previously we have described the origin of ionized metallic species produced by the impacts of meteorites at mesospheric heights. The presence of the corresponding spectral features can be used as indicators of the presence of ionospheres, as has recently been shown by Pallé et al. (2009) when analysing the transmission spectrum of the Earth atmosphere. Weak absorption lines corresponding to Ca+ at 393.4, 396.8 nm (H and K lines), 849.8, 854.2 and 866.2 nm (the near-infrared triplet) were detected. Other more abundant species, as Mg+, can be detectable at shorter wavelengths in the UV.

The H_3^+ molecule is generated by the ionization of molecular hydrogen by cosmic rays. Its transitions give rise to spectral features in the near infrared (Gottfried et al. 2003).

Molecular hydrogen was first observed in the Lyman and Werner bands (117–165 nm) in Jupiter (Broadfoot et al. 1979; Clarke et al. 1980). Emission lines of H_3^+ in the infrared were detected in the auroral regions of Jupiter at $2\,\mu$m (Drossart et al. 1989) and the polar ionosphere in the $4\,\mu$m band (Oka and Geballe 1990). Maps of the emission in Jupiter's auroral zones have been published by

4.7 The Ionosphere

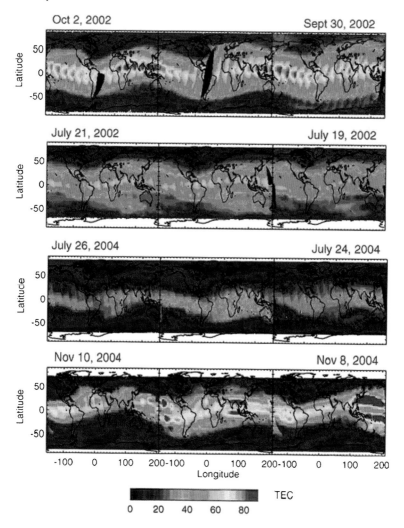

Fig. 4.21 Temporal and spatial changes of the total electron content of the atmosphere as observed by the GUVI experiment onboard the TIMED satellite. Credit: NASA/The Johns Hopkins University Applied Physics Laboratory (JHU/APL)

Lellouch (2006). Geballe et al. (1993) also detected the 2 μm emission in polar regions of Saturn although two orders of magnitude weaker than in Jupiter. Shokolov (1981) reported some emission of this ion in the upper layers of Venus.

The molecule H_3^+ is an efficient mechanism of cooling in the upper layers of planets, as the extended atmosphere of HD 209458b (Koskinen et al. 2007; Troutman, 2007). Laughlin et al. (2008) have established an upper limit for this emission in the giant exoplanet Tau Boo. A positive detection of 4 μm emission toward the star HD141569A (Brittain and Rettig 2002) was later not verified by Goto et al. (2005).

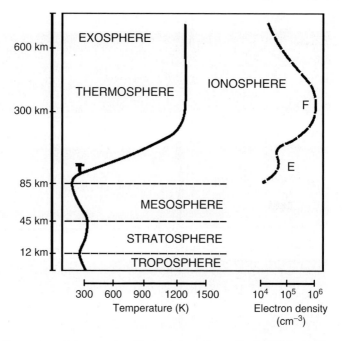

Fig. 4.22 Comparison of the neutral and ionized atmospheres. Credit: Wikipedia Commons

4.7.3 Lightnings

Lightning can cause ionosphere perturbations in the D-region. Wilson (1925) proposed a mechanism by which electrical discharge from lightning storms could propagate from clouds to the ionosphere, but empirical verification (Davis and Johnson 2005) of this did not come until much later.

Observations from space allow electric discharges to be recorded from the top of storm clouds toward the upper atmospheric layers. They are called Red Sprites, Elves and Blue Jets. The first images of a sprite were accidentally captured in a video on 6 July 1989 (Franz et al. 1990). Beginning in 1990, about twenty images have been obtained from the space shuttle[15] (Boeck et al. 1992; Vaughan 1994). Sprites occur when a potent stroke creates an intense electrostatic field above the storm cloud from which it emanates. A synthetic spectrum has been calculated by Milikh et al. (1998).

Elves are also originated in lightning at 85–95 km but extend into wings up to 300 km (Nagano et al. 2003). They were discovered in 1992 by a low-light video camera on the Space Shuttle. Blue jets emerge from the top of the thundercloud, extending up in narrow cones and disappearing at altitudes around 40 km. They were first discovered by Alaska scientists onboard a NASA research jet in 1994

---

[15] An important experiment was carried out during the last flight of the Columbia in 2003, led by the astronaut Ilan Ramon (1954–2003).

4.8 The Magnetosphere

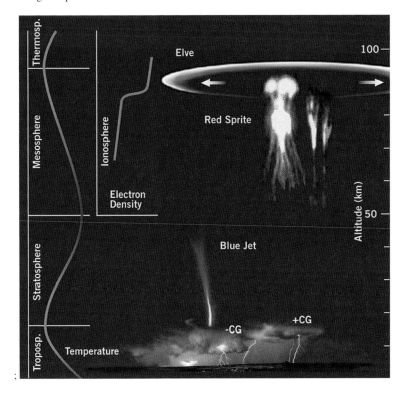

Fig. 4.23 Altitude separates blue jets, red sprites and elves. Credit: Danish National Space Center

(Pasko 2008). Aramyan et al. (2008) have proposed that the radiation of red sprites and blue jets is due to the superluminescence[16] of highly excited atomic oxygen induced by an acoustic wave.

Figure 4.23 shows the location of these phenomena (denominated jointly Transient Luminous Events, TLE), which last less than one second and may be an essential element of the Earth's global electrical circuit (Siingh 2007; Rycroft et al. 2008).

4.8 The Magnetosphere

4.8.1 Description

The view of the Earth in high-energy radiation (UV, X-rays) would be completely different without the presence of a magnetic field produced by a dynamo process

---

[16] Superluminescence is the same as amplified spontaneous emission: the emission of luminescence that experiences significant optical gain within the emitting device, and therefore can be relatively intense.

Fig. 4.24 The Magnetosphere and its interaction with the solar wind. Courtesy: GSFC/NASA

in the outer core. Its major part resembles the field of a bar magnet ('dipole field') inclined by about 10° to the rotation axis of Earth, but more complex parts ('higher harmonics') also exist, as first shown by C. Gauss.

The magnetosphere of Earth is a region in space whose shape is primarily determined by the distortion of Earth's internal magnetic field and by the solar wind plasma and the interplanetary magnetic field (IMF). On the side facing the Sun, the distance to its boundary (which can vary) is about 70,000 km (10–12 Earth radii or R_E).[17] The boundary of the magnetosphere ('magnetopause') is roughly bullet-shaped, about 15 R_E abreast of Earth and on the night side (in the 'magnetotail' or 'geotail') approaching a cylinder with a radius 20–25 R_E. The tail region stretches well past 200 R_E (see Fig. 4.24 for a diagram). If the Earth had no global magnetosphere, the solar wind would impact on our atmosphere and gradually erode it away.

4.8.2 Radiation Belts

The Van Allen radiation belt is a torus of energetic charged particles around Earth, held in place by the Earth's magnetic field (Fig. 4.25).[18] The inner radiation belt extends one Earth radius and consists of very energetic protons (>100 MeV), a by-product of collision by cosmic rays with atoms of the atmosphere. The outer

---

[17] All distances here are from the Earth's centre.

[18] J. Van Allen (1914–2006) made the first public announcement of the discovery at a special joint assembly of the National Academy of Sciences and the American Physical Society on 1 May 1958.

4.8 The Magnetosphere

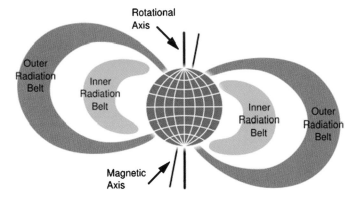

Fig. 4.25 The two radiation belts of our planet

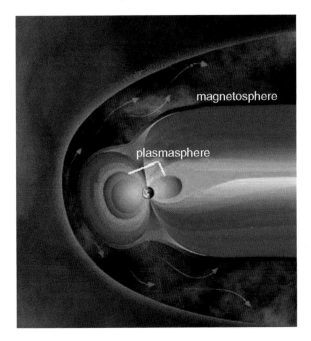

Fig. 4.26 The plasmasphere

belt is situated at an altitude of about 31,000–65,000 km (1–10 Earth radii) and is composed of high energy electrons (0.1–10 MeV). They are injected from the geomagnetic tail following geomagnetic storms.

The plasmasphere is a torus of cool (low-energy ~1 eV), dense (tens to thousands of particles per cubic centimetre) plasma that occupies the inner magnetosphere and is located above the ionosphere (Fig. 4.26). It is populated by the outflow of ionospheric plasma (hydrogen ions) along mid- and low-latitude magnetic field lines and co-rotates with the Earth. The plasmasphere extends out to as little as 2–3 Earth

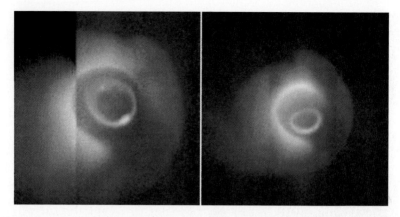

Fig. 4.27 The Earth's plasmasphere as viewed with the EUV instrument on board the IMAGE spacecraft. The Earth is at the centre of the images. The Sun is to the upper left. The view is toward Earth's north pole. The 30.4 nm emission from the plasmaspheric helium ions appears in false colour as a pale blue cloud surrounding the Earth. The two images were acquired at different times during the same orbit on 24 May 2000. In the first image, the spacecraft is in the evening sector and has not yet passed over the north pole; in the second image, it has crossed the pole and is at or near apogee (7.2 Earth radii) on the morning side. Credit: B. Sandel and T. Forrester, University of Arizona

radii and, under quiet conditions on the evening side, perhaps more than 6 Earth radii (Carpenter 1963; Lemaire and Gringauz (1998); De Keyser et al. 2009). Figure 4.27 shows this region as seen in EUV radiation.

Nilsson et al. (2008) have studied the outflow of oxygen ions at high altitude above the polar cap using data collected by CLUSTER quartet of satellites. They discovered that these ions were being accelerated in the direction of the Earth's magnetic field.[19] The magnetosphere is not only filled with solar energetic particles, but also with particles coming from the Earth's atmosphere.

4.8.3 Aurorae

A bright discrete arc near midnight is evident in Fig. 4.27 (lefthand image), the aurora. This phenomenon is caused by the collision of charged particles[20] (e.g. electrons), found in the magnetosphere, with atoms in the Earth's upper atmosphere (at altitudes above 80 km) and constitutes the clearest manifestation of the Earth's coupling to the solar wind via the magnetosphere. Light emitted by the aurora tends

---

[19] H. Nilsson commented in an ESA news item that, compared to the Earth's stock of oxygen, the amount escaping is negligible. However, in the far future, when the Sun's luminosity increases, the oxygen escape may become significant.

[20] Unlike the airglow, excited by solar photons.

Fig. 4.28 Aurora Borealis and lights in Finland, Russia, Estonia and Latvia are featured in this digital still picture taken by the Expedition 11 crew aboard the International Space Station. If it were daylight, parts of the Eastern Baltic Sea would be visible. The station was over a point on Earth located at 50.6 degrees north latitude and 15.1 degrees east longitude at the time. The cluster of stars to the lower right of the thin crescent Moon is the Praesepe Cluster in Cancer. Just to the right of that is the planet Saturn. Courtesy: NASA

to be dominated by emissions from atomic oxygen, resulting in a greenish glow (at a wavelength of 557.7 nm) and – especially at lower energy levels and at higher altitudes – the dark-red glow (at 630.0 nm of wavelength).

It differs from the airglow in three main aspects: (a) high latitude vs. global extension, (b) highly structured appearance vs. relatively uniform and (c) the source: solar wind instead of solar radiation for the airglow.

Its easy visibility to the naked eye has clearly influenced the mythology of many civilizations and its historical records have also been used as a proxy of solar activity in the past (see Chap. 6 of Vaquero and Vázquez 2009).

Figures 4.28 and 4.29 show spectacular views of the aurora photographed from the International Space Station. In the latter we can see the different heights of formation of the green and red aurora.

4.9 Radio Emission of the Earth and Other Planets

The radio engineer Karl G. Jansky (1905–1950), while working at Bell Telephone Laboratories, in 1932, was the first to detect radio noise from the region near the centre of the Milky Way, during an experiment to locate distant sources of terrestrial

Fig. 4.29 Aurora photographed from the ISS with a 58 mm lens on 2 February 2003. Green aurora extends upward from the Earth's airglow layer. When the red aurora is present, it usually extends above the green aurora. Courtesy: NASA

radio interference (Jansky 1933). The distribution of this galactic radio emission was mapped by Grote Reber (1911–2002), using a 9.5 m (31 ft) paraboloid that he built in his backyard in Wheaton, Illinois. He also discovered the long sought-after radio emission from the Sun (Reber 1944).

Two conditions are necessary for there to be significant planetary radioemission: a magnetosphere and the existence of a storage of charged particles, the radiation belts. The interaction of the latter with the magnetic fields gives rise to the emission or absorption of radiation. For rapidly moving particles (relativistic), the radiation occurs over a large range of frequencies, and is called the synchrotron radiation. When such charged particles (electrons and protons) move through a magnetic field, their paths are changed. The particles are accelerated and start to move in spirals around magnetic field lines towards either the south or the north pole. Charged particles that are accelerated emit radiation that depends on the energy and the type of the charged particles.

For slowly moving particles (non-relativistic) this happens at a single frequency, the cyclotron[21] frequency

$$f_c = (1/2p)eB/m_e,$$

---

[21] Both names are in reference to laboratory accelerators.

4.9 Radio Emission of the Earth and Other Planets

where e is the electron charge, B the magnetic field strength and m_e the electron mass. Both types of radiation also receive the name of non-thermal radiation in contrast to thermal radiation (black-body) dependent on the temperature.

Burke and Franklin (1955) discovered the first planetary source of radioemission in the Solar System: Jupiter. The radio signal was described as having the appearance of short random bursts of static resembling the thunderstorm interference on a broadcast receiver. It is the only planetary body that has a synchrotron radioemission at decimetric wavelengths (Radhakrishnan and Roberts 1960; Kloosterman et al. 2005) together with a strong cyclotron radiation (McCulloch 1968). An important source of particles is the volcanic activity of its satellite Io.

The cyclotron radioemission is centred in the kilometres wavelength and is generally associated with aurora. It has also been discovered in Saturn during the Voyager (Kaiser et al. 1981) and Cassini (Gurnett et al. 2005) missions. Similar findings have been made in Uranus (Warwick et al. 1986) and Neptune (Sawyer et al. 1990).

The radioemission of the Earth was discovered by Benediktov et al. (1965)[22] and Dunckel et al. (1970) using data centred in kilometric wavelengths. Later observations with the IMP and Hawkeye experiments[23] revealed the variability of this radiation generated between the plasmapause and the magnetopause.

Figure 4.30 shows the radio emission of the different planets having a magnetosphere, where we can contrast the importance of the Earth emission.

For a comparative study of planetary radioemission see Rucker et al. (1988). Desch and Kaiser (1984) established an empirical 'radiometric' Bode's law for the planets of the Solar System

$$P_{rad} = \varepsilon P_{SW}^\alpha,$$

where $\alpha \sim 1$, ε ranges from 10^{-6} to 10^{-3} and P_{SW} is the solar wind power incident on the planetary magnetosphere. P_{SW} depends on the solar wind density, the solar wind speed and the radius of the planetary magnetosphere.

Winglee et al. (1986) were probably the first to suggest the possibility to detect cyclotron radioemission from magnetized exoplanets. Radioemission from an exoplanet can be a direct method for its detection, also demonstrating the existence of a magnetosphere (Bastian et al. 2000). The upper frequency limit of this radiation gives the maximum strength of the magnetic field near the surface of the planet. This radioemission is likely modulated by the rotation period of the planet.

Radiosources at the Earth are located along magnetic field lines, which map down to discrete UV and X-ray sources (Huff et al. 1988).

---

[22] Published 1968 in english in the Cosmic Research Journal (vol. 3, p. 492).

[23] The Interplanetary Monitoring Platform (IMP) was a project of the Goddard Space Flight Center to investigate the interplanetary plasma and magnetic field. The instrument was included in eight spacecrafts launched from 1963 to 1973. It was followed by the Hawkeye satellite (Explorer 52) launched in 1974.

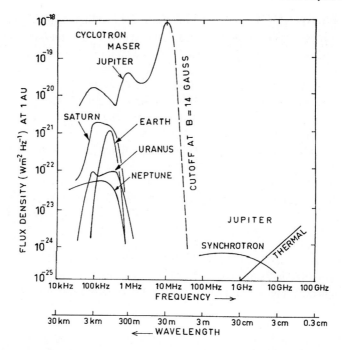

Fig. 4.30 Flux densities of solar system planets normalized to a distance of 1 AU. Source: Bastian et al. (2000) Fig. 1. Reproduced by permission of the American Astronomical Society

4.10 The Earth in X-Rays

The Sun is the strongest source of X-rays in the Solar System. However, planets are also emitters in this range. As in other magnetized planets, aurorae are the main source of terrestrial X-rays. Recently, the Chandra Observatory has observed our planet from 120,000 km with X-ray detectors. Bhardwaj et al. (2007) discovered low energy (0.1–10 keV) X-rays generated during auroral activity, shown as the bright arcs in this sample of images (Fig. 4.31).

Approximately 1% of the auroral input energy is transformed into electron cyclotron radioemission (Gurnett 1974). Auroral radio sources typically map to auroral optical sources (Huff et al. 1988).

Non-aurora X-ray emission is produced by scattering of solar X-rays. The outer layers of the Earth's atmosphere reflect part of the incident X-ray photons coming from cosmic sources. Churazov et al. (2008) have calculated the X-ray albedo for the Earth.

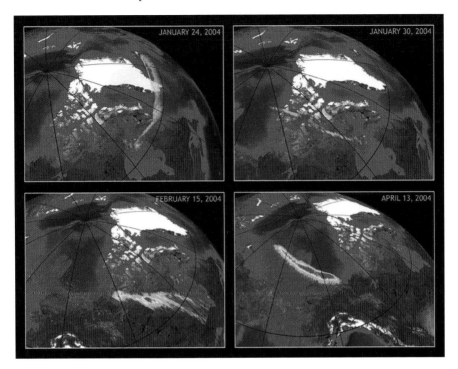

Fig. 4.31 X-rays radiation from aurorae observed by the Chandra observatory. The images – seen here superimposed on a simulated image of the Earth – are from approximately 20 min scans during which Chandra was pointed at a fixed point in the sky, while the Earth's motion carried the auroral region through the field of view. The colour code of the X-ray arcs represent the X-ray brightness, with maximum values shown in *red*. Credit: NASA/MSFC/CXC/A. Bhardwaj and R. Elsner/MSFC/CXC/A.Bhardwaj and R.Elsner

4.11 The Earth's Gamma Ray Emission

This high energy radiation is emitted by the upper atmosphere following the interaction between the cosmic rays and their main constituents. A cascade of secondary particles (muons, pions, positrons and electrons) and the Cherenkov light[24] are produced. In this way, the atmosphere blocks this harmful radiation from reaching the surface.

The first measurements of this radiation were done in 1972 and 1973 by the NASA SAS-II satellite (Thompson et al. 1981) but had very limited photon statistics. The Compton Gamma-Ray Observatory (GCRO) was active from 1991 to 2000, orbiting at an average altitude of 420 km. From this distance, the Earth appears as a disk with an angular diameter of 140 degrees. Combining exposures of

---

[24] Electromagnetic radiation which is emitted when a charged particle, such as a proton, passes through an insulator at a speed greater than the speed of light in that medium.

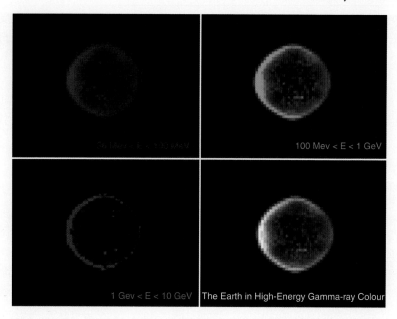

Fig. 4.32 Gamma ray image of the Earth in different energy ranges. Courtesy: D. Petry, NASA Goddard Space Flight Center

our planet taken by the EGRET instrument during the complete CGRO mission, it was possible to produce a γ ray picture of our planet (Fig. 4.32). Brightest near the edge and faint near the centre, the picture indicates that the gamma rays are coming from high in the Earth's atmosphere.

4.12 The Outer Layers of the Early Earth

The composition of the atmosphere of the Early Earth was different from that of the present time. Therefore, the intrinsic spectrum was also different, dominated by carbon dioxide, methane and the absence of oxygen. Moreover, other external factors were also different.

In previous chapters we have written on the stronger levels of solar ionizing radiation (solar wind, X-UV rays) during the early times of our planet. Apart from possible implications on life, this has clearly affected the photochemistry processes occurring in the upper layers of the atmosphere (cf. Zahnle and Walker 1982; Ayres 1997; Ribas et al. 2005). Kulikov et al. (2007) provide an excellent review on this matter, also with references on the early atmospheres of Venus and Mars.

The enhanced XUV radiation will produce warmer and extended exospheres (see Fig. 4.33). The base of the exosphere is located at the height where the mean free path of the photons is equal to the local scale height, $H = kT_{exo}mg^{-1}$, where m is

4.12 The Outer Layers of the Early Earth

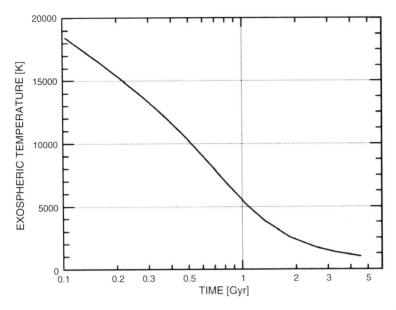

Fig. 4.33 Time evolution of the exospheric temperature based on Earth's present atmospheric composition. Source: Kulikov et al. (2007)

the mass of the main atmospheric species. The thermal escape parameter is defined as the ratio between the potential energy of a particle and the thermal energy kT_{exo} (see Güdel 2007)

$$X = \frac{GM_E m}{kT_{exo} r},$$

where r is the distance to the centre of the Earth. If $X < 1.5$ or analogously $T_{exo} > 2GM_E m/3kr$, the exosphere becomes unstable. One first consequence should be the blow-off of hydrogen. The H_2 concentration of the early atmosphere was determined by balancing volcanic outgassing and escape to the space.

The strength of the solar wind was no doubt larger at that time, although the quantitative estimation is still far from clear (see Wood et al. 2005; Holzwarth and Jardine 2007). Nevertheless, the non-thermal processes of erosion of the early Earth's atmosphere were probably larger than those due to the thermal escape, although this could also depend on the strength of the Earth's magnetic field.

The expected high rates of evaporation were probably reduced by the presence of an early atmosphere containing high amounts of CO_2 cooling the base of the exosphere (Tian et al. 2005). This claim has been contradicted by Catling (2006).[25] McGovern (1969) studied the structure of the upper atmosphere for a methane-dominated atmosphere.

---

[25] See subsequent response of Tian et al. in the same volume.

In absence of the ozone layer, absorption of radiation by other atmospheric components was essential to determine the amount of XUV radiation received at the surface of the Early Earth. Cnossen et al.'s (2007) calculations indicate that changes in the assumed CO_2 concentration play a major role. In any case, radiation levels on the Archaean Earth were several orders of magnitude higher in the wavelength range below 200 nm than current levels in this range. That means that any form of life that might have been present at Earth's surface 4–3.5 Ga ago must have been exposed to much higher quantities of damaging radiation than at present. Terrestrial life will be the topic of the next chapter.

References

Appleton, E.V.: Geophysical influences on the transmission of wireless waves. Proc. Phys. Soc. Lond. **37**, D16–D22 (1924)

Aramyan, A.R., Galechyan, G.A., Harutyunyan, G.G.: Superluminescence of atomic oxygen in the upper atmosphere. Laser Phys. **19**, 835–841 (2008)

Assman, R.: Über die existenz eines wärmeren Luftstromes in der Höhe von 10 bis 15 km. Proc. Roy. Prussian Acad. Sci. **24**, 1–10 (1902)

Ayres, T.R.: Evolution of the solar ionizing flux. J. Geophys. Res. **102**, 1641–1652 (1997)

Bastian, T.S., Dulk, G.A., Leblanc, Y.: A Search for Radio Emission from Extrasolar Planets. Astrophys. J. **545**, 1058–1063 (2000)

Bauer, S.J.: Physics of Planetary Ionospheres. Springer (1973)

Bedinger, J.F., Manring, E.: Emission from Sodium Vapor Ejected into the Earth's Atmosphere at Night. J. Geophys. Res. **62**, 162 (1957)

Bhardwaj, A., Elsner, R.F., Randall Gladstone, G., Cravens, T.E., Lisse, C.M., Dennerl, K., Branduardi-Raymont, G., Wargelin, B.J., Hunter Waite, J., Robertson, I., Østgaard, N., Beiersdorfer, P., Snowden, S.L., Kharchenko, V.: X-rays from solar system objects. Planet. Space Sci. **55**, 1135–1189 (2007)

Bilitza, D., Reinisch, B.W.: International Reference Ionosphere 2007: Improvements and new parameters. Adv. Space Res. **42**, 599–609 (2008)

Boeck, W.L., Vaughan Jr., O.H., Blakeslee, R., Vonnegut, B., Brook, M.: Lightning induced brightening in the airglow layer. Geophys. Res. Lett. **19**, 99–102 (1992)

Brasseur, G., Solomon, S.: Aeronomy of the Middle Atmosphere. D. Reidel, Dordrecht (1994)

Breit, G., Tuve, M.A.: A Test of the Existence of the Conducting Layer. Phys. Rev. **28**, 554–575 (1926)

Brittain, S.D., Rettig, T.W.: CO and H3+ in the protoplanetary disk around the star HD141569. Nature **418**, 57–59 (2002)

Burch, J.L.: IMAGE mission overview. Space Sci. Rev. **91**, 1–14 (2000)

Burke, B.F., Franklin, K.L.: Observations of a variable radio source associaetd with the planet Jupiter. J. Geophys. Res. **60**, 213–217 (1955)

Carpenter, D.L.: Whistler Evidence of a 'Knee' in the Magnetospheric Ionization Density Profile. J. Geophys. Res. **68**, 1675–1682 (1963)

Carruthers, G.R., Page, T., Meier, R.R.: Apollo 16 Lyman alpha imagery of the hydrogen geocorona. J. Geophys. Res. **81**, 1664–1672 (1976)

Catling, D.C.: Comment on "A Hydrogen-Rich Early Earth Atmosphere". Science **311**, 38 (2006)

Chakrabarti, S., Paresce, F., Bowyer, S., Kimble, R., Kumar, S.: The extreme ultraviolet day airglow. J. Geophys. Res. **88**, 4898–4904 (1983)

Chamberlain, J.W.: The Ultraviolet Airglow Spectrum. Astrophys. J. **121**, 277–286 (1955)

Chamberlain, J.W.: Physics of the Aurora and Airglow. Academic Press, London (1961)

References

Chamberlain, J.W.: Physics of the Aurora and Airglow. American Geophysical Union (1995)
Chapman, S.: A theory of upper-atmosphere ozone. Mem. Roy. Meteorol. Soc. **3**, 103–125 (1930)
Chapman, S.: The absorption and dissociative or ionizing effect of monochromatic radiation in an atmosphere on a rotating earth. Proc. Phys. Soc. **43**, 26–45 (1931)
Chapman, S.: Notes on Atmospheric Sodium. Astrophys. J. **90**, 309–316 (1939)
Christie, M.: The Ozone layer. Cambridge University Press, Cambridge (2000)
Churazov, E., Sazonov, S., Sunyaev, R., Revnivtsev, M.: Earth X-ray albedo for cosmic X-ray background radiation in the 1-1000 keV band. Mon. Not. Roy. Astron. Soc. **385**, 719–727 (2008)
Cnossen, I., Sanz-Forcada, J., Favata, F., Witasse, O., Zegers, T., Arnold, N.F.: Habitat of early life: Solar X-ray and UV radiation at Earth's surface 4-3.5 billion years ago. J. Geophys. Res. E **112**(11), (2008); (2007)
Craven, J.D., Nicholas, A.C., Frank, L.A., Strickland, D.J., Immel, T.J.: Variations in the FUV dayglow after intense auroral activity. Geophys. Res. Lett. **21**, 2793–2796 (1994)
Davis, C.J., Johnson, C.G.: Lightning-induced intensification of the ionospheric sporadic E-layer. Nature **435**, 799–801 (2005)
de Keyser, J., Carpenter, D.L., Darrouzet, F., Gallagher, D.L., Tu, J.: CLUSTER and IMAGE: New Ways to Study the Earth's Plasmasphere. Space Sci. Rev. **145**, 7–53 (2009)
Desch, M.D., Kaiser, M.L.: Predictions for Uranus from a radiometric Bode's law. Nature **310**, 755–757 (1984)
Dessler, A.E.: The Chemistry and Physics of Atmospheric Ozone. Academic Press, London (2000)
Dickinson, R.E.: Effects of solar electromagnetic radiation on the terrestrial environment. In: Physics of the Sun, pp. 155–191. D. Reidel (1986)
Dobson, G.: Forty years research on atmospheric ozone at Oxford. Appl. Optic. **7**, 387–345 (1968)
Drossart, P., Maillard, J.P., Caldwell, J., Kim, S.J., Watson, J.K.G., Majewski, W.A., Tennyson, J., Miller, S., Atreya, S.K., Clarke, J.T., Waite, J.H., Wagener, R.: Detection of H3(+) on Jupiter. Nature **340**, 539–541 (1989)
Dunckel, N., Ficklin, B., Rorden, L., Helliwell, R.A.: Low-frequency noise observed in the distant magnetosphere with OGO 1. J. Geophys. Res. **75**, 1854–1862 (1970)
Feldman, P.D., McNutt, D.P.: Far infrared nightglow emission from atomic oxygen. J. Geophys. Res. **74**, 4791–4793 (1969)
Feldman, P.D., Sahnow, D.J., Kruk, J.W., Murphy, E.M., Moos, H.W.: High-resolution FUV spectroscopy of the terrestrial day airglow with the Far Ultraviolet Spectroscopic Explorer. J. Geophys. Res. **106**, 8119–8130 (2001)
Fowler, A., Strutt, R.J.: Absorption Bands of Atmospheric Ozone in the Spectra of Sun and Stars. Roy. Soc. Lond. Proc. A **93**, 577–586 (1917)
Franz, R.C., Nemzek, R.J., Winckler, J.R.: Television Image of a Large Upward Electrical Discharge Above a Thunderstorm System. Science **249**, 48–51 (1990)
Geballe, T.R., Jagod, M.F., Oka, T.: Detection of H3(+) infrared emission lines in Saturn. Astrophys. J. **408**, L109–L112 (1993)
Giménez, A., Sabau-Graziati, L.: The Spanish MINISAT-01 mission. Memor. Soc. Astronom. Ital. **67**, 563–568 (1996)
Goto, M., Geballe, T.R., McCall, B.J., Usuda, T., Suto, H., Terada, H., Kobayashi, N., Oka, T.: Search for H3+ in HD 141569A. Astrophys. J. **629**, 865–872 (2005)
Gottfried, J.L., McCall, B.J., Oka, T.: Near-infrared spectroscopy of H3+ above the barrier of linearity. J. Chem. Phys. **118**, 10,890–10,899 (2003)
Greer, R.G.H., Murtagh, D.P., McDade, I.C., Dickinson, P.H.G., Thomas, L., Nishhizumi, K.: ETON 1 – A data base pertinent to the study of energy transfer in the oxygen nightglow. Planet. Space Sci. **34**, 771–788 (1986)
Güdel, M.: The Sun in Time: Activity and Environment. Living Rev. Sol. Phys. **4** (2007)
Gurnett, D.A.: The earth as a radio source: terrestrial kilometric radiation. J. Geophys. Res. **79**, 4227–4238 (1974)
Gurnett, D.A., Kurth, W.S., Hospodarsky, G.B., Persoon, A.M., Averkamp, T.F., Cecconi, B., Lecacheux, A., Zarka, P., Canu, P., Cornilleau-Wehrlin, N., Galopeau, P., Roux,

A., Harvey, C., Louarn, P., Bostrom, R., Gustafsson, G., Wahlund, J.E., Desch, M.D., Farrell, W.M., Kaiser, M.L., Goetz, K., Kellogg, P.J., Fischer, G., Ladreiter, H.P., Rucker, H., Alleyne, H., Pedersen, A.: Radio and Plasma Wave Observations at Saturn from Cassini's Approach and First Orbit. Science **307**, 1255–1259 (2005)

Hargreaves, J.K.: The upper atmosphere and Solar-terrestrial relations. Van Nostrand Reinhold (1979)

Harris, M.J.: A new coupled terrestrial mesosphere-thermosphere general circulation model: Studies of dynamic, energetic, and photochemical coupling in the middle and upper atmosphere. PhD Thesis, University of London (2001)

Heppner, J.P., Meredith, L.H.: Nightglow Emission Altitude From Rocket Measurements. J. Geophys. Res. **63**, 51–65 (1958)

Herzberg, G.: Molecular Spectra and Molecular Structure. I.- Spectra of Diatomic Molecules. Prentice Hall Inc., New York (1939)

Herzberg, G.: Molecular Spectra and Molecular Structure. II.- Infrared and Raman Spectra of Polyatomic Molecules. Van Nostrand Co., New York (1945)

Herzberg, G.: Molecular Spectra and Molecular Structure. III.- Electronic Spectra and Electronic Structure of Polyatomic Molecules. Van Nostrand Co., New York (1966)

Holzwarth, V., Jardine, M.: Theoretical mass loss rates of cool main-sequence stars. Astron. Astrophys. **463**, 11–21 (2007)

Huff, R.L., Calvert, W., Craven, J.D., Frank, L.A., Gurnett, D.A.: Mapping of auroral kilometric radiation sources to the aurora. J. Geophys. Res. **93**, 11,445–11,454 (1988)

Hulburt, E.O.: Ionization in the Upper Atmosphere of the Earth. Phys. Rev. **31**, 1018–1037 (1928)

Hunsucker, R.D., Hargreaves, J.K.: The High-latitude ionosphere and its Effects on Radio Propagation. Cambridge University Press, Cambridge (2002)

Jansky, K.G.: Electrical disturbances apparently of extraterrestrial origin. Proc. IRE **21**, 1387–1398 (1933)

Kaiser, M.L., Desch, M.D., Lecacheux, A.: Saturnian kilometric radiation - Statistical properties and beam geometry. Nature **292**, 731–733 (1981)

Kelley, M.C.: The Earth's Ionosphere: Plasma Physics and Electrodynamics. Academic Press, London (1989)

Khomich, V.Y., Semenov, A.I., Shefov, N.N.: Airglow as an indicator of Upper atmospheric structure and dynamics. Springer (2008)

Kloosterman, J.L., Dunn, D.E., de Pater, I.: Jupiter's Synchrotron Radiation Mapped with the Very Large Array from 1981 to 1998. Astrophys. J. Suppl. **161**, 520–550 (2005)

Koomen, M.J., Scolnik, R., Tousey, R.: Measurements of the night airglow from a rocket. Astron. J. **61**, 182–182 (1956)

Kopp, E.: On the abundance of metal ions in the lower ionosphere. J. Geophys. Res. **102**, 9667–9674 (1997)

Koskinen, T.T., Aylward, A.D., Miller, S.: A stability limit for the atmospheres of giant extrasolar planets. Nature **450**, 845–848 (2007)

Kulikov, Y.N., Lammer, H., Lichtenegger, H.I.M., Penz, T., Breuer, D., Spohn, T., Lundin, R., Biernat, H.K.: A Comparative Study of the Influence of the Active Young Sun on the Early Atmospheres of Earth, Venus, and Mars. Space Sci. Rev. **129**, 207–243 (2007)

López-Moreno, J.J., Morales, C., Gómez, J.F., Trapero, J., Bowyer, S., Edelstein, J., Lampton, M., Korpela, E.J.: EURD observations of EUV nightime airglow lines. Geophys. Res. Lett. **25**, 2937–2940 (1998)

Laughlin, L., Troutman, M.R., Brittain, S., Rettig, T.W.: Investigation of H3+ Emission from Exoplanet Tau Boo. In: American Astronomical Society Meeting Abstracts, vol. 212, p. 10.11 (2008)

Lellouch, E.: Spectro-imaging observations of H3+ on Jupiter. Phil. Trans. Roy. Soc. Lond. A **364**, 3139–3146 (2006)

Lemaire, J.F., Gringauz, K.I.: The Earth's Plasmasphere. Cambridge University Press, London (1998)

Lindemann, F.: On the Solar Wind. Phil. Mag. **38**, 674 (1919)

References

Lindemann, F., Dobson, G.: A theory of Meteors, and the Density and Temperature of the Outer Atmosphere to which it Leads. Proc. Roy. Soc. Lond. Ser. A **102**, 411–437 (1922)

Link, R., Gladstone, G.R., Chakrabarti, S., McConnell, J.C.: A reanalysis of rocket measurements of the ultraviolet dayglow. J. Geophys. Res. **93**, 14,631–14,648 (1988)

Llewellyn, E.J., Solheim, B.H.: The excitation of the infrared atmospheric oxygen bands in the nightglow. Planet. Space Sci. **26**, 533–538 (1978)

McLennan, J.C., Shrum, G.M.: On the origin of the auroral green line 5577 - and other associated with aurora borealis. Proc. Roy. Soc. Lond. A **108**, 501 (1925)

McCormack, B.M. (ed.): The radiating atmosphere (1971)

McCulloch, P.M.: Interpretation of the radio rotation period of Jupiter in terms of the cyclotron theory. Aust. J. Phys. **21**, 409–413 (1968)

McGovern, W.E.: The Primitive Earth: Thermal Models of the Upper Atmosphere for a Methane-Dominated Environment. J. Atmos. Sci. **26**, 623–635 (1969)

Meier, R.R., Warren, H.P., Nicholas, A.C., Bishop, J., Huba, J.D., Drob, D.P., Lean, J.L., Picone, J.M., Mariska, J.T., Joyce, G., Judge, D.L., Thonnard, S.E., Dymond, K.F., Budzien, S.A.: Ionospheric and dayglow responses to the radiative phase of the Bastille Day flare. Geophys. Res. Lett. **29**, 99–1 (2002)

Milikh, G., Valdivia, J.A., Papadopoulos, K.: Spectrum of red sprites. J. Atmos. Sol. Terr. Phys. **60**, 907–915 (1998)

Mlynczak, M.G., Martin-Torres, F.J., Marshall, B.T., Thompson, R.E., Williams, J., Turpin, T., Kratz, D.P., Russell, J.M., Woods, T., Gordley, L.L.: Evidence for a solar cycle influence on the infrared energy budget and radiative cooling of the thermosphere. J. Geophys. Res. A **112**(11), 12,302 (2007)

Müller-Wodarg, I.: Planetary upper atmospheres, pp. 331–353. Imperial College Press, London (2005)

Nagano, I., Yagitani, S., Miyamura, K., Makino, S.: Full-wave analysis of elves created by lightning-generated electromagnetic pulses. J. Atmos. Sol. Terr. Phys. **65**, 615–625 (2003)

Nagy, A.F., Cravens, T.E.: Solar System Ionospheres, AGU Geophysical Monograph Series, Vol. 130, pp. 39 (2002)

Newcomb, S.: A Rude Attempt to Determine the Total Light of all the Stars. Astrophys. J. **14**, 297–312 (1901)

Nilsson, H., Waara, M., Marghitu, O., Yamauchi, M., Lundin, R., Rème, H., Sauvaud, J.A., Dandouras, I., Lucek, E., Kistler, L.M., Klecker, B., Carlson, C.W., Bavassano-Cattaneo, M.B., Korth, A.: An assessment of the role of the centrifugal acceleration mechanism in high altitude polar cap oxygen ion outflow. Ann. Geophys. **26**, 145–157 (2008)

Offerman, D., Conway, R.R.: Mission studies the composition of Earth's middle atmosphere. EOS Trans. **76**, 337 (1995)

Oka, T., Geballe, T.R.: Observations of the 4 micron fundamental band of H3(+) in Jupiter. Astrophys. J. **351**, L53–L56 (1990)

Osterbrock, D.E., Fulbright, J.P., Cosby, P.C., Barlow, T.A.: Faint OH (nu' = 10), ^{17}OH, and ^{18}OH Emission Lines in the Spectrum of the Night Airglow. Publ. Astron. Soc. Pac. **110**, 1499–1510 (1998)

Pallé, E., Zapatero Osorio, M.R., Barrena, R., Montañés Rodríguez, P., Martin, E.L.: Earth's transmission spectrum from lunar eclipse observations. Nature **459**, 814–816 (2009)

Pasko, V.P.: Blue jets and gigantic jets: transient luminous events between thunderstorm tops and the lower ionosphere. Plasma Phys. Contr. Fusion **50**(12), 124,050 (2008)

Radhakrishnan, V., Roberts, J.A.: Polarization and Angular Extent of the 960-Mc/sec Radiation from Jupiter. Phys. Rev. Lett. **4**, 493–494 (1960)

Reber, G.: Cosmic Static. Astrophys. J. **100**, 279–287 (1944)

Ribas, I., Guinan, E.F., Güdel, M., Audard, M.: Evolution of the Solar Activity over Time and Effects on Planetary Atmospheres. I. High-Energy Irradiances (1-1700 Å). Astrophys. J. **622**, 680–694 (2005)

Richmond, A.D.: The Ionosphere. In: The Solar Wind and the Earth, Reidel, D. pp. 123–140 (1987)

Rishbeth, H., Garriott, O.K.: Introduction to Ionospheric Physics. Academic Press, (1969)

Roach, F.E., Gordon, J.L.: The Light of the Night Sky. D. Reidel (1973)
Rodgers, C.D., Walshaw, C.D.: The computation of infrared cooling rate in planetary atmospheres. Q. J. Roy. Meteorol. Soc. **92**, 67–92 (2006)
Rucker, H.O., Bauer, S.J., Pedersen, B.M. (eds.): Planetary radio emissions II. Aust. Acad. Sci. (1988)
Rycroft, M.J., Harrison, R.G., Nicoll, K.A., Mareev, E.A.: An Overview of Earth's Global Electric Circuit and Atmospheric Conductivity. Space Sci. Rev. **137**, 83–105 (2008)
Sawyer, C., Warwick, J.W., Romig, J.H.: Smooth radio emission and a new emission at Neptune. Geophys. Res. Lett. **17**, 1645–1648 (1990)
Scharringhausen, M., Aikin, A.C., Burrows, J.P., Sinnhuber, M.: Space borne measurements of mesospheric magnesium species - a retrieval algorithm and preliminary profiles. Atmos. Chem. Phys. **8**, 1963–1983 (2008)
Shokolov, V.S.: H3+ in the Upper Atmosphere of Venus. Astronomicheskij Tsirkulyar **1174**, 3 (1981)
Siingh, D., Gopalakrishnan, V., Singh, R.P., Kamra, A.K., Singh, S., Pant, V., Singh, R., Singh, A.K.: The atmospheric global electric circuit: An overview. Atmos. Res. **84**, 91–110 (2007)
Slanger, T.G., Wolven, B.C.: Airglow Processes in Planetary Atmospheres. In: Atmospheres in the Solar System: Comparative Aeronomy, pp. 77– (2002)
Slipher, V.M.: Emissions of the spectrum of the night sky. Popular Astron. **37**, 327–328 (1929)
Stewart, B.: On the Cause of the Solar-Diurnal Variations of Terrestrial Magnetism . Proc. Phys. Soc. Lond. **8**, 38–49 (1886)
Strickland, D.J., Lean, J.L., Meier, R.R., Christensen, A.B., Paxton, L.J., Morrison, D., Craven, J.D., Walterscheid, R.L., Judge, D.L., McMullin, D.R.: Solar EUV irradiance variability derived from terrestrial far ultraviolet dayglow observations. Geophys. Res. Lett. **31**, 3801 (2004)
Strutt, R.: The aurora line in the spectrum of the nightsky. Proc. Roy. Soc. A **100**, 366 (1922)
Teisserenc de Bort, L.P.: Variations de la température de l'air libre dans la zone comprise entre 8 et 13 km altitude. Compt. Rendus Acad. Sci. Paris **134**, 987–989 (1902)
Thompson, D.J., Simpson, G.A., Ozel, M.E.: SAS 2 observations of the earth albedo gamma radiation above 35 MeV. J. Geophys. Res. **86**, 1265–1270 (1981)
Tian, F., Toon, O.B., Pavlov, A.A., De Sterck, H.: A Hydrogen-Rich Early Earth Atmosphere. Science **308**, 1014–1017 (2005)
Troutman, M.R.: Searching for H+3 in the atmosphere of the exoplanet HD 209458b. Master's thesis, Clemson University, USA (2007)
Vaquero, J.M., Vázquez, M.: The Sun recorded through History. Springer, Berlin (2009)
Vaughan, O.H.: NASA shuttle lightning research: observations of nocturnal thunderstorms and lightning displays as seen during recent space shuttle missions. In: J. Wang, P.B. Hays (eds.) Proc. SPIE Vol. 2266, p. 395-403, Optical Spectroscopic Techniques and Instrumentation for Atmospheric and Space Research, Presented at the Society of Photo-Optical Instrumentation Engineers (SPIE) Conference, vol. 2266, pp. 395–403 (1994)
Vázquez, M., Hanslmeier, A.: Ultraviolet Radiation in the Solar System, Astrophysics and Space Science Library, vol. 331. Springer (2005)
Warwick, J.W., Evans, D.R., Romig, J.H., Sawyer, C.B., Desch, M.D., Kaiser, M.L., Alexander, J.K., Gulkis, S., Poynter, R.L.: Voyager 2 radio observations of Uranus. Science **233**, 102–106 (1986)
Werner, S., Keller, H.U., Korth, A., Lauche, H.: UVIS/HDAC Lyman-α observations of the geocorona during Cassini's Earth swingby compared to model predictions. Adv. Space Res. **34**, 1647–1649 (2004)
Whipple, F.: The Propagation to great Distances of Airwaves from Gunfire. Progress of the Investigation during 1931. Q. J. Roy. Meteorol. Soc. **63**, 471–478 (1932)
Wilson, C.T.: The electric field of a thundercloud and some of its effects. Proc. Phys. Soc. Lond. **37**, 32–37 (1925)

References

Winglee, R.M., Dulk, G.A., Bastian, T.S.: A search for cyclotron maser radiation from substellar and planet-like companions of nearby stars. Astrophys. J. **309**, L59–L62 (1986)

Wood, B.E., Müller, H.R., Zank, G.P., Linsky, J.L., Redfield, S.: New Mass-Loss Measurements from Astrospheric Lyα Absorption. Astrophys. J. **628**, L143–L146 (2005)

Yntema, Y.L.: On the brightness of the sky and total amount of starlight. Publications Kapteyn Astronomical Laboratory Groningen **22**, 1–55 (1909)

Zahnle, K.J., Walker, J.C.G.: The evolution of solar ultraviolet luminosity. Rev. Geophys. Space Phys. **20**, 280–292 (1982)

Chapter 5
Biosignatures and the Search for Life on Earth

Although many extrasolar planets have been discovered, so far none of them is comparable to Earth: basically, Earth is the only life-hosting planet in the Universe that we currently know of. However, it is very likely that we will eventually reach the technological level required for the detection of other Earth-like planets. It is reasonable to expect that by analysing their spectral signatures, we will be able to approach the fundamental question of how common life is in the universe. These spectral signatures are the features that only living systems leave on the environment and are readily distinguishable from the planetary geology; their presence or absence requires life. Referred to as *biosignatures*, they can be detected either directly by simple detection of living organisms or indirectly through the detection of chemical compounds or physical structures that their metabolism has left on the environment. Obviously, for the moment the Earth is our only available laboratory to search for, and to define, biosignatures. Thus, for these biosignatures to be the least biased by our particular understanding of life, we need to adopt a non-geocentric approach, and to find general physical features, common to all living systems, which are present in the planetary signal and detectable through remote sensing.

5.1 The Physical Concept of Life

> Whether there are some general characteristics which would apply not only to life on this planet with its very special set of physical conditions, but to life of any kind, is an interesting but so far purely theoretical question. I once discussed it with Einstein, and he concluded that any generalized description of life would have to include many things that we only call life in a somewhat poetical fashion. John Desmond Bernal (1901–1971) (1949).

There is no unique definition of life that simultaneously satisfies astronomers, biologists, geologists and philosophers. Nevertheless, if we propose to detect life on Earth-like planets and beyond, we should be able to define it, or at least define some of its physical, measurable properties. This is not an easy task, and the problems associated with the identification of potential extraterrestrial life encountered during the Viking field experiments in Mars (Levin and Straat 1977) or the analysis of the ALH84001 Martian meteorite (McKay et al. 1996) indicate the complexity of the problem.

Life can be understood as the highest level of complexity in the universe, resulting as a consequence of the chemical evolution of the elements. This evolution started with the formation of hydrogen and helium in the Big Bang, then heavier elements such as carbon came about as a result of supernova explosions, and eventually the complex organic compounds found in living forms arose. Each of these steps represents a major advance in complexity and development. Since life has been present on Earth for about 80% of its history, it is often assumed that life on Earth started relatively early and became more complex, organized and conscious. Chaisson (1998) connects the emergence of life and the increase of its complexity with the expansion of the Universe (time arrow).

From a thermodynamic point of view, living systems can be studied as open systems that take energy from the environment. The second law of thermodynamics predicts that the total amount of disorder (or entropy) in the universe always increases with time (Carnot 1824; Clausius 1865; Thomson 1851). However, living systems seem to challenge this law. The food chain of life on Earth is based on plants, which trap solar radiation from the Sun and turn it into highly structured sugars through the process of photosynthesis (Deisenhofer and Michel 1989). Photosynthesis dissipates solar energy and in this way terrestrial plants seem to create order, apparently defying the second law. It was Schrödinger (1944) who first explained that the only way for organisms (or living systems) to remain alive is 'by continually drawing from its environment negative entropy', that is Schrödinger proposed that the level of complexity (order) in a body can only increase provided that the amount of disorder in its surroundings increases by a greater amount. In this book, he introduced the idea of an 'aperiodic crystal' that could contain genetic information in its configuration of covalent chemical bonds, an idea that helped stimulate the search for the genetic molecule: DNA.

For living organisms, this negative entropy is achieved through the biochemical process of absorbing energy from the environment and expelling waste byproducts, referred to as the metabolism. Organisms need their metabolism to maintain a state of disequilibrium in the surrounding media, and to do this they require an external and inexhaustible energy source, such as solar radiation in the case of plants (see Lineweaver and Egan 2008; Schneider and Sagan 2006 for a review and a monograph on this topic). The entropy calculations must be done for the full system formed by the living being and the environment.

This ability to apparently challenge the second law is not a characteristic exclusive to life. Illya Prigogine (1917–2003) and co-workers developed the theory that some systems are able to sustain order remaining in a condition of disequilibrium as long as there is a continuous source of energy (Nicolis and Prigogine 1977, 1989; Prigogine and Stengers 1984). These systems are based on the existence of gradients that generate energy fluxes. Referred to as 'dissipative structures' because they dissipate when the energy source is not available, these systems would be capable

5.1 The Physical Concept of Life

Fig. 5.1 Several natural systems exhibiting a rich level of self-organization

of self-organization. Luisi (2006) describes different approaches to the topic of life definition, including the concept of emergence.[1]

Many self-organizing systems exist in nature (see Fig. 5.1), such as crystallization, convection Benard cells, cyclones etc.; life is simply the most sophisticated of them, with a high capacity for transmitting information. An overview of this topic can be found in Heylighen (2003). Self-organization is definitely an inherent quality of living organisms, as they contain complex dissipative structures within them; it has also been related to evolution (Eigen and Schuster 1979; Eigen et al. 1981; Eigen and Winkler 1993).

Another important contribution to the study of the role of self-organization in natural selection was given by Kauffman (1991, 1993), who proposed that evolution is not a mere sequence of accidents or random events, as Darwinism maintains, rather there is self-organization in natural selection. If life were bound to arise, not as an incalculable improbable accident, but as an expected fulfillment of the natural order, then we truly are at home in the universe (Kauffman 1993).

In the Earth's atmosphere, oxygen can coexist with nitrogen and other highly reactive gases in a state of deep chemical disequilibrium owing to the presence of living organisms. This disequilibrium in the environment, caused by living organisms, can provide a detectable biosignature. A planet's atmosphere, the most accessible area to remote sensing, may offer reasonable biosignatures, because it may reveal negative entropy associated with chemical disequilibria. The Earth's atmosphere, for instance, contains the hydrogen-rich gases, and also oxygen, which

---

[1] Emergence is the way complex systems and patterns arise out of a multiplicity of relatively simple interactions.

Table 5.3 Critical temperature and pressure for cosmically common volatiles (from Bains 2004)

Liquid	Critical temperature (K)	Critical pressure (atmospheres)
Ammonia	405.6	111.5
Argon	150.9	48.5
Carbon dioxide	304.5	72.9
Ethane	305.2	48.2
Hydrogen	33.26	12.8
Methane	190.9	45.8
Neon	44.4	26.3
Nitrogen	126.1	33.5
Water	647.4	218

The phase of a substance also changes with the pressure. Table 5.3 shows critical temperatures[3] and pressures[4] for a set of common volatiles.[5]

Water has also a large dipolar momentum as compared to the alcohols. This favors the dissociation of ionizable groups leading to ionic groups, which can form additional H-bonds with water molecules, thus improving their solubility. Organic molecules can be divided into two families: *hydrocarbons* and *CHON-containing* molecules. When brought into the presence of liquid water, hydrocarbons try to escape the water molecules while CHONs have affinity for water (Brack 1993, 2002). Using this property of water, soaps probably played an important role in the formation of the membranes of the first biological cells. Charged atoms at one end of a soap molecule – the hydrophilic end – dissolve in water. The rest of the molecule is hydrophobic in the form of a long chain of organic products like oil. In the primordial soup, hydrophobic organic compounds clumped together naturally. Some of these were hydrocarbons similar to soap molecules, with charged atoms at one end. Like soaps, they packed together to build thin-skinned bubbles in which the charged hydrophilic atoms formed the outer surface and the hydrocarbon tails pointed inwards, mixing with the organic material crumpled together (Schopf 1999).

Other important properties of the water molecule essential for life, all derived from its high dipolar momentum, are the following:

- Water conducts heat more easily than any liquid except mercury.
- It has a neutral pH. Pure water is neither acidic nor basic, although its pH changes when other substances are dissolved in it.
- It has a high surface tension, that is it is elastic and it does not disaggregate but forms drops. Water at 20°C has a surface tension of 72.8 dynes cm^{-1} – compared to 22.3 for ethyl alcohol and 465 for mercury. Because of its surface tension, the

---

[3] Critical temperature of a substance is the temperature at and above which vapour of the substance cannot be liquified, no matter how much pressure is applied.

[4] Critical pressure of a substance is the pressure required to liquefy a gas at its critical temperature.

[5] Substances that readily vaporize at relatively low temperatures.

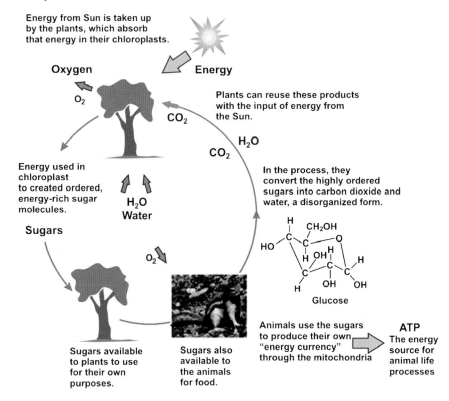

Fig. 5.3 The energy cycle for life is fueled by the Sun. The main end product for plants and animals is the production of highly energetic molecules like ATP. These molecules store enough immediately available energy to allow plants and animals to do their necessary work (Source: http://hyperphysics.phy-astr.gsu.edu/hbase/biology/enercyc.html)

5.3.3.2 Chemical Energy

Life is based on redox chemical reactions, in which atoms change their state of oxidation. Oxidation describes the loss of electrons by a molecule, atom or ion. Reduction describes the gain of electrons by a molecule, atom or ion.[6]

In the oxidation process, the organic material is broken down to obtain energy. The oxidant removes electrons from another substance and is thus reduced by itself. And because it 'accepts' electrons it is also called an electron acceptor.

$$\text{Reduction}: \text{Oxidant} + \text{Electrons} \longrightarrow \text{Product}.$$

---

[6] Several mnemonics are used to remember the concepts, for example, 'LEO the lion says GER' – Losing Electrons is Oxidation, Gaining Electrons is Reduction.

Some of the typical redox reactions used by terrestrial organisms to produce energy (the first ones known as methanogenesis) are

$$4H_2 + CO_2 \longrightarrow CH_4 + 2H_2O \; (1.4\,eV, 474.28\,kJ/mol)$$
$$H_2 + 2Fe(III) \longrightarrow 2H + +2Fe^{+3} \; (1.6\,eV, 148.6\,kJ/mol)$$

and, typical in hydrothermal vents and endolithic life, is the reaction

$$S + 6Fe^{+2} + 4H_2O \longrightarrow HSO_4 + 6Fe^{+2} + 7H^+.$$

During the process of reduction, new organic elements (*chemoautotrophs* and *chemoheterotrophs*) are created. The reductant transfers electrons to another substance and is thus oxidized by itself. And because it 'donates' electrons it is also called an electron donor.

$$\text{Oxidation : Reductants} \longrightarrow \text{Product} + \text{Electrons}.$$

A net input of reducing power is also required for life, since the compounds on which life is based are on average relatively reduced. In the case of autotrophs, electrons are needed to reduce the CO_2. In the case of heterotrophs, the need for electrons is smaller since some building blocks are taken directly.

If the source of electrons is an organic compound, we speak of an *organotroph*, and if it is an inorganic compound then we speak of a *lithotroph*. Phototrophs can use a variety of compounds as electron donors in photosynthesis (H_2O, H_2S, Fe^{+2} and H_2).

In chemotrophs, the nature of the electron acceptor in the energy generating reaction is also a source of metabolic diversity. The most versatile oxidant is oxygen (O_2), but it is not the only one. Alternative oxidants are nitrate, ferric iron, manganese, arsenate, sulfate and CO_2. For chemoorganotrophs the oxidation of organic matter is called respiration, where aerobic respiration uses O_2 as electron acceptor and anaerobic respirations use other alternatives. If the terminal electron acceptor is an organic molecule, the process is called fermentation.

Organic matter respiration is driven by the following general reaction. Some particular examples are given in Table 5.4.

$$\text{Reduced Organic Matter} + \text{Oxidant} \longrightarrow CO_2 + \text{Reduced Oxidant}.$$

The transition from anaerobic to aerobic respiration needs a long period of time and the creation of an oxygen atmosphere. Biological processes were mainly responsible, but the contribution of some abiotic processes to build present-Earth atmosphere cannot be disregarded.

All terrestrial cells use energy in nearly the same way. They use the same molecule, called ATP (adenosine triphosphate), to store and release energy. Once ATP is produced, it can be used to provide energy for any cellular reaction. Each time a cell draws energy from a molecule of ATP, it leaves a closely related

Table 5.4 Aerobic and anaerobic modes of organic matter respiration

Mode of organic matter respiration	Oxidant	Reduced oxidant	Energy (kJ mol^{-1})
Aerobic oxidation	O_2	H_2O	3,190
Manganese reduction	MnO_2	Mn^{2+}	3,090
Nitrate reduction	HNO_3	N_2	3,030
Iron reduction	Fe_2O^3	Fe^{+2}	1,410
	FeOOH	Fe^{+2}	1,330
Sulphate reduction	SO_4^{2-}	S_2^-	380
Methanogenesis	CO_2	CH_4	350

by-product, called ADP (adenosine diphosphate), which can be easily turned back into ATP. The fact that all life uses the same molecule (ATP) for energy storage suggests evidence of a common origin. However, there are no reasons why extraterrestrial organisms should not select other biomolecules for this process.

Table 5.7 summarizes the potential hosts for life on the solar system.

5.4 Biosignatures on Present Earth

A biosignature is an indicator of life or a fingerprint left by life in the environment. Many biosignatures are local, such as fossils, chemical compounds or the remnants of living organisms in geological materials. But other features produced by life can be found at planetary scale in the terrestrial atmosphere or surface (Table 5.5).

In this section, some *in-the-field* biosignatures are mentioned, but we focus on the latter case, on life fingerprints in the globally integrated reflected and emitted electromagnetic spectrum of Earth. This is so because, in the search for other Earth-like planets, only global biosignatures that do not require in situ measurements can be detected and studied remotely using powerful enough instrumentation. See Botta et al. (2008) and Selsis et al. (2008) for a summary of present techniques for life detection.

These global-scale biosignatures in our planet are either the atmospheric by-product of metabolism, ground-based biological pigments or a combination of features (or absence of features) that coexist due only to the existence of life. The most interesting cases are described in the following sections.

5.4.1 Spectral Biosignatures in the Atmosphere

As we show in Chap. 3, there are enormous differences between the atmospheric spectrum of Earth, Venus and Mars, even though the three of them are rocky planets with comparable masses (Fig. 5.4). We can use this empirical fact to develop techniques to find life signatures in the spectra of terrestrial exoplanets.

Table 5.5 Life plausibility categories

Category	Definition	Examples
I	Demonstrable presence of liquid water, readily available energy and organic compounds	Earth
II	Evidence for the past or present existence of liquid water, availability of energy and inference of organic compounds	Mars, Europa
III	Physically extreme conditions, but with evidence of energy sources and complex chemistry on Earth possibly suitable for life forms unknown on Earth	Venus, Titan, Triton, Enceladus
IV	Persistence of life very different from on Earth conceivable in isolated habitats or reasonable inference of past conditions suitable for the origin of life prior to the development of conditions so harsh as to make its perseverance at present unlikely but conceivable in isolated habitats	Mercury, Jupiter
V	Conditions so unfavourable for life by any reasonable definition that its origin or persistence cannot be rated a realistic probability	Sun, Moon

Adapted from Schulze-Makuch and Irwin (2004)

5.4.1.1 Atmospheric Carbon dioxide, Water Vapour and Ozone: The Triple Fingerprint

Oxygen is a very reactive gas, and its presence in a planetary atmosphere usually indicates a continuous generation via geological or biological effects. On Earth, large amounts of atmospheric molecular oxygen are generated by oxygenic photosynthesis, followed by burial of organic carbon in marine sediments (Cloud 1972 ; Walker 1977; Holland 1978, 1984, 2002).

However, O_2 can also be generated non-biologically in several ways (Kasting 1997; Schindler and Kasting 2000). One of the major sources is the photo-dissociation of water vapour in the stratosphere, which produces hydrogen escape to space and the oxidation of the atmosphere. This situation would occur in a runaway-greenhouse planet with a high average surface temperature, like an early Venus.

High atmospheric oxygen concentrations can also be caused by the inhibition of oxygen sinks. This second case would occur, for instance, in a Mars-like planet, where oxygen does not react with reduced minerals (endothermic reactions) due to the low temperature and there is no volcanic outgassing of H_2, which could reduce the oxygen amount (Kasting et al. 1997).

The above two cases occur just in the limiting region where liquid water is impossible on the planetary surface. However, Kasting (1997) shows that the detection of large amounts of atmospheric oxygen in an Earth-like planet whose surface

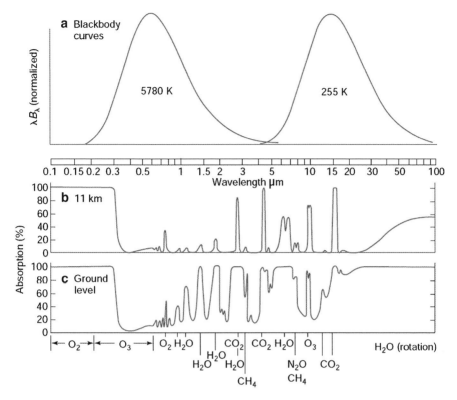

Fig. 5.4 *Upper panel* (**a**) the blackbody curve for the solar effective temperature (*left*) and for the Earth effective temperature (*right*). Note how the maximum emission for the Sun is at about 0.6 μm, while for the Earth is at 15 μm. *Lower panels* show the most important molecular absorption bands in the terrestrial atmosphere at the ground level (**b**) and at 11 km (**c**)

temperature is adequate to harbour liquid water is a strong indicator of continuous production through biological activity (photosynthesis). Such a planet would show the signs of water vapour in its atmosphere, besides oxygen and ozone.

O_2 has several absorption bands, and the most important ones located in the optical spectral range are α (578.8–583.4 nm), B (687.5 nm) and A (745.0–785.0 nm). Because of the magnitude of the stellar flux in the optical region, the detection of oxygen in an exoplanet atmosphere is not simple, unless the star is eclipsed. Ozone is a photolytic product formed in the stratosphere by the ultraviolet rays (see Chap. 4).

Their analysis shows that the O_3 column density is not a linear tracer of the atmospheric O_2 content, although its massive detection in a planetary atmosphere would indicate a certain content of atmospheric O_2. If the planet is detected within the circumstellar habitable zone of a star (see following sections), this could also be a strong indication of biological activity. One additional difficulty in detecting the ozone feature in the mid-infrared is the presence of other absorption bands, such as NH_3 and PH_3, in the region between 9 and 11 μm.

Segura et al. (2007) showed that even if CO_2 and O_2 may have abiotic sources, their simultaneous detection in a planet with abundant liquid water provides a biosignature. This is considered as the fundamental biosignature for the detection of exobiospheres similar to Earth's and is known as the *triple fingerprint* (CO_2, O_2 and H_2O), as first proposed by Selsis et al. (2002).

However, we must stress that oxygen (ozone) was not always present along the history of a really habitable planet such as the Earth (see Chap. 4). Jones (2008) summarizes the possibilities for life in absence of oxygen.

- Plenty of O_2 but too little stellar UV to form O_3.
- Efficient removal of O_3.
- Anoxygenic photosynthesis
- A planet too young for oxygen photosynthesis to feature in the spectrum of the planetary atmosphere.
- A biosphere beneath the planet's surface.

5.4.1.2 Other Atmospheric Biosignatures

The detection of O_2 or O_3 is definitely a better biomarker when associated with a reducing atmospheric component, due to the departure from the thermodynamical disequilibrium that such a situation would imply and following the ideas explained in Sect. 5.1.

Some important reducing biogenic trace gases in the Earth's atmosphere are nitrous oxide (N_2O), methane (CH_4), H_2, CO, CH_3Cl and freons. They are considered to be excellent biosignatures, although the detection of their spectral bands requires high spectral resolution. They are created in biogenic processes as follows:

1. N_2O is primarily generated by denitrification of agricultural soils by microorganisms (Stein and Yung 2003).
2. CH_4 is produced by methanogenic bacteria that live in anaerobic environments, such as intestines of ruminants and flooded soils in rice paddies (Segura et al. 2005). However, it also has an abiotic source: it emanates from mid-ocean ridge vents (Kelley et al. 2001), although its biological sources outweigh abiotic sources by a factor of 300 (Kasting and Catling 2003). More recently, Kelley et al. (2005) have measured an increase in the abiotic methane concentration in the atmosphere by a factor of 10.

    These two gases, N_2O and CH_4, however, are mainly confined to the troposphere, their abundances decline rapidly above 12–15 km (Kasting 1997), and their detection for an Earth-like planet is extremely difficult, in particular in a planet with abundant high clouds.
3. H_2 is produced under certain conditions by unicellular green algae during illumination (Gaffron 1939 and Gaffron and Rubin; 1942). H_2 production has been found to be very common throughout the procaryotic and eucaryotic kingdoms and it can be generated biologically through different pathways. The most common ones are the photolysis of water and the fermentation of small molecules

Table 5.6 Biotic and abiotic sources of some atmospheric molecules

Molecule	Biotic	Abiotic
N_2O	Denitrification agricultural soils Microorganisms	
CH_4	Methanogenic bacteria	Mid-ocean ridge vents
H_2	Unicellular green algae Photolysis H_2O Fermentation	Photolysis by UV
CH_3Cl	Tropical planets Fungii rotten woods	Ocean Biomass burning

into H_2 and CO_2. However, the photolysis by UV light of different atmospheric components to produce H_2 is also produced abiotically (Sleep and Bird 2008).

4. Methyl chloride (CH_3Cl) is the most abundant halocarbon in the atmosphere. Its main biological sources on Earth are some common tropical plants, certain types of ferns and Dipterocarpaceae that have been shown to contribute significantly to its abundance in the global atmosphere (Yokouchi et al. 2002), and fungal activity in rotten woods (Harper 1985). But CH_3Cl also has two important abiotic sources: the ocean (Singh and Kanakidou 1993) and biomass burning (Lobert et al. 1991).

We should remark that the absence of atmospheric or even surface biosignatures is not a proof that the planet is abiotic. A particular metabolism or biological pigment might just not be detectable through remote sensing in that planet (Table 5.6).

5.4.2 Chlorophyll and Other Spectral Biosignatures of the Planetary Surface: The Red Edge

Photosynthesis is closely linked to several light-harvesting molecules, chlorophyll-a (Chl-a) being the dominant one on Earth. Chl-a behaves like an antenna, helping plants to collect solar energy. In the process of photosynthesis, light-harvesting chlorophyll-protein carries out a selective collection of photons, with high absorption in the blue and red and lower absorption in the green, which gives its colour to plants (Papageorgiou and Govindjee 2005).

The idea of detecting chlorophyll in other planets is not new. Liais (1865) already speculated that dark albedo regions in Mars were due to vegetation and not due to water. In the science fiction novel *The War of the Worlds* (1898), H.G. Wells (1866–1946) already speculated about the idea of a different kind of chlorophyll on Mars (Fig. 5.13). Arcichovsky (1912) suggested looking for chlorophyll in the Earthshine spectrum. G.A. Tikhov (1875–1960) in his book *Astrobotany* (1949) presented the results of several expeditions to take spectra of vegetation under extreme conditions[7] (Briot et al. 2004; Omarov and Tashenov 2005).

---

[7] He also observed Mars using the 30-inch refractor at Pulkovo Observatory.

Besides its reflectance in the green, the chlorophyll spectrum also shows strong reflectance in the near-infrared. This feature, known as *red edge*, is remarkable, since it is four times brighter than the green reflectivity in the visible spectral region (Clark 1999) – this means that if we were able to see in the near infrared we would not see vegetation green, but in that new infrared colour – and it is detectable through remote sensing, as shown in Fig. 5.5. Remote sensing data have been used to study the vegetation condition and seasonal vegetation dynamics (Reed and Bradley 2006) in order to predict the impact of climate change on ecosystems. These results shows that all healthy vegetation is chemically similar and exhibit green and infrared reflectance enhancement, and also that there is a positive correlation between temperature and chlorophyll content (Almond et al. 2007).

In cellular thermodynamics, the red-edge feature plays the role of releasing energy from the leaf interior, preventing it from overheating. It is caused by the refraction between leaf mesophyll cell walls and air spaces within the leaf, as shown in Fig. 5.6.

The red edge has been used to describe the variation in leaf and canopy chlorophyll concentration, because it is sensitive to plant health conditions, species and incident sunlight. The leaves substantially reduce their absorbance of incident radiation during the hot periods of the year by changing their moisture and hence dissolved salt contents. At these times, the light intensity required for saturation of photosynthesis is low and a reduction in the radiation absorbed by the leaves therefore results in a greater water-use efficiency (Mooney et al. 1977).

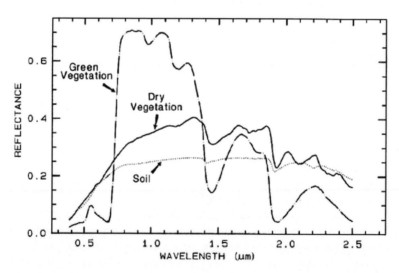

Fig. 5.5 Reflectance spectra of photosynthetic (*green*) vegetation, non-photosynthetic (*dry*) vegetation and soil. The green vegetation has absorptions short of 1 μm due to chlorophyll. Those at wavelengths greater than 0.9 μm are dominated by liquid water. The dry vegetation shows absorptions dominated by cellulose, and also lignin and nitrogen. These absorptions must also be present in the green vegetation, but can be detected only weakly in the presence of stronger water bands. The soil spectrum shows a weak signature at 2.2 μm due to montmorillonite. Adpated from Clark (1999) Fig. 1–18. Copyright: John Wiley Sons

5.4 Biosignatures on Present Earth

Fig. 5.6 Plant leaves are the primary photosynthesizing organs affecting planetary biogeochemical cycles. Leaf structure is closely associated with its photosynthetic function. They must permit carbon dioxide access to the photosynthetic cells but impede water from diffusing out. The oxygen that is a waste product of photosynthesis must be allowed to escape from the leaf. It is well established that the reflectance and transmission spectra of leaves is a function of both the concentration of light absorbing compounds (chlorophyll, water, dry matter, etc.) and the internal scattering of light that is not absorbed or absorbed less efficiently. Courtesy: Ron Neumeyer

On Earth, vegetation may extend over large areas of the planetary surface, allowing the direct detection of its pigments spectral signal from space. For this reason the red edge is used by satellites to identify vegetated areas, and from that, quantify the vegetative cover on the Earth's surface. The concept of leaf area index (LAI), the ratio of total upper leaf surface of vegetation divided by the surface area of the land on which the vegetation grows,[8] is frequently used to determine chlorophyll contents from satellite images (Fig. 5.7).

Another interesting concept is the relationship between the red and near-infrared (NIR) reflected energy to the amount of vegetation present on the ground (Colwell 1974). The amount of red and NIR radiation reflected from a plant canopy and reaching the space varies with solar irradiance, atmospheric conditions and canopy background, structure and composition. One cannot use a simple measure of reflected energy to quantify plant biophysical parameters to monitor vegetation on a global scale.

---

[8] LAI is a dimensionless value, typically ranging from 0 for bare ground to 6 for a dense forest.

Fig. 5.7 The MODIS on the Terra and Aqua satellites collects global leaf area index (LAI) data on a daily basis. This map, created from Terra data, shows the LAI for the month of April 2008, expressed in terms of square metres of leaf area per square metre of ground area

Several vegetation indexes have been defined as the arithmetic combination of molecular bands related to spectral characteristics of plants, in order to classify them through remote sensing (phonologic monitoring). The most commonly used is the normalized difference vegetation index (NDVI), given by the following expression, which cancels out a large portion of noise due to its ratio property:

$$\text{NDVI} = \frac{R_{NIR} - R_{RED}}{R_{NIR} + R_{RED}},$$

where R_{RED} and R_{NIR} stand for the spectral reflectance measurements acquired in the red (0.54–0.68 μm) and near-infrared (0.7–1.1 μm) regions. These reflectances are themselves ratios of that reflected over the incoming radiation in each spectral band individually, here they vary between 0 and 1 (Carlson and Ripley 1997).

Figure 5.8 gives another view of the global photosynthesis, where the NDVI measured on land and the oceanic chlorophyll production[9] are combined.

Two additional concepts are also used by satellite monitoring, the photosynthetically active radiation incident on a plant canopy (PAR) and the fraction of photosynthetically active radiation absorbed by a plant canopy (FPAR). The lat-

---

[9] The concentration of microscopic marine plants, called phytoplankton, can be derived from satellite observation and quantification of ocean colour. This is due to the fact that the colour in most of the world's oceans in the visible light region (wavelengths of 400–700 nm) varies with the concentration of chlorophyll and other plant pigments present in the water, that is the more the phytoplankton present, the greater the concentration of plant pigments and the greener the water.

5.4 Biosignatures on Present Earth

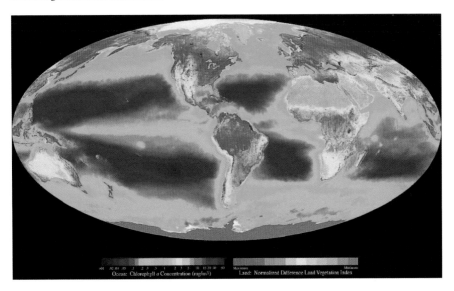

Fig. 5.8 This composite image gives an indication of the magnitude and distribution of global primary production, both oceanic (mg m^{-3} chlorophyll a) and terrestrial (NDVI). Provided by the SeaWiFS Project, NASA/Goddard Space Flight Center and ORBIMAGE. Data corresponding to the period September 1997–August 1998

ter excludes the fraction of incident PAR reflected from the canopy and the fraction absorbed by the soil surface, but includes the portion of PAR which is reflected by the soil and absorbed by the canopy on the way back to space.

While the remote sensing detection of atmospheric components will give an indirect indication of the presence of biological material, the detection of surface signals, like the red edge, will give a direct confirmation of the existence not only of exobiological organisms, but of an advanced degree of complexity and evolution.

Apart from satellite monitoring data, detection of the red-edge signal on Earth was reported by Sagan et al. (1993) from spectroscopic observations of the Galileo mission when pointing toward our planet in 1990, covering a relatively small, cloud-free green surface area (Fig. 5.9). The Mars Express and Venus Express missions took similar data in July 2003 and October 2008, respectively (Grinspoon et al. 2008).

The red edge has also been studied through the earthshine technique in the visible range (Arnold et al. 2002; Woolf et al. 2002; Montañés-Rodríguez et al. 2005; Seager et al. 2005; Hamdani et al. 2006; Montañés-Rodríguez et al. 2006). Several important implications have been attained from these studies, in relation to the use of the red edge signal as a biomarker in exoplanets.

First, variability in the slope of the red edge has been detected and measured, showing values that oscillate between 0 and 11%. This variation can partially be attributed to seasonal or geographical differences between the observations. But, although this signal is a reliable chlorophyll indicator for spatially resolved areas, in

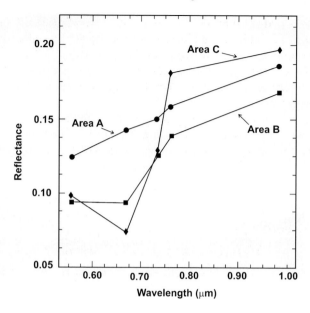

Fig. 5.9 Representative spectrophotmetric data taken by the Galileo single-launch Jupiter orbiter during its close approach to Earth in December 1990. The spectra correspond to three areas of land surface centred over 72°W, 34°S at 20–30 km resolution. A gently sloping spectrum (*circles*, Area A) is consistent with any of the several types of rock or soil. An intermediate spectrum (*squares*, Area B) shows some evidence of an absorption band near 0.67 μm. Substantial areas on the surface have an unusual spectrum (*diamonds*, Area C) with a strong absorption in the red band and a steed band edge just beyond 0.7 μm. This spectrum is inconsistent with all likely rock and soil types and is plausibly associated with photosynthetic pigments. From Sagan et al. 1993, Fig. 3. Reprinted by permission from Macmillan Publishers Ltd. Nature Vol. 365, p. 715. Copyright (1993)

the case of globally integrated planetary observations it remains, in general, an ambiguous biomarker, because other surface and atmospheric compounds have spectral features that contribute to enhance or to reduce the vegetation index.

Clouds, and low clouds in particular, have demonstrated to contribute with a positive slope to the red-edge index (Tinetti et al. 2006a, b). The typical cloud cover for Earth is about 60%, thus vegetation is significantly obscured by clouds in our planet. The sand of the Sahara desert, an area usually uncovered by clouds, also shows a positive slope (Arnold 2008). Furthermore, there are minerals that show enhancements in their reflectance spectra that are spectrally coincidental with the red edge (Seager et al. 2005). Other relatively darker surface components, such as oceans, non-vegetated land areas and snow or ice, cause a decline in the red-edge intensity (Tinetti et al. 2006a).

Considering this ambiguity, the measured change in reflectance in the red edge spectral region could be attributed only to vegetation after carefully analyzing the real distribution of land, oceans and real cloud cover from satellite data during earthshine observations (Montañés-Rodríguez et al. 2006). However, despite the discouraging results, the models indicate that an unambiguous detection on an

5.4 Biosignatures on Present Earth

extrasolar planet could occur at certain times of its orbit (Montañés-Rodríguez et al. 2006). As the planet orbits around its star and rotates, the vegetation signal should become more dominant at certain geometrical configurations: when the planet is in the desired phase range, typically only about 10% of the planet's sunlit surface is visible to the observer, which means that the planet's signal will be faintest compared to its parent star.

Figure 5.10a compares the solar spectral flux (at the top and bottom of the atmosphere) with the absorption spectra of different photosynthetic pigments (See Kiang (2007a) for further details and for data sources). The chlorophyll's red edge

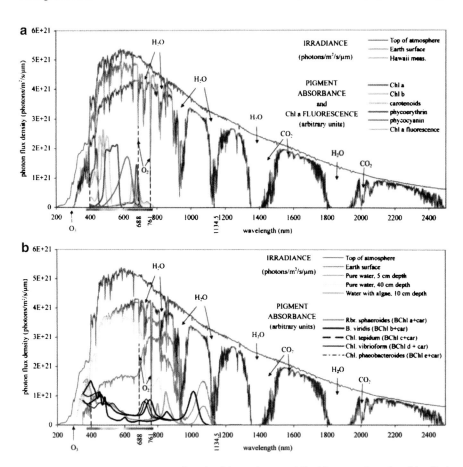

Fig. 5.10 (a) Solar spectral photon flux densities at the top of Earth's atmosphere (*top blue line*) and at the Earth's surface (*dark red line*), and estimated absorbtion spectra of photosynthetic pigments of plants and algae (*brightly colored lines below*). *Green lines* show the two different kinds of chlorophyll common in plants. In addition, measurements from buoys at Hawaii (NOAA) are plotted to show how varying atmospheric transparency can affect the incident light spectrum. (b) Solar spectral photon flux densities at the top of the Earth's atmosphere, at the Earth's surface, at 5 cm deep in pure water, at 10 cm deep in water with an arbitrary concentration of brown algae, and algae and bacteria pigment absorbance spectra. (See Kiang (2007a) for further details and for data sources). The publisher for this copyrighted material is Mary Ann Liebert, Inc. publishers

detected on Earth is associated only with vegetation. The chlorophyll signal is also detected in marine phytoplankton, especially in tropical areas.[10] However, the phytoplankton's signal could hardly be detected through remote sensing in the global Earth spectrum. These single-cell aquatic plants can only use the solar radiation that reach the first 10 cm of the water column,[11] but this column is more than enough to obscure the signal of the red edge (Fig. 5.10b).

Although chlorophyll-a is the pigment most commonly used by surface terrestrial photosynthetic life, it is not the only one. Light-harvesting pigments are categorized into three groups: the chlorophylls, the carotenoids[12] and the phycobilioproteins [13] (Fig. 5.11). The latter two are not as commonly used on Earth but they could be dominant on other Earth-like planets, depending on the spectrum of light of the parent star. They not only have different colours, but they also absorb at different wavelengths.

Halobacteria use for photosynthesis a light-sensitive molecule called retinal that absorbs green light and reflects back red and violet light,[14] the combination of which

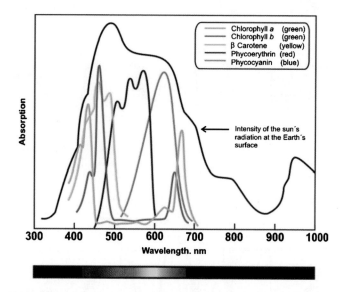

Fig. 5.11 Absorption spectrum of several plant pigments: the chlorophylls, the carotenoids and the phycobilioproteins (Phycoerythrin and Phycocyanin)

---

[10] Through photosynthesis, phytoplankton are responsible for much of the oxygen present in the Earth's atmosphere – half of the total amount produced by all plant life.

[11] Water is highly transmitting in the visible and highly transmitting in the near infrared.

[12] Because of their absorption region, carotenoids appear red and yellow and provide most of the red and yellow colours present in fruits and flowers.

[13] Accessory pigments to chlorophyll, present in cyanobacteria and red algae.

[14] This pigment is similar to sensory rhodopsin, the pigment that humans and other animals use for vision.

5.4 Biosignatures on Present Earth

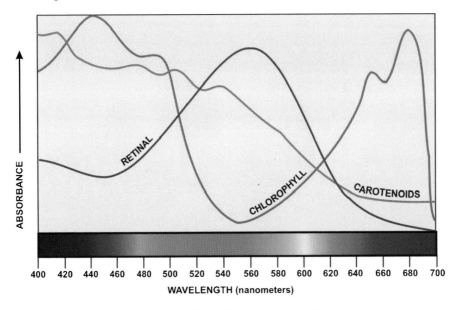

Fig. 5.12 The retinal pigment absorbs light of a wavelength that human perceive as green, but reflects light at the red and, to a lesser extent, the violet ends of the spectrum, a pattern that yields a purple appearance (*purple line*). By contrast, photosynthetic chlorophyll pigments absorb indigo and red and reflect green (*green line*). This mirror image relation suggest that chlorophyll evolved to exploit parts of the spectrum left used by the purple pigment. The carotenoid pigments (*orange line*) shield haloarchaea from high-energy violet and UV light waves but reflect lower-energy orangish-red colors. The absorbance spectra have been scaled for comparison. Adapted from DasSarma (2007), Fig. 4. Reproduced by permission of American Scientist

appear purple (Fig. 5.12). Primitive microbes that used retinal to harness the sun's energy might have dominated early Earth tinting with a purple colour the first biological hotspots (Sparks et al. 2006; DasSarma 2007).

Chlorophylls are the pigments of the cellular reaction centre. Chl-a is the one predominantly used, but oxygenic photosynthesis is also preformed in the near infrared by cyanobacteria that live in environments with little visible light. Their dominant light-harvesting pigment is Chl-d, which has its major peak absorbance at ∼720 nm.

Kiang et al. (2007b) showed how spectral characteristics of photosynthesis could be different from what we know for Earth when one looks at the terrestrial planets orbiting around F, K and M stars. Figure 5.13 represents how possible vegetation in an exoplanet orbiting a star different from the Sun could harvest and reflect different wavelengths other than chlorophyll. Pigments in Earth-sized planets orbiting stars somewhat brighter than the Sun could absorb blue (450 nm) and reflect yellow, orange, red, or a combination of these colours. For stars cooler than the Sun (M spectral type), evolution might favor photosynthetic pigments to pick up the full range of visible and infrared light. With little light reflected, plants might look blank to human eyes (see also Kiang 2008).

It has also been argued that the red edge spectral position could be shifted for other Earth-like planets with a different parent star (Tinetti et al. 2006a, b; Kiang

Fig. 5.13 This is an illustration of what plants may look like on different planets. Credit: Caltech illustration by Doug Cummings

et al. 2007a, b). However, it should essentially be located in a spectral region without major absorption bands to carry out its function of regulating the cellular heat exchanged. This would make it more easily detectable through remote sensing.

5.4.3 Chirality and Polarization as Biosignatures

The term chiral (handedness) of a molecule can be described as follows: if one takes the geometrical structure of a molecule, creates its image on a mirror and then try to match the two structures by means of translations and/or rotations, there are only two possible outcomes, either they are identical, in which case the molecule is achiral, or they are not identical, then the molecule is chiral.

Many biologically active molecules are chiral, including amino acids and sugars, and most of these compounds must be homochiral to function (Thiemann et al. 2001). Chiral molecules produce a typical response when they interact with solar light (which is not polarized): they induce circular polarization of light, which, following a classical description, can be interpreted as an electromagnetic wave that oscillates in only one plane of space. This plane rotates to the left or the right around the direction of propagation of the light. The induction of left or right handed circular polarization by a certain compound is called circular dichroism and is illustrated in Fig. 5.14.

Fig. 5.14 Circular dichroism in glucose

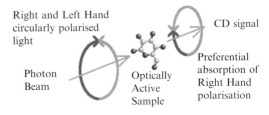

Living systems on Earth use only L-amino acids (left-handed circular polarized light) in proteins and D-sugars (right-handed circular polarized light) in nucleic acids (Gleiser et al. 2008). An existing preferred direction of circular polarization on the incident radiation may have led to the selection of a dominant direction (L-amino acids) (Meierhenrich et al. 2005; Barron 2007).

The search for chirality by means of identifying circular polarization through remote sensing has been proposed as a biomarker (Thaler et al. 2006). But it has to be used with care because the identification of circular polarization in itself is not an unambiguous indication of the existence of chiral molecules since it can be induced in several other natural ways, for instance there are minerals that induce polarization. In fact, the highest known circular polarization in the solar system has been found in Mercury (Kemp et al. 1971) and has been attributed to surface crystals and minerals.

Interestingly, our own senses are able to recognize chirality. For example, our taste buds sense the two chiral forms of a compound as different (Goraieb et al. 2007).

5.5 Biosignatures on Early-Earth

In Chap. 2, we described the evolutionary steps of life on Earth. Most of the time, our planet was inhabited only by unicellular organisms, which were able to interact with, and transform, their environments. Only recently did multicellular life appear.

The spectra shown in the previous sections is not a permanent property of our planet. Life has a strong influence on the Earth's atmosphere, and consequently early eras will have different life footprints in the atmosphere. In the early Archean period, for instance, methanogens were already producing CH_4, and could have generated a detectable methane-rich atmosphere (Schindler and Kasting 2000). Studies of ancient sediments demonstrate that O_2 was present in only trace amounts before 2.5 Ga (Kaufman et al. 2007; Farquhar et al. 2004)

With the advent of cyanobacteria and the subsequent rise in O_2 concentration (Battistuzzi et al. 2004; Bern and Goldberg 2005), even larger changes in atmospheric composition took place. Sleep and Bird (2008) recently published a review on the evolution of ecology during this rise of O_2 in the Earth's atmosphere.

Kaltenegger et al. (2007) characterized the evolution of the Earth's atmosphere and surface in order to model the observable spectra of an Earth-like planet through

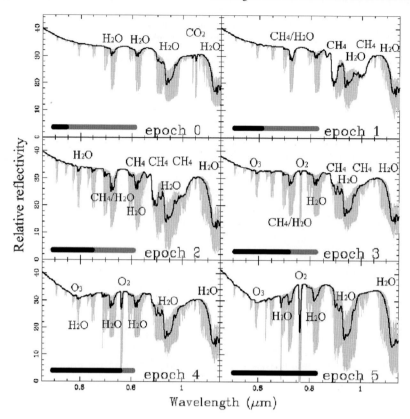

Fig. 5.15 Visible and near-infrared spectra of an Earth-like planet at six geological epochs. The spectral features change considerably as the planet evolves from a CO_2-rich (epoch 0) to a CO_2/CH_4-rich atmosphere (epoch 3) to a present-day atmosphere (epoch 5). The *black lines* show spectral resolution of 70. Adapted from Kaltenegger et al. (2007), Fig. 9. Reproduced by permission of the American Astronomical Society

its geological history. Kaltenegger et al. (2007) chose six epochs that exhibit a wide range in atmospheric abundances, ranging from a CO_2-rich early atmosphere 4 Ga ago, to a CO_2/CH_4-rich atmosphere around 2 Ga ago, to a present-day atmosphere (see Chap. 2 for a review of the Earth's evolution).

In the visible range (Fig. 5.15), the most pronounced historical changes were the deepening of the atmospheric water vapour bands and the appearance of O_2 and O_3 spectroscopic signatures. On the other hand, the signature of methane became increasingly small. In the infrared range (Fig. 5.16), O_2 also appeared in epoch 3 and onwards, and the signature of CO_2 and H_2O became easier to identify. As in the visible range, the CH_4 signatures disappeared with time. Surface features, as the red chlorophyll edge, became relevant only for the present Earth, at 0.44 Ga ago, with the appearance of land plants (Kaltenegger et al. 2007).

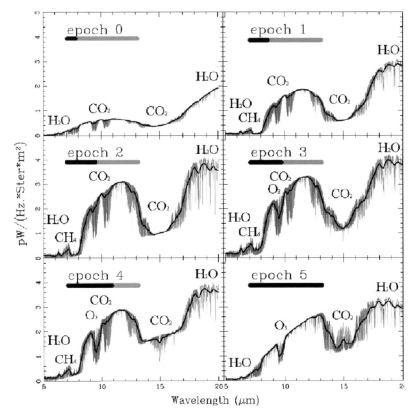

Fig. 5.16 Thermal-infrared spectra of an Earth-like planet at six geological epochs. The spectral features change considerably as the planet evolves from a CO_2-rich (epoch 0) to a CO_2/CH_4-rich atmosphere (epoch 3) to a present-day atmosphere (epoch 5). The *black lines* show spectral resolution of 20. Adapted from Kaltenegger et al. (2007), Fig. 10. Reproduced by permission of the American Astronomical Society

5.6 Life in the Universe

After studying life on our planet, it is appropriate to establish criteria for finding life elsewhere and to select suitable targets. Let us start with the study of habitability conditions, keeping in mind that the criteria could depend on the complexity of the living being that is considered.

5.6.1 Circumstellar Habitable Zone

The circumstellar habitable zone is related to the presence of liquid water as a necessary condition for the origin and the evolution of life as we know it. W. Whewell

(1794–1866) in his 1853 *Of the plurality of worlds*[15] remarked in clear reference to water: '[Earth] is situated just in that region of the System, where the existence of matter, both in a solid, a fluid, and a gaseous condition, is possible'. A.R. Wallace (1823–1913) in his 1903 book *Man's Place in the Universe*[16] established the conditions for the existence of life on Earth: (1) a regular heat supply, resulting in a limited range of temperatures, (2) a sufficient amount of solar light and heat, (3) water in great abundance and (4) alterations of day and night.

The works of Huang (1960), Dole (1970) and Hart (1979) established the modern concept of a habitable zone around a star. For recent reviews on this topic see Kasting and Catling (2003), Gaidos et al. (2005), Vázquez (2005) and Kasting et al. (2008).

A planet is considered to be within the habitable zone around a star if there is sufficient temperature and pressure in its atmosphere to sustain stable liquid water at the planetary surface. As seen in Chap. 2, the temperature would depend on the luminosity of the parent star, L_S, and on the star–planet distance, d, with the amount of greenhouse gases, g, and the planetary albedo, a, as additional factors characterizing this region. In short, the surface temperature of the planet, T, is expressed as

$$T^4 = \frac{F_S(1-a)}{4\sigma(1-g)} = \frac{L_S(1-a)}{4\sigma\, d^2(1-g)}.$$

In Fig. 5.17 a simple sketch of the habitable zone around different spectral types is shown, taking into account only the distance to the star.

We can now briefly analyse the different factors influencing habitability.

5.6.1.1 Stellar Constraints

The luminosity of a star is related with its effective temperature, T_{ef}, according the expression

$$L_S = 4\pi R_S^2 F_S = 4\pi R_S^2\, \sigma T_{ef}^4,$$

where F_S and R_S are the observed radiation flux and the radius of the star, respectively.

The stellar spectra is the best way to determine the stellar temperature and proceed to a classification. The standard Morgan–Keenan system has seven categories, each of which is associated with a temperature (Morgan and Keenan 1973 and Table 5.7). By analogy with the heating of solid objects which change color from a dull red to blue-white as they increase in temperature, the spectral type is telling us about the temperature of the star (just as with colour indexes but with more detail). Note that the hotter stars are sometimes termed 'early-type' and the cooler

---

[15] Reedited in 2001 by University of Chicago Press with new introductory material by M. Ruse.

[16] Published in volume 55 of The Independent (New York) and simultaneously in Fortnightly Review (London). It was reprinted in 2008 by BiblioLife.

5.6 Life in the Universe

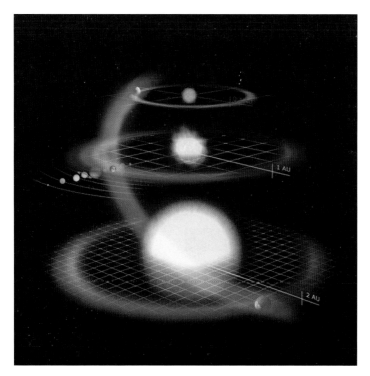

Fig. 5.17 A depiction of the habitable zone around three different stars. Depending on the stellar type, habitable planets will be at different distances from the parent star, so that the planet's surface temperature is capable of sustaining liquid water. The range of orbits around each star that allows liquid water is known as the habitable zone. Credit: ESA

Table 5.7 The MK spectral classification scheme

Spectral type	Temp [K]	Colour	Dominant absorption lines
O	>20,000	Hottest blue stars	Ionized helium, strong ultraviolet
B	20–10,000	Hot blue stars	Neutral helium dominates (HeI)
A	10–7,000	Blue/blue–white stars	Neutral hydrogen (H) dominates
F	7–6,000	White stars	CaII, neutral H weaker, other metals
G	6–5,000	Yellow stars	CaII , neutral metals (e.g. iron–FeI)
K	5,000–3,500	Orange–red stars	Neutral metals, molecular bands appear
M	3,500–2,000	Coolest red stars	Molecular bands, neutral metals

ones 'late-type'. From hottest to coolest, these categories are O, B, A, F, G, K and M. These categories are each divided into ten numbered subclasses from 0 to 9. Representative spectra for spectral types O5 through M5 are shown in Fig. 5.18.

In Fig. 5.18 we can see the relationship between stellar type and the elements that create each type of spectrum. A dominant characteristic of spectral class A stars is the presence of strong hydrogen lines, but ionized helium lines are present only in the class O stars. Since helium ionizes only at high temperatures, this tells us that

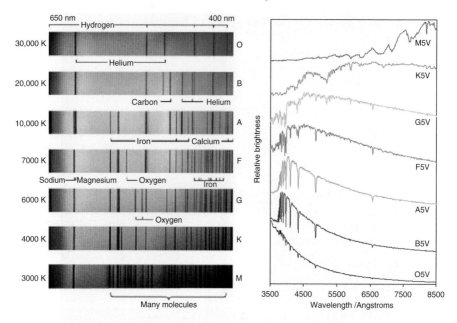

Fig. 5.18 A compilation of actual spectra from stars of different spectral types. Each spectrum is shifted vertically by an arbitrary amount to separate it from its neighbours (data from Pickles 1998)

class O stars must have very high surface temperatures. On the other hand, spectral lines associated with molecules are found only for spectral classes K and M. This is because these correspond to low surface temperatures, and molecules can only hold together in stars with relatively low surface temperatures.

The brightness of the star is a way to determine empirically its luminosity. Historically this was expressed by the concept of apparent magnitude.[17] The difference of magnitude between two stars is given by

$$m_1 - m_2 = -2.5 \log \frac{F_{S,1}}{F_{S,2}}.$$

Absolute magnitude is the apparent magnitude an object would have if it were at a standard luminosity distance (10 parsecs) away from the observer, in the absence of astronomical extinction. It allows the true brightnesses of objects to be compared without regard to distance. Bolometric magnitude, M_{bol}, is luminosity expressed in

---

[17] The brighter the object appears, the lower the value of its magnitude. In 1856, N. Pogson (1829–1891) formalized the system by defining a typical first magnitude star as a star that is 100 times as bright as a typical sixth magnitude star. Pogson's scale was originally fixed by assigning Polaris a magnitude of 2. Astronomers later discovered that Polaris is slightly variable, and so they first switched to Vega as the standard reference star, and then switched to using tabulated zero points for the measured fluxes.

magnitude units; it takes into account energy radiated at all wavelengths, whether observed or not:

$$\frac{L_S}{L_{Sun}} = 10^{-0.4\,(M_{bol}-M_{bol,Sun})}.$$

In 1911, Ejnar Hertzprung (1873–1967) plotted the first diagram of the relative magnitudes of stars in a cluster vs. their spectral types. Two years later Henry Russell (1877–1957), working independently, produced a plot of the absolute magnitude of nearby stars with well-determined distances against their spectral types. The resulting diagram has come to be known as the Hertzprung–Russell or HR diagram. It is one of the most useful diagrams in astronomy and is fundamental to our understanding of stars (see Fig. 5.19 for an example).

The main points to note from Fig. 5.19 are that 90% of stars lie on a narrow diagonal band running from top left (bright and hot) to bottom right (faint and cool). This is called the main sequence. The Sun lies approximately in the middle of the

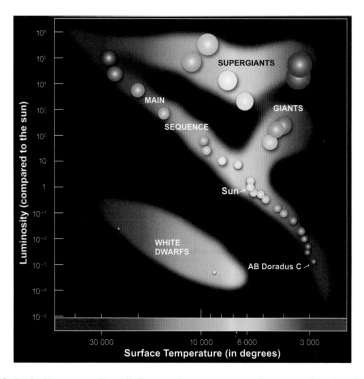

Fig. 5.19 In the Hertzprung–Russell diagram, the temperatures of stars are plotted against luminosities. The position of a star in the diagram provides information about its present stage and its mass. Stars that burn hydrogen into helium lie on the diagonal branch, the so-called main sequence. Red dwarfs like AB Doradus C lie in the cool and faint corner. When a star exhausts all the hydrogen, it leaves the main sequence and becomes a red giant or a supergiant, depending on its mass. Stars with the mass of the Sun which have burnt all their fuel evolve finally into a white dwarf (*left low corner*). Credit: European Southern Observatory (ESO)

main sequence. Giants and supergiants are much more luminous for a given temperature, lying above the main sequence and generally at the cooler end (the right) of the HR diagram. Far below the main sequence are the faint white dwarf stars. This diagram is explained in terms of stellar evolution, based on the energy consumption in the stellar core.

The main sequence is the most stable phase where the energy is obtained by burning hydrogen into helium. The time of permanence in this phase, T_{MS}, is related with the mass of the star, M_S, (Scalo and Miller 1979)

$$\log T_{MS} = 9.87 - 3.79 \log M_S + 1.07 (\log M_S)^2.$$

For main sequence stars the luminosity of a star is related with its mass through a simple power law ratio of the type

$$L_S \propto M_S^3.$$

The relationship between the radius, R_S, and the mass of a main sequence star is given approximately[18] by

$$R_S \propto M_S^{0.8}.$$

In summary, habitable planets are expected around stars in the main sequence. Massive stars are less numerous in the Universe and stay too short to allow life in orbiting planets to develop. Low-mass stars are more numerous and live long, but they present some problems that we describe in some detail below. In the middle we have 'solar-like' stars, neither 'too hot' nor 'too cold' and living just long enough. This spectral range likely accounts for about 5–10% of stars in our galaxy. Superficial liquid water may exist on planets orbiting these spectral types at a distance that does not induce tidal lock (see Sect. 5.6 and Chap. 2 for the adequate formulation).

5.6.1.2 M Stars and Tidal Locking

Red M dwarfs make up between 70 and 90% of all the stars in the Galaxy and have long main sequence lifetimes, significantly larger than the age of the Universe. However, astronomers for many years ruled out red dwarfs as potential abodes for life, because of their small size (from $0.1 - 0.6 M_S$) implicating that their nuclear reactions proceed exceptionally slowly, and thus they emit very little light, from 3% produced by the Sun to as little as 0.01%. Any planet in orbit around a red dwarf would have to huddle very close to its parent star to attain Earth-like surface temperatures; from 0.3 AU (just inside the orbit of Mercury) for a star like Lacaille 8760 to as little as 0.032 AU (such a world would have a year lasting just 6.3 days) for a star like Proxima Centauri. At those distances, the star's gravity would cause tidal lock. The daylight side of the planet would eternally face the star, while the nighttime

---

[18] Empirically determined in well-detached binary systems.

side would eternally face away from it. The only way potential life could avoid either an inferno or an utter deep freeze would be if the planet had an atmosphere thick enough to transfer the star's heat from the day side to the night side. But it was long assumed that such a thick atmosphere would prevent sunlight from reaching the surface in the first place, preventing photosynthesis.

Studies have shown that a planet's atmosphere need only be 15% thicker than Earth's for the star's heat to be effectively carried to the night side (Joshi et al. 1997). This is well within the levels required for photosynthesis, though water would still remain frozen on the dark side in some of the models. Heath et al. (1999) have shown that seawater too could be effectively circulated without freezing solid if the ocean basins were deep enough to allow free flow beneath the night side's ice cap. So, a planet with deep enough sea basins and a thick enough atmosphere could, at least potentially, harbor life in a red dwarf system.

Mere size is not the only factor in making red dwarfs potentially unsuitable for life, however. On a red dwarf planet, photosynthesis on the night side would be impossible, since it would never see the Sun. On the day side, because the Sun does not rise or set, areas in the shadows of mountains would remain so forever, making photosynthesis difficult. Photosynthesis as we understand it would be further complicated by the fact that a red dwarf produces most of its radiation in the infrared, and on our planet the process depends on visible light. Raven and Cockell (2006) have studied the efficiency of photosynthesis for different sources of light and absorption by the water in the planetary oceans (Raven 2007).

Tides may also place important constraints on planetary habitability. Barnes et al. (2010) have defined a 'tidal habitable zone' (THZ) for a range of stellar and planetary masses, which complements the classical habitable zone bases based on the existence of liquid water on the planetary surface. The tidal heating must be substantial enough to contribute, together with other internal energy sources, to the start of plate tectonics, but not so large as to cause Io-like volcanism.

Further, while the odds of finding a planet in the habitable zone around any specific red dwarf are slim, the total amount of habitable zone around all red dwarfs combined is equal to the total amount around Sun-like stars. For summary papers on this topic see Tarter et al. (2007) and Scalo et al. (2007).

5.6.1.3 Planetary Constraints

The limits of the circumstellar habitable zone are also conditioned by the properties of the planet.

The inner boundary of the continuously habitable zone for an Earth-like planet is determined by the conditions that inevitably lead to a runaway greenhouse effect in the early stages of planet history. This would occur when the planet is too close to the star and too much water enters the atmosphere. A moist greenhouse effect was also proposed by Kasting et al. (1993). In that scenario, water gets into the stratosphere, where it is dissociated by solar UV radiation, and the resulting H atoms at the top of the atmosphere are lost to space.

The outer boundary of the habitable zone is defined by the conditions that lead to runaway glaciation after an oxidizing atmosphere develops. This would happen when the planet is too far from the star and too much water freezes, leading to runway glaciations, or with the condensation of CO_2 clouds, which would limit the potential amount of greenhouse warming.

Kasting et al. (1993) calculated the inner and outer boundaries at 0.95 and 1.37 AU, respectively. Most recently, Franck et al. (2001) found these limits to be between 0.95 and 1.2 AU. Their habitable zone is defined by surface temperature boundaries of 0°C and 100 °C, but they restricted the CO_2 partial pressure to above 10–5 bar to ensure that conditions are suitable for biological productivity via photosynthesis. According to their model, the extinction of life for an Earth-like planet around a star between 0.6 and 1.1 M_S is caused by planetary geodynamics, at the age of 6.5 Ga.

5.6.1.4 The Continuously Habitable Zone

On the basis of the terrestrial experience, we assume that a long time must elapse before multicellular life appears. This means that the planet must remain habitable for a long time, at least a few Gigayears. It is defined a continuously habitable zone (CHZ) as the region in which a planet could remain habitable for a long period of time, suitable for the development of multicellular life.

Several time-varying factors determine the possibility of keeping the habitability conditions (liquid water on the planetary surface). They were summarized in Fig. 2.22, and in short are a balance between the increase of stellar luminosity and the changes in the concentration of greenhouse gases in the atmosphere.

The Potsdam group have developed a geodynamical model to estimate the time of a planet in the CHZ (Franck et al. 2000; Bounama et al. 2007). For planets around massive stars ($M > 2$ M_S), the main limitation is given by the star. For those planets around less massive stars, the limit is given by the duration of tectonic activity on the planet, T_{TP}, depending on the planetary mass through the expression (Fogg 1992)

$$T_{TP} = 5.1 (M_P/M_E)^{0.71} \text{Ga}.$$

5.6.2 Additional Constraints for Habitability

5.6.2.1 Short-term Stellar Variability

Changes in luminosity are common to all stars but the amplitude of such fluctuations covers a broad range. Most stars are relatively stable, but a significant minority of variable stars often experience sudden and intense increases in luminosity and, consequently, the amount of energy radiated toward bodies in orbit. These are considered poor candidates for hosting life-bearing planets as their unpredictability and energy output changes would impact negatively on organisms. Living things adapted to a particular temperature range would likely be unable to stand too great a deviation.

These changes in stellar brightness are driven by variations of the different types of energy available at the star. Short-term changes in main sequence solar-like stars are associated to the magnetic activity driven by a dynamo process in the stellar convection zone. However, the measured changes across a cycle of solar activity, 11 years, on the sun are benign, roughly 0.1%. There is strong (though not undisputed) evidence that larger changes in solar flux ($\sim$0.25%) have had significant effects on the Earth's climate well within the historical era; the Little Ice Age in the 1600s, for instance, might have been caused by a relatively long-term decline in the Sun's irradiance (Eddy 1976). However, for M stars these changes are stronger and their effects on life could not be negligible.

5.6.2.2 Ultraviolet and Ionizing Radiation

Ionizing radiation[19] has the potential to produce direct lethal effects on living organisms (Sinha and Häder 2002; Fernandez-Capetillo 2005). Ultraviolet radiation is also especially damaging for DNA and moreover can produce important photochemistry reactions in a planetary atmosphere.

Table 5.8 summarizes the main sources of UV and ionizing radiation on Earth and on other exoplanets (see also Vázquez and Hanslmeier 2005).

Main sequence stars emit enough high frequency ultraviolet radiation to initiate important atmospheric dynamics such as ozone formation (Kasting et al. 1997). Segura et al. (2003) have studied the UV environment of Earth-like planets orbiting F, G and K stars, finding that high O_2 planets orbiting K2V and F2V stars are better protected from surface UV radiation than is modern Earth. Buccino et al. 2006 studied the limits of an ultraviolet habitable zone considering the mentioned damaging effects and also those essential for several biogenesis processes.

Table 5.8 Sources of UV and ionizing radiation on Earth and other potential exoplanets

Source	Radiation	Time scale
Stellar photosphere	Ultraviolet	Ga
Stellar flares	Ultraviolet, X-rays	Days
Coronal mass ejections	Stellar wind	Days
Planet radioactivity	γ rays	Ga
Radiation belts	High-energy particles	
Artificial sources (Earth)		
Supernova explosions	Cosmic rays	Months
γ ray bursts	Cosmic rays	Hours

---

[19] It consists of subatomic particles or photons that have enough energy to detach electrons from atoms and molecules such as DNA.

The first effect to be considered is related with the photochemistry of the planetary atmosphere. Ionizing energy bursts would continually strip the planets of their protective covering. For example, the Venus atmospheric oxygen escape during the passage of several coronal mass ejections from the Sun is well documented (Luhmann et al. 2008).

Grenfell et al. (2007) have shown that high levels of cosmic rays could affect the strength of biomarkers such as ozone or methane when comparing planets orbiting M stars with our Earth.

Scalo and Wheeler (2002) suggested that the stochastic ionizing radiation could be an accelerating factor in the evolution of life through direct mutational enhancement or sterilization. In fact, primitive bacteria that evolved in a strong UV environment (see Chaps. 2 and 4) have probably developed a survival procedure.

For high-energy photons or particles, the greatest mutational lesions are those leading to lethality, but the situation is different for UV mutagenesis. The mutation doubling dose for microorganisms due to UV radiation may be smaller than the lethal dose. Flares in the youngest M stars could provide stochastic bursts of ionizing radiation to accelerate life evolution in nearby planets (Smith et al. 2004).

Other constraints are given by the stability of the planetary orbit, explained in some detail in Chap. 8.

5.6.3 Galactic Habitable Zone

The concept of the galactic habitable zone (GHZ) is not precisely defined. Basically, it is the region in the Milky Way with the necessary conditions to form Earth-like planets and with a sufficiently calm environment, over several Ga, to allow the development of multicellular life, which took about 4 Ga for Earth.

The Milky Way is a common spiral disk galaxy, although the largest one in its vicinity. Its structure is represented in Fig. 5.20 (see also Fig. 1.30). It is formed by the nuclear bulge and the Galactic Centre, the disk and the halo. The halo is a spherical distribution that contains the oldest stars in the galaxy, known as *Extreme*

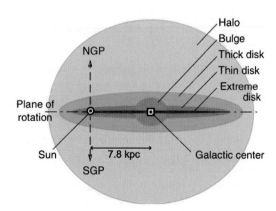

Fig. 5.20 Schematic (edge-on) view of the major components of the Milky Way Galaxy's overall structure. Adapted from Buser (2000) Fig. 1. Reprinted with permission of the AAAS

Population II stars. The globular clusters seen in the halo of the galaxy are the residues of the primordial star clusters existing when the galaxy was formed. These clusters are metal poor (by 'metal' is meant all elements heavier than helium). The elements up to Fe on the periodic table are produced during the nuclear reactions in the stellar core or during the collapse of this core and resulting supernova explosion. Therefore, the only existing material when Population II stars formed was basically hydrogen and helium. The *Population I* stars formed in the galactic disk from a material that already contained heavy elements from the first generation of supernovae. They are younger and have higher metallicity than the Population II stars.

The development and the successful evolution of complex life depends on the temporal and the spatial location within the galaxy. Sterilizing radiation and certain exterminator events are associated to some galactic regions. Nearby transient sources of ionizing radiation, for instance, including supernovas and gamma ray bursts, as well as comet impacts are threats to complex life. These events tend to increase close to the galactic centre.

Furthermore, an adequate abundance of heavy elements is a determining factor for the formation of terrestrial planets. With too little metallicity, Earth–mass planets are unable to form; with too much metallicity, the planetary system develops giant planets orbiting close to the parent star (Santos et al. 2003; Laws et al. 2003), which would destroy any Earth–mass planets.

Considering these facts, the GHZ is defined by Gonzalez et al. (2001) as an annular region lying in the plane of the galactic disk. The inner (closest to the galactic centre) limit is set by the supernova frequency and the outer limit is imposed by the galactic chemical evolution, in particular, the abundance of heavier elements.

Lineweaver et al. (2004) modelled the evolution of the galaxy and situated a GHZ between the crowded inner bulge and the barren outer Galaxy. This is shown in Fig. 5.21, which represents the evolution of the annular region with time. The green area contains the stars that can harbor life at the present time. This GHZ emerged about 8 Ga ago (inner white contour in Fig. 5.21) and expanded with time as metallicity spread outward in the Galaxy and the supernovae rate decreased. By comparing the age distribution on the right of Fig. 5.21 to the origin of the Sun, they found that ∼75% of the stars that may harbour complex life in the Galaxy are older than the Sun and that their average age is ∼1 Ga older than the Sun.

Recently, Blair et al. (2008) have measured formaldehyde (H_2CO) in molecular clouds in different regions of the galaxy, constraining the inner boundary of the galactic habitable zone.

5.7 Signatures of Technological Civilizations

The emergence of intelligence in our planet was an important feature of its history, although our perspective could be strongly biased. With this in mind, we try to establish the fingerprints that such intelligence could leave in the surface and atmosphere, which could possibly be observed remotely.

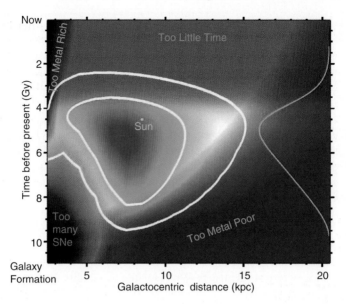

Fig. 5.21 The GHZ in the disk of the Milky Way based on the star formation rate, metallicity (*blue*), sufficient time for evolution (*grey*) and freedom from life-extinguishing supernova explosions (*red*). The *white* contours encompass 68% (*inner*) and 95% (*outer*) of the origins of stars, with the highest potential to be harbouring complex life today. The *green line* on the right is the age distribution of complex life (Lineweaver et al. 2004). Reprinted with permission from the AAAS

Our civilization is characterized by the growing production and the use of energy. According to the principles of thermodynamics, this inevitably leads to emission of residuals in some form. Some of these leakages could be observable from space.

5.7.1 Night Lights

Light pollution is light created by humans. About 30–60% of energy consumed in such lighting is unnecessary, and is sent to space. Figure 5.22 perfectly illustrates the uneven distribution of this emission of light to outer space.

The first world atlas of artificial night sky brightness (Fig. 5.22) from observations taken at sea level was published by Cinzano et al. (2001). This can be compared with a well know composite of hundreds of images taken by the orbiting DMSP[20] satellites (Fig. 5.23). About two-thirds of the world population and 99% of the population in the United States (excluding Alaska and Hawaii) and European Union

---

[20] Defense Meteorological Satellites Program. A program of the US Air Forces active since the 1970s.

5.7 Signatures of Technological Civilizations

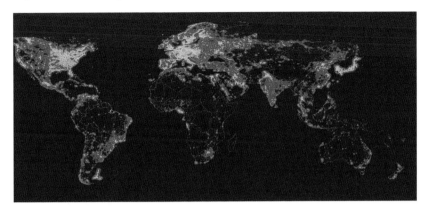

Fig. 5.22 Artificial night sky brightness at sea level. The map has been computed for the photometric astronomical V band at the zenith. Colours correspond to ratios between the artificial and natural sky brightness (<0.01 *black*, 0.01–0.11 *dark-gray*; 0.11–0.33 *blue* and 9–27 *red*). Adapted from Fig. 1 of Cinzano et al. (2001). Reproduced by permission of Blackwell-Wiley

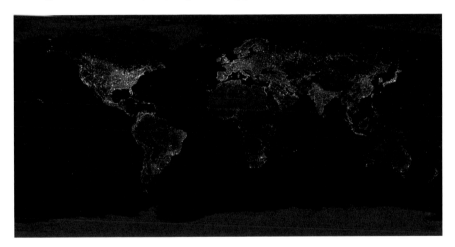

Fig. 5.23 Earth at night. Data courtesy Marc Imhoff of NASA GSFC and Christopher Elvidge of NOAA NGDC. Image by Craig Mayhew and Robert Simmon, NASA GSFC. Available from http://visibleearth.nasa.gov

live in areas where the night sky is above the threshold set for polluted status. Assuming average eye functionality, about one-fifth of the world population, more than two-thirds of the United States population and more than one half of the European Union population have already lost naked eye visibility of the Milky Way. Additionally, about one-tenth of the world population, more than 40% of the United States population and one sixth of the European Union population no longer view the heavens with the eye adapted to night vision, because of the sky brightness.

5.7.2 Spectral Features

Laws of sky protection put strict limits on the type of lights that can be used for outdoor lighting, concerning their power and orientation with respect to the ground. These limits imply that, after local midnight, most of the high-pressure sodium (HPS) and mercury lights must be extinguished, as well as all the discharge-tube illumination.

Pedani (2004, 2005) shows sky spectra in one of the ENO (European Northern Observatory) observatories, where the level of light pollution is limited by law[21] (Fig. 5.24). The most damaging lights are those that emit ultraviolet light, as light at these wavelengths is the most strongly dispersed in the atmosphere and has no value in terms of illumination.

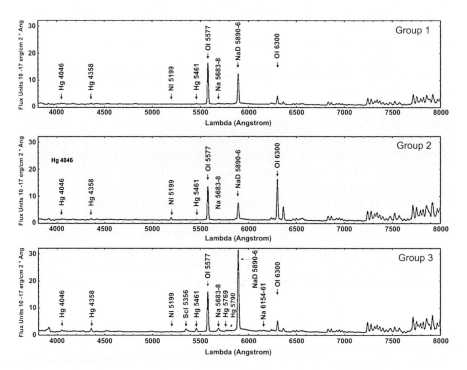

Fig. 5.24 The night-sky spectra at Roque de los Muchachos Observatory (ORM) from observations at the TNG. The Group 1 (4 h total exposure) is the average of eight spectra and best represents the average observing conditions at ORM. The Group 2 spectrum was taken towards the NW, the least light-polluted zone at ORM. The Group 3 spectrum was taken towards the most light-polluted region of sky at ORM, before midnight. The presence of thin clouds could explain the abnormally high fluxes of the light polluting lines. Adapted from Pedani et al. (2005) Fig. 1

---

[21] The IAC Sky Quality Protection Unit ensures adherence to the law.

Industrial activities produce some gases that are exclusively of anthropogenic origin. The Moderate Resolution Imaging Spectroradiometer (MODIS) aboard the Terra and Aqua satellites allows the monitoring of pollution from space.

5.7.3 Artificial Radioemission

Non-thermal electromagnetic radiation is mainly generated by humans in the telecommunications industry. At specific radio frequencies the Earth is now brighter than the Sun.

At radio wavelength, the signals of the ballistic missile early warnings systems (BMEWS), the Arecibo radiotelescope and TV and FM transmitters[22] dominate the electromagnetic spectra of the Earth and are larger than the natural emissions from the Earth or the Sun (Sullivan et al. 1978; Billingham and Tarter 1992). The bandwidths are narrow and so the power per unit bandwidth is large.[23] Figure 5.25 shows the Earth radio spectrum with the technologies available in the 1980s compared to the main natural sources of the solar system (Sun, magnetospheres and radiation belts).

The electron density of the ionosphere, N_e, determines the plasma frequency $f_p \sim 1$–5 MHz. Radio emissions from the sunlit Earth should be detectable only at $f > f_p$ (at night f_p should be lower). During the passage of the Galileo Spacecraft close to the Earth, Sagan et al. (1993) studied the radio spectrum of our planet. They detected narrow-band emissions near the range 4–5 MHz, which were identified as ground-based transmitters. The signals were observed only on the nightside, where f_p is sufficiently low to allow the radiowaves to escape through the ionosphere.

Traditionally, radio is considered the best range of the electromagnetic spectrum to search for and detect 'artificial' signatures emitted by an intelligent civilization. That might be an anthropogenic prejudice due to the fact that humans have chosen this specific technology for telecommunication processes, but there are certain advantages to the use of radio waves for interstellar communications, if such are ever to occur. For example, there is almost nothing in outer space able to block or absorb radio frequencies, and it is easy to pinpoint the origin of such waves with accuracy. It is also quite easy to encrypt information within.

Rotational modulation of the radio emission of Earth would inform a distant observer about the spatial distribution of the emitters and additionally about the distribution of the continents (Fig. 5.26). The radar of the US Naval Space Surveillance has a detectability range of leaking terrestrial signals to 60 light-years for an Arecibo-type antenna (Sullivan et al. 1978; Sullivan 1981) .

---

[22] The TV signal is not due to the modulation that produces the picture but to the very narrow bandwidth carrier (less than 1 Hz), which is the backbone of the signal.

[23] TV services are increasingly being delivered by other means such as cable that do not leak radio waves into the space.

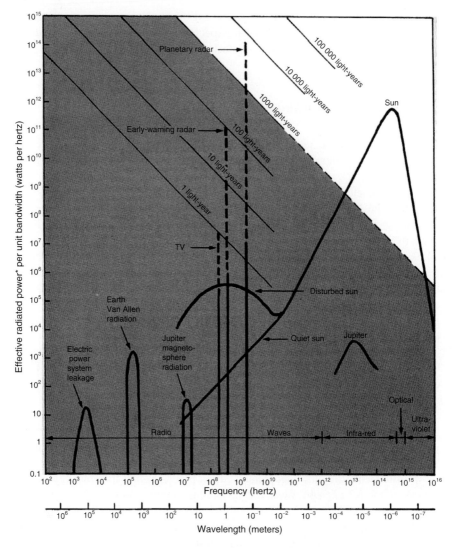

Fig. 5.25 The radio spectrum of the solar system seen at different distances. Adapted from Cosmic Search Vol. 2 No. 1, p. 30 (1980)

Billingham and Tarter (1992) estimated the maximum distance, R, at which radar signals could be detected

$$R = \sqrt{\frac{\text{EIRP}}{4\pi \phi_{\min}}},$$

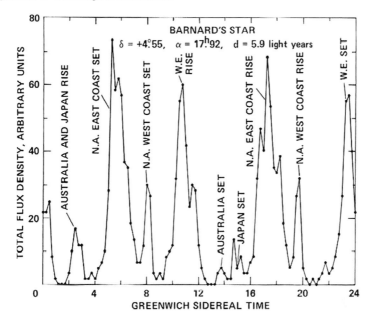

Fig. 5.26 Calculated flux density (summed over all frequencies) of TV radiation that would be measured over a sidereal day by an observer located in the direction of Barnard's star. The origin of the various peaks is indicated: 'rise' and 'set' refer, respectively, to the appearance at the western limb of a particular region on the rotating Earth (N.A., North America; W.E., Western Europe). Fig. 5 of Sullivan (1981)

where ϕ_{min} is the sensitivity of the detector (in watts per square metre), and EIRP (Equivalent Isotropically Radiated Power) is the product of the transmitted power and directive antenna gain.

5.7.4 Nuclear Explosions

Atmospheric nuclear explosions produce a unique signature: a short and intense flash lasting around 1 ms, followed by a much more prolonged and less intense emission of light lasting from a fraction of a second to several seconds.

Nuclear explosions in the atmosphere left radioactive contamination that can be identified. Chlorine-36 is produced in large amounts in such events. This isotope is produced naturally in the stratosphere when atoms are bombarded by cosmic rays, high-energy particles that streak through space from beyond our solar system. This so-called cosmogenic production of chlorine-36 has varied over time due to fluctuations in the strength of Earth's magnetic field. The worldwide concentration of chlorine-36 substantially increased during atmospheric nuclear weapon tests conducted in the Pacific between 1952 and 1962[24] (Elmore et al. 1982; Green et al. 2004; Delmas et al. 2004). Sr-90 is also created during nuclear explosions or in nuclear reactors in the process of fission of heavy nuclei, such as uranium-235 and plutonium-239.

On 9 July 1962, the US exploded a thermonuclear bomb at 400 km of altitude, called Starfish-Prime, above Johnston Island in the Pacific Ocean. As a consequence, an artificial aurora was produced lasting 7 min (Fig. 5.27) and the radiation belts were amplified. The subsequent ionization of the upper atmosphere also gave rise to an intense pulse of radioemission (McDiarmid et al. 1963; Dyal 2006). In the end, the Test Ban Treaty of 1963 prohibited nuclear explosions in the atmosphere, outer space and under water.

Elliot (1973) analyzed the X-ray emissions of the terrestrial nuclear explosions carried out in the early 1960s. He estimated the distance at which the Starfish test could be detected in X-rays from space. Assuming an energy of the explosion equivalent to 1.4 megatons and an isotropic propagation, he found a distance of 400 AU, about ten times the radius of Pluto's orbit.

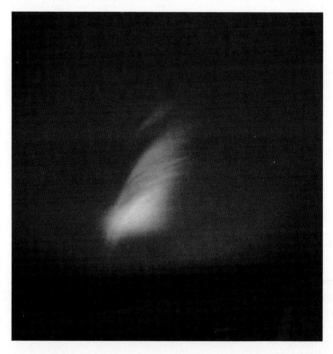

Fig. 5.27 The Starfish artificial aurora as seen from a KC-135 surveillance aircraft (July 9, 1962)

---

[24] Chlorine-36 is also potentially useful as a tracer to study movement of contaminants around large nuclear reactor complexes and near repositories for radioactive waste.

5.7.5 Extraterrestrial Pulses

A technological civilization could produce light pulses that last nanoseconds (Howard et al. 2004). The current technology on the Earth could generate a directed laser pulse that would outshine the broadband visible light of the Sun by four orders of magnitude. Ross (2000) describes a strategy for detecting such an extraterrestrial signal.

Shkhlovoski and Sagan (1966) suggested that alien civilizations could add short-lived isotopes to the atmospheres of their stars, to be detected by the presence of the corresponding absorption lines.

References

Alberts, B., Bray, D., Hopkin, K., Johnson, A., Lewis, J., Raff, M., Roberts, K., Walter, P.: Essential Cell Biology. Garland Science Publishing, London (2003)

Almond, S., Boyd, D.S., Curran, P.J., Dash, J.: The response of UK vegetation to elevated temperatures in 2006. Coupling ENVISAT MERIS terrestrial chlorophyll index (MTCI) and mean air temperature. In: Proceedings on the 2007 Annual Conference of the Remote sensing and Photogrammetry Society: Newcastle University (2007)

Arcichovsky, V.M.: Auf der Suche nach Chlorophyll auf den Planeten. Annales de l'Institut Polytechnique Don Cesarevitch Alexis at Novotcherkassk **17**, 195 (1912)

Arnold, L.: Earthshine Observation of Vegetation and Implication for Life Detection on Other Planets. A Review of 2001–2006 Works. Space Sci. Rev. **135**, 323–333 (2008)

Arnold, L., Gillet, S., Lardière, O., Riaud, P., Schneider, J.: A test for the search for life on extrasolar planets. Looking for the terrestrial vegetation signature in the Earthshine spectrum. Astron. Astrophys. **392**, 231–237 (2002)

Bains, W.: Many Chemistries Could Be Used to Build Living Systems. Astrobiology **4**, 137–167 (2004)

Ball, P.: Life's Matrix: A biography of water. University of California Press, California (2001)

Barnes, R., Jackson, B., Greenberg, R., Raymond, S.N.: Tidal Limits to Planetary Habitability. Proceedings "Pathways Towards Habitable Planets" Symposium., ASP Conference Series, Eds.: D. Gelino, V. Coude du Foresto, I. Ribas, (in press) (2010)

Barron, L.D.: Chirality and Life. Space Sci. Rev. **135**, 187–201 (2007)

Battistuzzi, F.U., Feijao, A., Hedges, S.: A genomic timescale of prokaryote evolution: insights into the origin of methanogenesis, phototrophy, and the colonization of land. Evol. Biol. **4**, 44 (2004)

Bennett, J., Shostak, S., Jakosky, B.: Life in the Universe. Addison Wesley (2003)

Bern, M., Goldberg, D.: Automatic selection of representative proteins for bacterial phylogeny. BMC Evol. Biol. **5**, 34 (2005)

Bernal, J.D.: The physical basis of life. Proc. Phys. Soc. B **62**, 597–618 (1949)

Billingham, J., Tarter, J.: Detection of the Earth with the SETI microwave observing system assumed to be operating out in the galaxy. Acta Astronautica **26**, 185–188 (1992)

Blair, S.K., Magnani, L., Brand, J., Wouterloot, J.G.A.: Formaldehyde in the Far Outer Galaxy: Constraining the Outer Boundary of the Galactic Habitable Zone. Astrobiology **8**, 59–73 (2008)

Botta, O., Bada, J.L., Gomez-Elvira, J., Javaux, E., Selsis, F., Summons, R.: "Strategies of Life Detection": Summary and Outlook. Space Sci. Rev. **135**, 371–380 (2008)

Bounama, C., von Bloh, W., Franck, S.: How Rare Is Complex Life in the Milky Way? Astrobiology **7**, 745–756 (2007)

Brack, A.: Liquid water and the origin of life. Orig. Life Evol. Biosph. **23**, 3–10 (1993)

Brack, A., Pillinger, C.T.: Life on Mars, Extremophiles **2**, 313–319 (1998)

Brack, A.: Water, the spring of life, pp. 79–88. Astrobiology. The quest for the conditions of life. G. Horneck, C. Baumstark-Khan (eds.). Physics and astronomy online library. Berlin: Springer (2002)

Briot, D., Schneider, J., Arnold, L.: G. A. Tikhov, and the Beginnings of Astrobiology. In: J. Beaulieu, A. Lecavelier Des Etangs, C. Terquem (eds.) Extrasolar Planets: Today and Tomorrow, Astronomical Society of the Pacific Conference Series, vol. 321, pp. 219–220 (2004)

Brook, M.A.: Silicon in Organic, Organometallic, and Polymer Chemistry. Wiley, New York, Toronto (2000)

Buccino, A.P., Lemarchand, G.A., Mauas, P.J.D.: Ultraviolet radiation constraints around the circumstellar habitable zones. Icarus **183**, 491–503 (2006)

Buser, R.: The formation and early evolution of the Milky Way galaxy. Science **287**, 69–74 (2000)

Carlson, T.N., Ripley, D.A.: On the relation between NDVI, fractional vegetation cover, and leaf area index. Rem. Sens. Environ. **62**, 241–252 (1997)

Carnot, S.: Réflexions sur la puissance motrice du feu et sur les machines propres à déveloper cette puissance. Bachelier, Paris (1824)

Chaisson, E.J.: The cosmic environment for the growth of complexity. BioSystems **46**, 13–19 (1998)

Chyba, C.F., Hand, K.P.: Astrobiology: The Study of the Living Universe. Annu. Rev. Astron. Astrophys. **43**, 31–74 (2005)

Cinzano, P., Falchi, F., Elvidge, C.D.: The first World Atlas of the artificial night sky brightness. Mon. Not. Roy. Astron. Soc. **328**, 689–707 (2001)

Clark, R.: Chapter 1: Spectroscopy of rocks and minerals, and principles of spectroscopy. In: A. Rencz (ed.) Manual of Remote Sensing: Remote Sensing for the Earth Sciences, vol. 3, pp. 3–58. Wiley, London (1999)

Clausius, R.: The Mechanical Theory of Heat with its Applications to the Steam Engine and Physical Properties of Bodies. John van Voorst, London (1865)

Cloud, P.E.: A working model of the primitive Earth. Am. J. Sci. **272**, 537–548 (1972)

Colwell, J.E.: Vegetation Canopy reflectance. Rem. Sens. Environ. **3**, 175–283 (1974)

DasSarma, S.: Extreme Microbes. Am. Sci. **95**, 224–231 (2007)

Deisenhofer, J., Michel, H.: The Photosynthetic Reaction Center from the Purple Bacterium Rhodopseudomonas viridis. Science **245**, 1463–1473 (1989)

Delmas, R.J., Beer, J., Synal, H.A., Muscheler, R., Petit, J.R., Pourchet, M.: Bomb-test ^{36}Cl measurements in Vostok snow (Antarctica) and the use of ^{36}Cl as a dating tool for deep ice cores. Tellus Series B Chem. Phys. Meteorol. B **56**, 492–498 (2004)

Des Marais, D.J., Nuth III., J.A., Allamandola, L.J., Boss, A.P., Farmer, J.D., Hoehler, T.M., Jakosky, B.M., Meadows, V.S., Pohorille, A., Runnegar, B., Spormann, A.M.: The NASA Astrobiology Roadmap. Astrobiology **8**, 715–730 (2008)

Dole, S.H.: Habitable planets for man. New York: American Elsevier, 2nd ed (1970)

Dyal, P.: Particle and field measurements of the Starfish diamagnetic cavity. J. Geophys. Res. **111**(10), 12,211 (2006)

Eddy, J.A.: The Maunder Minimum. Science **192**, 1189–1202 (1976)

Eigen, M., Gardiner, W., Schuster, P., Winkler-Oswatitsch, R.: The origin of genetic information. Sci. Am. **224**(4), 88–92, 96 (1981)

Eigen, M., Schuster, P.: The Hypercycle: A Principle of Natural Self-organization . Naturwissenschaften **65**, 341–369 (1979)

Eigen, M., Winkler, R.: Laws of the Game: How the principles govern chance. Princeton University Press, Princeton (1993)

Elliot, J.L.: X-ray pulses for interstellar communication. In: C. Sagan (ed.) Communication With Extraterrestrial Intelligence, pp. 398–402. The MIT Press (1973)

Elmore, D., Tubbs, L.E., Newman, D., Ma, X., Finkel, R., Beer, J., Oeschger, H., Andree, M.: ^{36}Cl bomb pulse measured in a shallow ice core from Dye3, Greenland. Nature **300**, 735–737 (1982)

Encrenaz, T.: Searching for water in the Universe. Springer Praxis (2007)

Falkowski, P.G.: The ocean's invisible forest. Sci. Am. **287**(2), 54–61 (2002)
Farquhar, J., Johnston, D.T., Calvin, C., Condie, K.: Implications of Sulfur Isotopes for the Evolution of Atmospheric Oxygen. In: S. Mackwell, E. Stansbery (eds.) Lunar and Planetary Institute Conference Abstracts, vol. 35, p. 1920 (2004)
Feinberg, G., Shapiro, R.: Life Beyond Earth - The Intelligent Earthling's Guide to Life in the Universe. William Morrow and Company (1980)
Fernández-Capetillo, O.: Radiation and DNA repair constraints for the development of a DNA-based life. In: M. Vázquez (ed.) Fundaments and Challenges in Astrobiology, pp. 85–98. Research SignPost, India (2005)
Field, C.B., Behrenfeld, M.J., Randerson, J.T., P., F.: Primary Production of the Biosphere: Integrating Terrestrial and Oceanic Components. Science **281**, 237–240 (1998)
Fogg, M.J.: An Estimate of the Prevalence of Biocompatible and Habitable Planets. J. Br. Interplanet. Soc. **45**, 3–12 (1992)
Franck, S., Block, A., von Bloh, W., Bounama, C., Schellnhuber, H.J., Svirezhev, Y.: Habitable zone for Earth-like planets in the solar system. Planet. Space Sci. **48**, 1099–1105 (2000)
Franck, S., von Bloh, W., Bounama, C., Steffen, M., Schönberner, D., Schellnhuber, H.J.: Limits of photosynthesis in extrasolar planetary systems for earth-like planets. Adv. Space Res. **28**, 695–700 (2001)
Gaffron, H.: Reduction of carbon dioxide with molecular hydrogen in green algae. Nature **143**, 204–205 (1939)
Gaffron, H., Rubin, J.: Fermentation and photochemical production of hydrogen in algae. J. Gen. Physiol. **26**, 219–240 (1942)
Gaidos, E., Deschenes, B., Dundon, L., Fagan, K., Menviel-Hessler, L., Moskovitz, N., Workman, M.: Beyond the Principle of Plentitude: A Review of Terrestrial Planet Habitability. Astrobiology **5**, 100–126 (2005)
Gargaud, M., Martin, H., Claeys, P.: Lectures in Astrobiology II. Springer (2007)
Gilmour, I., Sephton, M.A.: An Introduction to Astrobiology. Cambridge University Press (2003)
Gleiser, M., Thorarinson, J., Walker, S.I.: Punctuated Chirality. Orig.Life Evol. Biosph. **38**, 499–508 (2008)
Gonzalez, G., Brownlee, D., Ward, P.: The Galactic Habitable Zone: Galactic Chemical Evolution. Icarus **152**, 185–200 (2001)
Goraieb, K., Alexandrea, T.L., Buenoa, M.: X-ray spectrometry and chemometrics in sugar classification, correlation with degree of sweetness and specific rotation of polarized light. Anal. Chim. Acta **595**, 170–175 (2007)
Govender, M.G., Rootman, S.M., Ford, T.A.: An ab initio study of the properties of some hydride dimers. Cryst. Eng. **6**, 263–286 (2003)
Green, J.R., Cecil, L.D., Synal, H.A., Santos, J., Kreutz, K.J., Wake, C.P.: A high resolution record of chlorine-36 nuclear-weapons-tests fallout from Central Asia. Nucl. Instrum. Meth. Phys. Res. B **223**, 854–857 (2004)
Grenfell, J.L., Grießmeier, J.M., Patzer, B., Rauer, H., Segura, A., Stadelmann, A., Stracke, B., Titz, R., Von Paris, P.: Biomarker Response to Galactic Cosmic Ray-Induced NO_x And The Methane Greenhouse Effect in The Atmosphere of An Earth-Like Planet Orbiting An M Dwarf Star. Astrobiology **7**, 208–221 (2007)
Grinspoon, D.H., Williams, D.M., Piccioni, G., Bertaux, J., Moore, C.: Observing Earth from the Vantage Point of Venus Orbit. In: Bulletin of the American Astronomical Society, vol. 40, p. 386 (2008)
Hamdani, S., Arnold, L., Foellmi, C., Berthier, J., Billeres, M., Briot, D., François, P., Riaud, P., Schneider, J.: Biomarkers in disk-averaged near-UV to near-IR Earth spectra using Earthshine observations. Astron. Astrophys. **460**, 617–624 (2006)
Harper, D.B.: Halomethane from halide iona highly efficient fungal conversion of environmental significance. Nature **315**, 55–57 (1985)
Hart, M.H.: Habitable Zones about Main Sequence Stars. Icarus **37**, 351–357 (1979)
Heath, M.J., Doyle, L.R., Joshi, M.M., Haberle, R.M.: Habitability of Planets Around Red Dwarf Stars. Orig. Life Evol. Biosph. **29**, 405–424 (1999)

Heylighen, F.: The science of self-organization and adaptivity. In: in: Knowledge Management, Organizational Intelligence and Learning, and Complexity, in: The Encyclopedia of Life Support Systems, pp. 253–280. EOLSS Publishers Co. Ltd, Oxford (2003)

Holland, H.D.: The Chemistry of the Atmosphere and Oceans. Wiley, New York (1978)

Holland, H.D.: The Chemical Evolution of the Atmosphere and Oceans. Princeton University Press, Princeton (1984)

Holland, H.D.: Volcanic Gases, Black Smokers, and the Great Oxidation Event. Geochim. Cosmochim. Acta **66**(22), 3811–3826 (2002)

Horneck, G., Rettberg, P.: Complete course in astrobiology. Wiley-VCH, p. 413 (2007)

Howard, A.W., Horowitz, P., Wilkinson, D.T., Coldwell, C.M., Groth, E.J., Jarosik, N., Latham, D.W., Stefanik, R.P., Willman Jr., A.J., Wolff, J., Zajac, J.M.: Search for Nanosecond Optical Pulses from Nearby Solar-Type Stars. Astrophys J. **613**, 1270–1284 (2004)

Huang, S.S.: The Sizes of Habitable Planets. Publ. Astron. Soc. Pac. **72**, 489–493 (1960)

Jones, B.W.: Exoplanets search methods, discoveries, and prospects for astrobiology. Int. J. Astrobiology **7**, 279–292 (2008)

Joshi, M.M., Haberle, R.M., Reynolds, R.T.: Simulations of the Atmospheres of Synchronously Rotating Terrestrial Planets Orbiting M Dwarfs: Conditions for Atmospheric Collapse and the Implications for Habitability. Icarus **129**, 450–465 (1997)

Joyce, G.F.: Hydrothermal vents too hot? Nature **334**, 564 (1988)

Kaltenegger, L., Traub, W.A., Jucks, K.W.: Spectral Evolution of an Earth-like Planet. Astrophys. J. **658**, 598–616 (2007)

Kasting, J.F.: Habitable zones around low mass stars and the search for extraterrestrial life. Orig. Life Evol. Biosph. **27**, 291–307 (1997)

Kasting, J.F.: Habitable planets around the Sun and other stars. In: H. Deeg, J.A. Belmonte, A. Aparicio (eds.) Extrasolar Planets, pp. 217–244 (2008)

Kasting, J.F., Catling, D.: Evolution of a Habitable Planet. Annual Review of Astron. Astrophys. **41**, 429–463 (2003)

Kasting, J.F., Whitmire, D.P., Reynolds, R.T.: Habitable Zones around Main Sequence Stars. Icarus **101**, 108–128 (1993)

Kasting, J.F., Whittet, D.C.B., Sheldon, W.R.: Ultraviolet radiation from F and K stars and implications for planetary habitability. Orig. Life Evol. Biosph. **27**, 413–420 (1997)

Kauffman, S.: Antichaos and adaptation. Sci. Am. **265**(2), 78–84 (1991)

Kauffman, S.: The Origins of Order: Self-organization and Selection in Evolution. Oxford University Press, New York (1993)

Kaufman, A.J., Johnston, D.T., Farquhar, J., Masterson, A.L., Lyons, T.W., Bates, S., Anbar, A.D., Arnold, G.L., Garvin, J., Buick, R.: Late Archean Biospheric Oxygenation and Atmospheric Evolution. Science **317**, 1900– (2007)

Kelley, D.S., Karson, J.A., Blackman, D.K., Frh-Green, G., Butterfield, D., Lilley, M., Olson, E., Schrenk, M., Roe, K., Lebon, G., Rivizzigno, P., the AT3-60 Shipboard Party: An off-axis hydrothermal vent field near the Mid-Atlantic Ridge at 30°N. Nature **412**, 145–149 (2001)

Kelley, D.S., Karson, J.A., Frh-Green, G.L., Dana, R., Yoerger, D.R., Shank, T.M., Butterfield, D.A., Hayes, J.M., Schrenk, M.O., Olson, E.J., Proskurowski, G., Jakuba, M., Bradley, A., Larson, B., Ludwig, K., Glickson, D., Buckman, K., Bradley, A.S., Brazelton, W.J., Roe, K., Elend, M.J., Delacour, A., Bernasconi, S.M., Lilley, M.D., Baross, J.A., Summons, R.E., Sylva, S.P.: A serpentinite-hosted ecosystem: the Lost City hydrothermal vent field. Science **307**, 1428–1434 (2005)

Kemp, J.C., Wolstencroft, R.D.: Circular Polarization: Jupiter and Other Planets. Nature **232**, 165–168 (1971)

Kiang, N.: The Color of Plants on other worlds. Sci. Am., **298**, 48–55 (2008)

Kiang, N.Y., Segura, A., Tinetti, G., Govindjee, Blankenship, R.E., Cohen, M., Siefert, J., Crisp, D., Meadows, V.S.: Spectral Signatures of Photosynthesis. II. Coevolution with Other Stars And The Atmosphere on Extrasolar Worlds. Astrobiology **7**, 252–274 (2007)

Kiang, N.Y., Siefert, J., Govindjee, Blankenship, R.E.: Spectral Signatures of Photosynthesis. I. Review of Earth Organisms. Astrobiology **7**, 222–251 (2007)

Küppers, B.O.: Information and the Origin of Life. The MIT Press (1990)
Laws, C., Gonzalez, G., Walker, K.M., Tyagi, S., Dodsworth, J., Snider, K., Suntzeff, N.B.: Parent Stars of Extrasolar Planets. VII. New Abundance Analyses of 30 Systems. Astron. J. **125**, 2664–2677 (2003)
Levin, G.V., Straat, P.A.: Recent results from the Viking Labeled Release experiment on Mars. J. Geophys. Res. **82**, 4663–4667 (1977)
Liais, M.E.: Schreiben des Herrn Emm. Liais an den Herausgeber. Astron. Nachr. **65**, 11 (1865)
Lineweaver, C.H., Egan, C.A.: Life, gravity and the second law of thermodynamics. Phys. Life Rev. **5**, 225–242 (2008)
Lineweaver, C.H., Fenner, Y., Gibson, B.K.: The Galactic Habitable Zone and the Age Distribution of Complex Life in the Milky Way. Science **303**, 59–62 (2004)
Lobert, J.M., Scharffe, D.H., Hao, W.M., Kuhlbusch, T.A., Crutzen, P.J.: Experimental evaluation of biomass burning emissions, pp. 289–304. MIT Press, Cambridge, Mass (1991)
Lovelock, J.E., Margulis, L.: Atmospheric homeostasis by and for the biosphere: The Gaia hypothesis. Tellus **26**, 2–10 (1974)
Lovelock, J.E., Margulis, L.: Homeostatic tendencies of the Earth's atmosphere. Orig. Life **25**, 93–103 (1974)
Luhmann, J.G., Fedorov, A., Barabash, S., Carlsson, E., Futaana, Y., Zhang, T.L., Russell, C.T., Lyon, J.G., Ledvina, S.A., Brain, D.A.: Venus Express observations of atmospheric oxygen escape during the passage of several coronal mass ejections. J. Geophys. Res. **113**(12) (2008)
Luisi, P.L.: The emergence of life: From chemical origins to synthetic biology. Cambridge University Press (2006)
Lunine, J.: Astrobiology: A Multi-Disciplinary Approach. Benjamin Cummings (2005)
Marrin, W.: Universal Water. Inner Ocean (2002)
McDiarmid, I.B., Burrows, J.R., Budzinski, E.E., Rose, D.C.: Satellite measurements in the "Starfish" artificial radiation zone. Canad. J. Phys. **41**, 1332–1345 (1963)
McKay, D.S., Gibson, E.K., Thomas-Keprta, K.L., Vali, H., Romanek, C.S., Clemett, S.J., Chiller, X.D.F., Maechling, C.R., Zare, R.N.: Search for past life on Mars: Possible relic biogenic activity in martian meteorite ALH84001. Science **273**, 924–930 (1996)
Meierhenrich, U., Nahon, L., Alcaraz, C., Bredehoft, J., Hoffmann, S., Barbier, B., A., B.: Asymmetric Vacuum UV photolysis of the Amino Acid Leucine in the Solid State. Angew. Chem. Int. Ed. **44**, 5630–5634 (2005)
Miller, S.L., Orgel, L.E.: The origins of life on the Earth. Englewood Cliffs, NJ: Prentice-Hall (1974)
Montañés-Rodriguez, P., Pallé, E., Goode, P.R., Hickey, J., Koonin, S.E.: Globally Integrated Measurements of the Earth's Visible Spectral Albedo. Astrophys. J. **629**, 1175–1182 (2005)
Montañés-Rodriguez, P., Pallé, E., Goode, P.R., Martín-Torres, F.J.: Vegetation Signature in the Observed Globally Integrated Spectrum of Earth Considering Simultaneous Cloud Data: Applications for Extrasolar Planets. Astrophys. J. **651**, 544–552 (2006)
Mooney, H.A., Ehleringer, J., Bjorkman1, O.: The energy balance of leaves of the evergreen desert shrub. Oecologia **29**, 301–310 (1977)
Morgan, W.W., Keenan, P.C.: Spectral Classification. Annu. Rev. Astron. Astrophys. **11**, 29–50 (1973)
Nealson, K.H., Tsapin, A., Storrie-Lombardi, M.: Searching for life in the Universe: unconventional methods for an unconventional problem. Int. Microbiol. **5**, 223–230 (2002)
Nicolis, G., Prigogine, I.: Self-Organization in Non-Equilibrium Systems: From Dissipative Structures to Order Through Fluctuations. Wiley, New York (1977)
Nicolis, G., Prigogine, I.: Exploring complexity. W.H. Freeman, San Francisco (1989)
Omarov, T.B., Tashenov, B.T.: Tikhov's Astrobotany as a Prelude to Modern Astrobiology. In: R.B. Hoover, A.Y. Rozanov, R. Paepe (eds.) Perspectives in Astrobiology, pp. 86–87 (2005)
Pace, N.R.: The universal nature of biochemistry. Proc. Natl. Acad. Sci. **98**(3), 805–808 (2001)
Papageorgiou, G., Govindjee, G. (eds.): Chlorophyll a Fluorescence: A Signature of Photosynthesis. Springer, Berlin (2005)
Patai, S., Z., R.: The Chemistry of Organic Silicon Compounds. Wiley, Chichester, UK (1989)

Pedani, M.: Light pollution at the Roque de los Muchachos Observatory. New Astron. **9**, 641–650 (2004)

Pedani, M.: An updated view of the light pollution at the Roque de los Muchachos Observatory. ING Newlett. **9**, 28–31 (2005)

Pickles, A.J.: A Stellar Spectral Flux Library: 1150 - 25000 A (Pickles 1998). VizieR Online Data Catalog **611**, 863 (1998)

Plaxco, K.W., Gross, M.: Astrobiology: A brief introduction. Johns Hopkins University Press, Baltimore (2006)

Prigogine, I., Stengers, I.: Order out of chaos. Bantam Books, New York (1984)

Raven, J.: Astrobiology: Photosynthesis in watercolours. Nature **448**, 418 (2007)

Raven, J.A., Cockell, C.: Influence on Photosynthesis of Starlight, Moonlight, Planetlight, and Light Pollution (Reflections on Photosynthetically Active Radiation in the Universe). Astrobiology **6**, 668–675 (2006)

Reed, B.C., Bradley: Trend analysis of time series phenology of North America derived from satellite data. GISci. Rem. Sens. **43**, 24–38 (2006)

Ross, M.: Search Strategy for Detection of SETI Short Pulse Laser Signals. In: G. Lemarchand, K. Meech (eds.) Bioastronomy 99: A New Era in Bioastronomy, Astronomical Society of the Pacific Conference Series, vol. 213, pp. 541–544 (2000)

Sagan, C., Thompson, W.R., Carlson, R., Gurnett, D., Hord, C.: A Search for Life on Earth from the Galileo Spacecraft. Nature **365**, 715–718 (1993)

Santos, N.C., Israelian, G., Mayor, M., Rebolo, R., Udry, S.: Statistical properties of exoplanets. II. Metallicity, orbital parameters, and space velocities. Astron. Astrophys. **398**, 363–376 (2003)

Scalo, J., Kaltenegger, L., Segura, A.G., Fridlund, M., Ribas, I., Kulikov, Y.N., Grenfell, J.L., Rauer, H., Odert, P., Leitzinger, M., Selsis, F., Khodachenko, M.L., Eiroa, C., Kasting, J., Lammer, H.: M Stars as Targets for Terrestrial Exoplanet Searches And Biosignature Detection. Astrobiology **7**, 85–166 (2007)

Scalo, J., Wheeler, J.C.: Astrophysical and Astrobiological Implications of Gamma-Ray Burst Properties. Astrophys. J. **566**, 723–737 (2002)

Scalo, J.M., Miller, G.E.: Constraints on the evolution of peculiar red giants. II - Masses and space densities. Astrophys. J. **233**, 596–610 (1979)

Schindler, T.L., Kasting, J.F.: Synthetic spectra of simulated terrestrial atmospheres containing possible biomarker gases. Icarus **145**, 262–271 (2000)

Schneider, E.D., Sagan, D.: Into the Cool: Energy Flow, Thermodynamics, and Life. University of Chicago Press, Chicago (2006)

Schopf, J.W.: Cradle of Life. Princeton University Press, Princeton (1999)

Schroedinger, E.: What is life? The physical aspect of the living cell. Based on Lectures delivered under the auspices of the Institute at Trinity College, Dublin, in February 1943. University Press, Cambridge (1944)

Schulze-Makuch, D., Irwin, L.N.: Life in the Universe. Springer, Berlin (2004)

Seager, S., Turner, E.L., Schafer, J., Ford, E.B.: Vegetation's Red Edge: A Possible Spectroscopic Biosignature of Extraterrestrial Plants. Astrobiology **5**, 372–390 (2005)

Segura, A., Kasting, J.F., Meadows, V., Cohen, M., Scalo, J., Crisp, D., Butler, R.A.H., Tinetti, G.: Biosignatures from Earth-Like Planets Around M Dwarfs. Astrobiology **5**, 706–725 (2005)

Segura, A., Krelove, K., Kasting, J.F., Sommerlatt, D., Meadows, V., Crisp, D., Cohen, M., Mlawer, E.: Ozone Concentrations and Ultraviolet Fluxes on Earth-Like Planets Around Other Stars. Astrobiology **3**, 689–708 (2003)

Segura, A., Meadows, V.S., Kasting, J.F., Crisp, D., Cohen, M.: Abiotic formation of O_2 and O_3 in high-CO_2 terrestrial atmospheres. Astron. Astrophys. **472**, 665–679 (2007)

Selsis, F., Despois, D., Parisot, J.P.: Signature of life on exoplanets: Can Darwin produce false positive detections? Astron. Astrophy. **388**, 985–1003 (2002)

Selsis, F., Paillet, J., Allard, F.: Biomarkers of extrasolar planets and their observability. In: H. Deeg, J.A. Belmonte, A. Aparicio (eds.) Extrasolar Planets, pp. 245–268. Cambridge University Press, New York (2008)

Sertorio, L., Tinetti, G.: Available energy for life on a planet with or without stellar radiation. Nuovo Cimento C Geophys. Space Phys. C **24**, 421–443 (2001)

Shklovsky, J.S., Sagan, C.: Intelligent life in the universe. San Francisco: Holden-Day (1966)

Singh, H.B., Kanakidou, M.: An investigation of the atmospheric sources and sinks of methyl bromide. Geophys. Res. Lett. **20**, 133–139 (1993)

Sinha, R.P., Häder, D.P.: UV-induced DNA damage and repair: a review. Photochemical and Photobiological Sciences **1**, 225–236 (2002)

Sleep, N.H., Bird, D.K.: Evolutionary ecology during the rise of dioxygen in the Earth's atmosphere. Phil. Trans. Roy. Soc. Lond. B Biol. Sci. **363**, 2651–2664 (2008)

Smith, D.S., Scalo, J., Wheeler, J.C.: Importance of Biologically Active Aurora-like Ultraviolet Emission: Stochastic Irradiation of Earth and Mars by Flares and Explosions. Orig. Life Evol. Biosph. **34**, 513–532 (2004)

Sparks, W.B., DasSarma, S., Reid, I.N.: Evolutionary Competition Between Primitive Photosynthetic Systems: Existence of an early purple Earth? Bull. Am. Astron. Soc. **38**, 901 (2006)

Stein, L.Y., Yung, Y.L.: Production, isotopic composition, and atmospheric fate of biologically produces nitrous oxide. Annu. Rev. Earth Planet. Sci. **31**, 329–356 (2003)

Sullivan, W.T.: Eavesdropping Mode and Radio Leakage from Earth. In: J. Billingham (ed.) Life in the Universe, pp. 377–390. NASA CP-2156 (1981)

Sullivan III, W.T., Brown, S., Wetherill, C.: Eavesdropping - The radio signature of the earth. Science **199**, 377–388 (1978)

Tarter, J.C., Backus, P.R., Mancinelli, R.L., Aurnou, J.M., Backman, D.E., Basri, G.S., Boss, A.P., Clarke, A., Deming, D., Doyle, L.R., Feigelson, E.D., Freund, F., Grinspoon, D.H., Haberle, R.M., Hauck II, S.A., Heath, M.J., Henry, T.J., Hollingsworth, J.L., Joshi, M.M., Kilston, S., Liu, M.C., Meikle, E., Reid, I.N., Rothschild, L.J., Scalo, J., Segura, A., Tang, C.M., Tiedje, J.M., Turnbull, M.C., Walkowicz, L.M., Weber, A.L., Young, R.E.: A Reappraisal of The Habitability of Planets around M Dwarf Stars. Astrobiology **7**, 30–65 (2007)

Thaler, T.L., Gibbs, P.R., Trebino, R.P., Bommarius, A.S.: Search for Extraterrestrial Life Using Chiral Molecules: Mandelate Racemase as a Test Case. Astrobiology **6**, 901–910 (2006)

Thiemann, W.H., Rosenbauer, H., Meierhenrich, U.J.: Conception of the 'Chirality-Experiment' on Esa's mission ROSETTA to comet 46p/Wirtanen. Adv. Space Res. **27**, 323–328 (2001)

Thomson, W.: On the Dynamical Theory of Heat, with numerical results deduced from Mr Joules equivalent of a Thermal Unit, and M. Regnaults Observations on Steam. Trans. Roy. Soc. Edinb. p. 174 (1851)

Tikhov, G.A.: Astrobotany. Alma Ata, Kazakhstan: Kazakhstan SSR Academy of Sciences Press (*in Russian*) (1949)

Tinetti, G., Meadows, V.S., Crisp, D., Fong, W., Fishbein, E., Turnbull, M., Bibring, J.P.: Detectability of Planetary Characteristics in Disk-Averaged Spectra. I: The Earth Model. Astrobiology **6**, 34–47 (2006)

Tinetti, G., Meadows, V.S., Crisp, D., Kiang, N.Y., Kahn, B.H., Fishbein, E., Velusamy, T., Turnbull, M.: Detectability of Planetary Characteristics in Disk-Averaged Spectra II: Synthetic Spectra and Light-Curves of Earth. Astrobiology **6**, 881–900 (2006)

Ulmschneider, P.: Intelligent Life in the Universe. Springer, Berlin (2002)

Vázquez, M.: Habitability of Planetary Systems. In: M. Vázquez (ed.) Fundaments and Challenges in Astrobiology, pp. 171–212. Research SignPost, India (2005)

Vázquez, M., Hanslmeier, A.: Ultraviolet Radiation in the Solar System. Springer, Berlin (2005)

Walker, J.C.: Evolution of the atmosphere. Macmillan, New York (1977)

Whewell, W.: Of a Plurality of Worlds. Sheldon, Blakeman Co., New York (1858)

Whitesides, G.M., Grzybowski, B.: Self-assembly at all scales. Science **295**, 5564 (2002)

Woolf, N.J., Smith, P.S., Traub, W.A., Jucks, K.W.: The Spectrum of Earthshine: A Pale Blue Dot Observed from the Ground. Astrophys. J. **574**, 430–433 (2002)

Yokouchi, Y., Ikeda, M., Inuzuka, Y., T., Y.: Strong emission of methyl chloride from tropical plants. Nature **416**, 163–165 (2002)

Chapter 6
Detecting Extrasolar Earth-like Planets

6.1 First Attempts to Discover Exoplanets

The first documented search for extrasolar planets was undertaken by the Dutch astronomer Peter van de Kamp (1901–1995) in the mid-twentieth century. He reported the discovery of an extrasolar planet around Barnard's Star (Van de Kamp 1963, 1971, 1975). This dim star, named after its discoverer, noted astronomer Edward E. Barnard (1857–1923), is the second closest to Sol after Alpha Centauri 3 and has the largest proper motion of all known stars (10.3 arcsec per year). It is located about 6 ly away in the northernmost part of the Constellation Ophiuchus, the Serpent Holder.

Van de Kamp (1982) studied photographic plates charting the proper motion of Barnard's Star and found a slight wobble in the star's path. He concluded that the wobble was caused by a planet of around 1.6 times the mass of Jupiter (M_J) in an eccentric orbit. Over the next couple of decades, Van de Kamp refined his calculations so that, by 1982, he asserted that there were, in fact, two planets in circular orbits around Barnard's Star with masses of 0.7 and 0.5 that of Jupiter. However, further studies attempting to verify Van de Kamp's work found only that there was either no wobble at all or a wobble caused by discrepancies in the methods used to produce the photographic plates (Gatewood and Eichhorn 1973). More recent observations with the Hubble Space Telescope have also failed to yield supporting evidence for a large Jupiter or brown dwarf-sized object around Barnard's star (Benedict et al. 1999), and high-precision radial velocity measurements (Kurster et al. 2003) set even more stringent upper mass limits on any large planets in orbit around this small star. Peter Van de Kamp died in 1995 still convinced of his findings.

In the late 1980s and 1990s, G.A. Walker and co-authors carried out the first attempts to discover exoplanets by using radial velocity measurements. They tried to detect the exoplanets' Doppler shifts by using laboratory lines as a reference (Campbell et al. 1988; Walker et al. 1992; Walker et al. 1995). Their results were negative, but it was probably due to their small sampling size and the fact that they did not consider variation at time scales smaller than 40 days.

It is not clear when the first of the known exoplanets was detected. In 1989, Latham et al. (1989) reported spectroscopic evidence for a probable brown dwarf companion to the solar-type star HD114762. This star presents periodic variations in radial velocity, which were attributed to the presence of an unseen companion. There are, however, some concerns about whether this companion is an exoplanet or it is, in fact, a brown dwarf. In spite of the fact that its minimum mass ($M\sin(i) = 11 M_J$) is smaller than the $13 M_J$ limit for brown dwarfs, there are indications that its inclination is low (i much less than $90°$; Hale 1995), so that its mass is finally likely to be above $13 M_J$ (Mazeh et al. 1996).

The first widely accepted discovery of an extrasolar planet was made by Wolszczan and Frail (1992). A terrestrial-sized object and an even smaller planet orbiting a pulsar star were detected by measuring the periodic variation in the pulse arrival time. Despite the close resemblance in terms of mass to the rocky planets in our own solar system, these are exotic objects, which are difficult to catalogue. A pulsar is a neutron star, the remnant of a massive star that exploded as a supernova. In the supernova event, any planetary system that the star might have possessed was necessarily destroyed, and any planetary objects orbiting the pulsar were probably formed afterwards (Lin et al. 1991). The formation mechanism of such late planets are presently unresolved, but some authors consider the detection of pulsar planets as encouraging in the sense that it might indicate that planet formation is probably a common rather than a rare phenomena.

There is not much doubt, however, that the discovery of an exoplanet around the star 51 Pegasi marked a milestone in exoplanet studies. 51 Peg b (Mayor and Queloz 1995) was the first exoplanet discovered around a main sequence star (a G5V star with a mass of $1.06\,M_{sun}$). The star has an apparent magnitude of 5.49, which makes it easily observable from Earth, with binoculars or even with the naked eye under very clear conditions. The planet has a mass similar to that of Jupiter and with an orbit closer to the parent star than Mercury is to the Sun. This discovery immediately challenged all the preconceptions scientists were harbouring about extrasolar planets as well as planetary formation theories. 51 Pegb was the first of a new class of exoplanets, known as 'Hot Jupiters', because of their large mass and proximity to the star.

The discovery of exoplanets was not entirely accepted by the scientific community without reticences (Gray 1997). However, the planetary nature of these new class of objects was soon confirmed by Charbonneau et al. (2000). They reported the discovery of a transit (a dimming in the star's brightness due to a planet passing in front) in the photometric light curve of the star HD 209458, which was previously known to have a planetary companion from radial velocity measurements. From the measurements, they concluded that the star's companion was indeed a gas giant with a radius of $1.27\,R_J$. Today, hundreds of exoplanets are being discovered and catalogued (Fig. 6.1), and the number continues to grow at an increasingly fast speed.

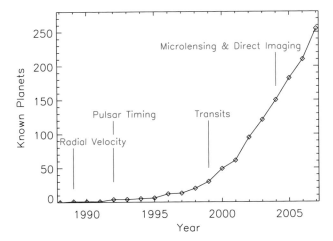

Fig. 6.1 Number of confirmed extrasolar planets discovered against time. Also marked are the years in which an extrasolar planet was found for the first time using different techniques. Data for the plot was obtained from the Extrasolar Planets Encyclopaedia (http://exoplanet.eu)

6.2 The Mass Limit: From Brown Dwarfs to Giant Planets

There have been numerous theoretical and observational studies dedicated to understanding the existence and nature of objects in the mass range between a star and a planet (Fig. 6.2). More than 40 years ago, Kumar (1963) and Hayashi and Nakano (1963) proposed that, under a certain mass limit, the thermonuclear fusion of the lighter hydrogen isotope would not occur, and degeneracy pressure would compensate the gravitational collapse and maintain the hydrostatic equilibrium. These kinds of objects were baptized by Tarter (1976) with the name of brown dwarfs.[1] Brown dwarfs, like extrasolar giant planets, whose masses are not sufficient to maintain stable deuterium combustion, are substellar objects. Recently, infrared observations with the Spitzer space telescope (Fig. 6.3) have allowed the direct imaging of two brown dwarfs orbiting stars (Luhman et al. 2007).

The first brown dwarfs (Rebolo et al. 1995; Nakajima et al. 1995) were discovered in the same year as the first extrasolar planet was discovered by radial velocity searches (Mayor and Queloz 1995). Teide 1 (Rebolo et al. 1995), a 55 M_J and 110-Ma-old brown dwarf in the Pleiades, was the first to be discovered. The presence of Lithium in its spectrum is a strong indicator that these objects are not low-mass stars (Magazzu et al. 1993; Rebolo et al. 1996). Nowadays, more than 500 substellar objects are known to be brown dwarf candidates. The fact that brown dwarfs are self-luminous and that they are not, in general, gravitationally tied to a more luminous star has increased their chances of detection as compared to extrasolar planets. Giant planets are also self-luminous, but the origin of their light is not nuclear fusion. But where is the limit between a planet and a star?

---

[1] Originally these objects were called black dwarfs (Jameson et al. 1983).

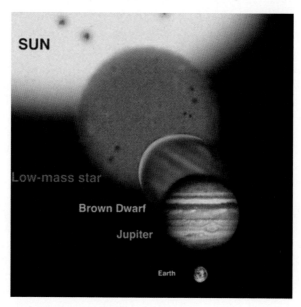

Fig. 6.2 A comparative drawing illustrating the relative sizes of the Sun, a low-mass M star, a brown dwarf, Jupiter and the Earth. Credit: Gemini Observatory, Artwork by J. Lomberg

Fig. 6.3 Two T brown dwarfs orbiting the stars HD 3651 (*left*) and HN Peg (*right*). Credit: NASA/JPL-Caltech/K. Luhman (Penn State University) and B. Patten (Harvard-Smithsonian)

The boundary between brown dwarfs and stars is defined by the hydrogen burning mass limit. For solar metallicity, the upper mass limit of substellar objects is theoretically established at 0.072 M_S (or about 75 M_J). For objects less massive than this limit, the internal temperature and pressure are not sufficiently intense to maintain nuclear processes converting hydrogen into helium (Chabrier and Baraffe 1997 and 2000).

6.2 The Mass Limit: From Brown Dwarfs to Giant Planets

The lower limit of the brown dwarf domain, however, is not so well-established and is subject to dispute. Several terms have been used to name planetary mass objects gravitationally isolated from any star that are observed in stellar clusters: free-floating planets, isolated planetary-mass objects (IPMOs), planemos, cluster planets or sub-brown dwarfs (Lucas and Roche 2000; Zapatero-Osorio et al. 2000; Boss 2001; Basri 2003). On the other hand, planetary mass objects detected with radial velocity, transits or microlensing, are named extrasolar giant planets. But it is not clear if these are two separate kinds of objects, different in their formation and evolution, or if these are different names for the same thing. From mass determinations, there is a clear overlap between brown dwarfs and exoplanets.

Some authors would like to define brown dwarf and IPMOs boundaries based on their different formation mechanism. This way brown dwarfs would be formed in a pseudo-stellar manner, through the fragmentation and collapse of cool molecular clouds, while exoplanets would form in stellar disks.

In practice, however, these criteria are difficult to implement, because we are still far from a deep understanding of the formation mechanisms of substellar objects, both planets and brown dwarfs (Pickett et al. 2000). Too many questions remain unanswered. For example, we do not know what is the minimum mass for the formation of very low mass objects in isolation, which would represent the bottom end of the Initial Mass Function (IMF). Photometric and spectroscopic searches suggest that the IMF extends further below the deuterium burning mass threshold at around 13 M_J. This is usually referred to as the 'planetary-mass' domain.

The least massive objects (Fig. 6.4) identified so far in young stellar clusters of Orion have masses around 3–8 M_J (Zapatero-Osorio et al. 2000; Lucas and Roche 2000). These objects are less massive than the minimum Jeans mass of 7–10 M_J and thus prompts us to refine the collapse-and-fragmentation models and/or to rethink possible formation mechanisms for such low-mass objects (see the review by Whitworth et al. (2007). Recently, several formation scenarios have been suggested, which include tidal interactions and ejection of low-mass objects from multiple systems before brown dwarfs and planetary-mass objects can accrete enough gas to become stars (Reipurth and Clarke 2001).

Other models suggest that brown dwarfs are formed in the same way as more massive hydrogen burning stars, that is, by the process of supersonic turbulent fragmentation (Padoan and Norlund 2005). In fact, some of the objects of the σ Orionis and similar stellar clusters are found to possess debris disks (see Fig. 6.5), which further supports the point that their formation mechanism might not be so different from low mass stars (Barrado y Navascués et al. 2001; Zapatero-Osorio et al. 2007). In fact, it is not a negligible possibility that some of these isolated planetary-mass objects might possess planets on their own.

A different criteria to classify substellar objects is through nuclear physics. The frontier between brown dwarfs and exoplanets is then limited by the minimum mass necessary for an auto-gravitating body to maintain the deuterium fusion reactions in its interior. For solar metallicity, the limiting mass is established at 13 M_J (4,100 M_E) (Saumon et al. 1996; Chabrier and Baraffe 2000).

Fig. 6.4 An image of the σOri region. Associated to this cluster, Zapatero-Osorio et al. (2002) discovered a free-floating methane dwarf, SOri70, which is likely the least massive planetary-mass body imaged to date outside the solar system. The multiple star, which is visible with the naked eye, is at the centre. The small white box indicates the position of the planet candidate, which is only 8.7 arcmin from the star and has a mass of about 3 M_J. The image was taken from the Digital Sky Survey and has a size of 23 × 22 square arcmin. The image inset shows the infrared image obtained using NICS at the Keck I telescope (Hawaii). Image credit: M. Zapatero-Osorio (IAC)

Substellar objects are not in thermodynamic equilibrium and are not capable of stabilizing their temperature. Thus, as they evolve, they cool and shrink. In Fig. 6.6, the evolution of the luminosity of very low mass stars, brown dwarfs and exoplanets is shown (Burrows et al. 1997). Already at an age of a few thousand million years, they become faint objects with extremely low luminosity and effective temperatures. Thus, it is very likely that the vast majority of brown dwarfs and isolated exoplanets in the solar neighbourhood remain undetected.

Because of their extremely low luminosity, the classical spectral series (O, B, A, F, G, K, M) has been extended to include the spectral types L (with temperatures in the range 1,300–2,000 K and metal hydrides and alkali metals in the spectra) and T (with temperatures between 700 and 1,300 K and showing methane in the spectra) (Martin et al. 1999; Kirkpatrick et al. 1999; Kirkpatrick 2005; Burgasser et al. 2002; Geballe et al. 2002). The most massive brown dwarfs start their lives as late spectral types M and later evolve to spectral types L and T. Young extrasolar planets exhibit spectral types L during the first few million years of their evolution.

6.2 The Mass Limit: From Brown Dwarfs to Giant Planets

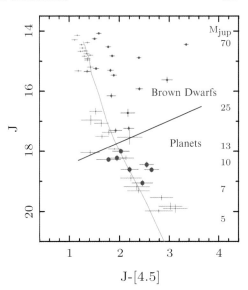

Fig. 6.5 Colour-magnitude diagram of low-mass σ Orionis member candidates. Objects with flux excesses at 5.8 and 8.0 μm (indicative of circumstellar disks) are plotted as large *filled circles*. Masses in Jovian units are given on the right hand side. Overplotted with a *solid line* is the 3-Myr isochrone from Baraffe et al. (2002). In blue is a line dividing the mass ranges between planetary mass objects and brown dwarfs. Adapted from Zapatero-Osorio et al. (2007)

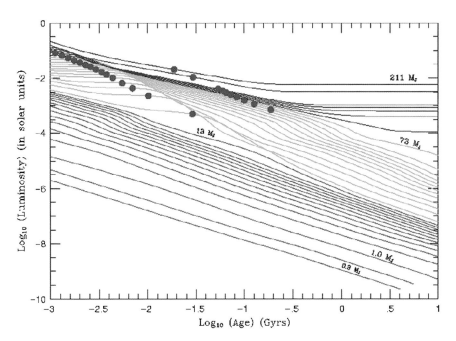

Fig. 6.6 Evolution of the luminosity of solar-metallicity M dwarfs and substellar objects vs. time after formation. The stars, brown dwarfs and planets are shown in *blue*, *green* and *red*, respectively. In this figure, brown dwarfs are designated as those objects that burn deuterium, while those that do not are designated as planets. The masses (in solar mass) label most of the curves, with the lowest three corresponding to the mass of Saturn, half the mass of Jupiter, and the mass of Jupiter. Adapted from Burrows et al. (1997). Reproduced by permission of the American Astronomical Society

Presently, there are dozens of planetary mass objects detected in isolation (Lucas and Roche 2000; Caballero et al. 2007) or orbiting brown dwarfs (Chauvin et al. 2004). Although the differences in physical and chemical properties between these objects and the giant planets tied to stars might be minimal, in the following sections we focus on the detection methodologies for conventional extrasolar planets, that is those planets orbiting around a star.

6.2.1 The Brown Dwarf Desert

According to the current formation models, vast numbers of brown dwarfs are thought to drift, like vagabonds, across interstellar space (Padoan and Norlund 2005). Observational studies of the initial mass functions (IMF) also seem to indicate a large number of expected brown dwarfs (Bouvier et al. 1998; Luhman et al. 2000; Bejar et al. 2001). However, according to radial velocity studies (Halbwachs et al. 2000), very few of the closest Sun-like stars have brown dwarf companions inside a radius of 5 AU. This under-population of brown dwarf companions has become known as the *brown dwarf desert*. Around low-mass stars, however, Close et al. (2002) found that there is likely no brown dwarf desert at wide separations.

If they are there, future space missions, such as Gaia, will be able to detect tens of thousands of brown dwarfs, drifting through both space and in orbit around other stars. These data will allow us to refine our theories and formation models.

6.3 The Detection of Earth-like Planets: A Complex Problem

The detection of extrasolar planets presents new challenges to modern astrophysics. Planets are faint and are located very close to a bright object, the parent star. Thus, their detection is extremely demanding.

6.3.1 Brightness Ratio

Perhaps the major challenge for the detection of exoplanets is their relatively low brightness compared to that of their parent star. In the visible range, planets do not shine but they do reflect some of the incident light. However, the starlight reflected by the planet is literally lost amidst the light from the parent star.

A planet orbiting around a star emitting a flux F_* acquires by reflection a brightness given by F_{rfl}. The ratio of both fluxes is given by

$$\frac{F_{rfl}}{F_*} = \frac{a_p}{4} \left(\frac{R_p}{D}\right)^2 \phi(t),$$

Fig. 6.7 A comparison of the emitted and reflected flux of the Earth (*black line*) to that emitted by the Sun (*red line*). In the visible, the brightness ratio between the Sun and the Earth is larger than in the infrared where there is a strong black body emission component from the Earth

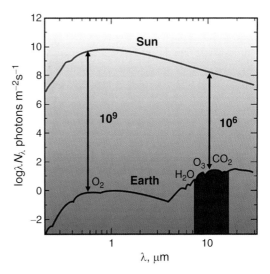

where a_p is the planetary albedo, D is the distance between the planet and the star, R_p is the radius of the planet and $\phi(t)$ is the orbital phase factor given by

$$\phi(t) = 1 - (\sin i)\left(\sin\left(\frac{2\pi t}{P}\right)\right),$$

where i is the inclination of the planetary orbit with respect to the sky plane and P the period of the planet around the star.

In the Earth's case, the light reflected from the planet's atmosphere and surface is 10^{-10} times fainter than the one emitted from the Sun (see Fig. 6.7).

But the visible region of the spectra is, arguably, not the best suited for exoplanet searches. In the infrared range, where thermal emission from the planet becomes significant and the star's emitted radiation is dimmer, the contrast is smaller. The ratio between the thermal flux of the planet (centred in the IR range), F_{th}, and the stellar flux is given by the expression

$$\frac{F_{th}}{F_*} = \frac{R_p^2}{2a},$$

where a is the semi-major axis of the planet's orbit.

Figure 6.8 illustrates the advantage of observing in the infrared owing to the better contrast with respect to the visible. Still the Earth is a million times (10^{-6}) fainter than the Sun.

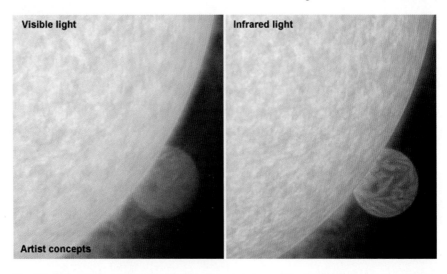

Fig. 6.8 This artist's concept shows what a fiery hot star and its close-knit planetary companion might look like close up in visible light and infrared. In infrared, the star is less blinding, and its planet perks up with its own glow. Image courtesy of NASA

6.3.2 Angular Distance

The second problem faced by extrasolar planet observers is that this faint source of light appears to be very close to the parent star when observed from astronomical distances.

The angular distance, θ, between two astronomical objects is usually expressed in arcseconds (a unit of angular distance equal to a 60th of a degree), and it is a function of the orbital distance between the two objects divided by the stellar distance to the observer. Following Fig. 6.9, where the distances and angles involved between the observer, the planet and its parent star are shown, the angular separation between the planet and the star is given by

$$\theta = \frac{a}{d},$$

where a is the star–planet separation and d is the distance of that system to Earth. Usually, $d \gg a$ so that the apparent angular separation θ is a very small number.

These really small angular separations are very difficult to resolve with current observing techniques and instrumentation. For an illustration, Fig. 6.10 represents two images of the brown dwarf Gliese 229B, obtained with an adaptive optics coronagraph system at the Palomar Observatory 60-in. telescope and with the Hubble Space Telescope. The brown dwarf is 7 arcsec to the lower right of the companion star, Gliese 229. The improvement in the Hubble imaging is evident and represented at that moment the state of the art in angular resolution. Currently, some

6.4 Methods for the Detection of Exoplanets

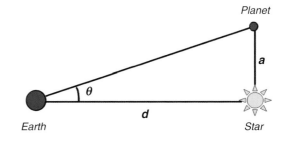

Fig. 6.9 Distances and angles involved in observing an extrasolar planet

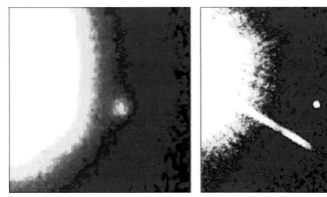

Fig. 6.10 Image of the brown dwarf Gliese 229B, obtained with an adaptive optics system at the Palomar Observatory 60-inch telescope (*left*) and with the Hubble Space Telescope (*right*). The brown dwarf is 7 arcsec to the lower right of the companion star, Gliese 229 (courtesy of Tadashi Nakajima)

ground-based instrumentation such as NAOS/CONICA (NACO) in the VLT telescope or FASTCAM in the WHT offer slightly better resolution in the infrared and optical ranges, respectively.

The star/brown dwarf brightness ratio is ∼5,000, and the distance between the two objects corresponds roughly to the Sun–Pluto separation. However, a Jupiter-mass planet seen from a distance of 10 pc would be 14 times closer to its parent star and roughly 200,000 times dimmer than Gliese 229B. In the case of the Earth, these numbers would be even smaller.

6.4 Methods for the Detection of Exoplanets

Theoretically, there is a large number of ways with which one could search for extrasolar planets. Perryman (2000) made an extensive review of the used and proposed methodologies which are synthesized in Fig. 6.11. To date, these exoplanet detection techniques have not been preceded by major theoretical advances, but rather by a strong development of well-known observational methodologies to their capability

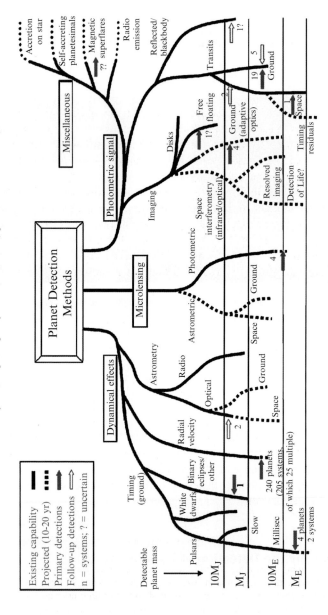

Fig. 6.11 Detection methods for extrasolar planets. The lower extent of the lines indicates, roughly, the detectable masses that are in principle within the reach of present measurements (*solid lines*), and those that might be expected within the next 10–20 years (*dashed*). The (logarithmic) mass scale is shown at left. The miscellaneous signatures to the upper right are less well quantified in mass terms. *Solid arrows* indicate (original) detections according to approximate mass, while *open arrows* indicate further measurements of previously detected systems. The "?" indicate uncertain or unconfirmed detections. The figure takes no account of the number of planets that may be detectable by each method. Courtesy of M. Perryman.

6.4.1 Indirect Detection of Exoplanets

6.4.1.1 Astrometry

Astrometry is an astronomical technique for recording the position of the celestial objects in the sky. In the search for extrasolar planets, astrometry is used to look for the periodic wobble that a planet induces in the position of its parent star. If the system is oriented face-on and the orbiting planet is massive enough, this small motion of the star can be detected by astrometry.

The effect is illustrated in Fig. 6.12. As the planet moves through its orbit (red dots), the star revolves around the system's centre of mass, called the barycentre (the black cross). For planetary systems similar to ours, the star's mass is so large compared to the planet's that the barycentre will most likely lie within the star itself, and the path in which it moves (blue dots) will be a much smaller circle. Astrometry, therefore, relies on very precise measurements of a star's position with time.

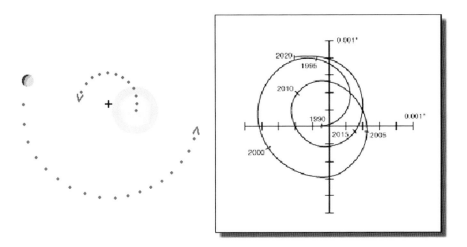

Fig. 6.12 *Left*: A drawing illustrating the movement induced on the star by an orbiting planet. *Right*: An image showing the Sun's motion (wobble) around the centre of mass due to Jupiter, over the course of 30 years. The scale of the motion is shown in angular units as if seen from 10 parsecs (32.6 light years) away. The end of each axis is 0.001″ or 0.001 arcsec (1 arcsec = 1/3,600 of a degree). This motion is too small to be detected for stars at present (even the Hubble Space Telescope only sees stars with a size of about 0.1 arcsec)

The astrometric signal, δ, of a planet with mass m_{pl} orbiting a star with mass m_* at a distance d in a circular orbit of period P is given by

$$\delta = 3\frac{m_p}{m_E}\left(\frac{m_*}{m_{Sun}}\right)^{-2/3}\left(\frac{P}{yr}\right)^{2/3}\left(\frac{d}{pc}\right)^{-1},$$

where δ is expressed in microarcseconds. The minimum detectable planet mass gets smaller in inverse proportion to the planet's distance from the star. Thus, using astrometry techniques, there is enough information to solve the orbital elements without the $\sin i$ ambiguity (Perryman and Hainaut 2005). This is very useful for mass determinations of planets detected by other methods, such as radial velocity. Astrometry is also applicable to all types of stars and more sensitive to planets with large orbital semi-major axes, but in this latter case it requires long periods of observation.

Astrometric quests for extrasolar planets started decades ago (Gatewood 1976; Gatewood et al. 1980), but so far no planets have been detected by this method. The real astrometric movements of the Sun due to Jupiter are plotted in Fig. 6.12. To detect an Earth-like planet orbiting a solar-like star with a period of 1 year and observed at 5 pc of distance, we would need a resolution of 0.6 µas, far beyond our present capabilities (Sozzetti 2005), and it would require sending instrumentation to space.

But efforts to reach the necessary precision continue. From the ground, the Keck and VLTI telescopes are being equipped to measure angles as small as 10–20 µas, leading to a minimum detectable mass in a 1 AU orbit of 66 M_E for a solar-mass star at 10 pc. ESA's next astrometric mission, Gaia, is designed to be the most precise astrometric observatory to date and will survey a thousand million stars after its launch, scheduled for 2011. Among other scientific goals, Gaia is expected to find between 10,000 and 50,000 gas giant planets beyond our Solar System by means of astrometry (Lindegren and Perryman 1996).

6.4.1.2 Radial Velocity

We have seen how a planet orbiting around its parent star exerts a pull that causes the star to 'wobble' around the centre of mass between the two objects. If the system is oriented nearly edge-on to the Earth, when the star moves towards Earth, the wavelengths of the spectral lines in the light it emits move towards the blue end of the spectrum. When the star travels away from Earth, the opposite happens, and the wavelengths move towards the red part of the spectrum. For this reason, radial velocity measurements are also known as Doppler spectroscopy.

A telescope outfitted with a precise spectrometer can measure these small shifts in the spectrum. The periodic variation of the star's radial velocity, ΔV, induced by the presence of a planet, can be deduced from the application of Kepler's laws. We have

$$\frac{(m_p \sin i)^3}{(m_* + m_p)} = \frac{P}{2\pi G} \Delta V^3 (1-e^2)^{3/2},$$

where m_p and m_* are the masses of the star and the planet, respectively, e is the orbital eccentricity and P is the orbital period. If $m_{pl} \ll m_*$, we have

$$m_p \sin i = \left(\frac{P}{2\pi G}\right)^{1/3} \Delta V (1-e^2) \, m_*^{2/3}$$

expressing the planetary mass in Jupiter masses, ΔV in m s^{-1}, and the period P in years, we have,

$$m_p \sin(i) = 3.5 \times 10^{-2} \Delta V \, P.$$

The vast majority of the exoplanets discovered so far were found using radial velocity measurements, including the so-called *Hot Jupiters* (Mayor and Queloz 1995; Butler et al. 1996; Butler et al. 1997; Cochran et al. 1997) and *Super–Earths* (Mayor et al. 2008; see Chap. 7 for details). Because the motion of the star caused by orbiting planets is so small, radial velocity measurements are best suited for finding large planets in tight orbits. The minimum detectable planet mass increases as the square root of the planet's orbital size.

This method can be used for main-sequence stars of spectral types mid-F through M. Stars hotter and more massive than mid-F rotate faster, pulsate, are generally more active and have less spectral structure, thus making it more difficult to measure their Doppler shift.

The technique is further limited because it will never be able to detect small, Earth-sized worlds, around solar type stars. With the best spectroscopes, astronomers can confidently detect motions of about 1 m s^{-1}, although precision of the order of tenths of centimetre per second are becoming more and more frequent (Mayor et al. 2008). However, Earth forces only the Sun to move at 10 cm s^{-1}. Thus, for an Earth-like planet ($M_p \sim 0.03 \, M_J$) orbiting a solar-like star with a period of 1 year, we need a resolution in velocity lower than the present capabilities, and even if a spectroscope could be made to detect such small velocities, for solar-type stars, the boiling of the star's gaseous surface could mask the effect of the planet (Saar and Donahue 1997; Hatzes 2002). It is possible, however, that this technique could be successfully applied in the future to detect Earth-size planets around M stars, for example.

6.4.1.3 Pulsar Timing

At the end of their lives, stars with masses of around 15–30 times the mass of our Sun explode in a supernova and leave behind a remnant called a neutron star. Less massive stars leave behind white dwarfs, while more massive stars lead to the formation of black holes (Israelian et al. 1999). Sometimes the magnetic field and spin axis of the neutron star are not aligned. These special neutron stars are called pulsars (Gold, 1968). If the geometry is right, as the star spins, a 'beacon' of radio waves,

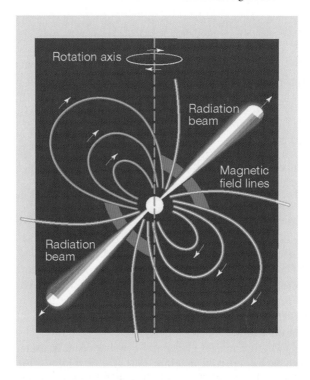

Fig. 6.13 The pulsar's magnetic field forms a magnetic dipole (like a bar magnet) with radio waves emitted from the magnetic poles. The magnetic poles are rarely aligned with the rotation axis so, like in a lighthouse, the radio waves appear as periodic pulses rather than a continuous beam. The gradual slowing down of the pulsar as it loses radiation energy can be used to estimate the pulsar's age (Gaensler and Frail, 2000). Adapted from Seiradakis et al. (2000). Reprinted by permission from Macmillan Publishers Ltd. Nature Vol. 406, p. 139. Copyright (2000)

streaming from the magnetic poles, hit the Earth at regular intervals, usually ranging between 2 ms and a couple of seconds (Fig. 6.13). Pulsars' signals are extremely regular, with a typical rate of change of only a second per ten million years (Davies et al. 1969). Because of this intrinsic regularity, small anomalies in the timing of pulsars can betray the existence of planets orbiting around them. Planets with masses similar to the Earth's or greater can be detected. The very first confirmed extrasolar planets were found using this method (Wolszczan and Frail 1992; Wolszczan 1994), around the pulsar PSR B1257 + 12, located 980 ly from the Sun. Three planets with masses estimated as 0.025, 4.3 and 3.9 M_E orbit the pulsar (Fig. 6.14).

The discovery of planets around pulsars was yet another surprise, unanticipated by theoretical models. At present we have no theory as to exactly how these planets form or their physical properties and composition. However, it is currently thought that any planets found around pulsars are likely to have formed after the supernova event, as any planets in the system before the star became nova would likely have been destroyed.

Fig. 6.14 A comparison of orbits and sizes of the inner solar system planets Mercury, Venus and Earth (*top*), and the three planets detected around the pulsar PSR B1257 + 12 (*bottom*). The Sun, the pulsar and the planets are not drawn to scale

Some authors have proposed that if some of the supernova explosion ejecta fails to escape, it may fall back onto the neutron star or it may possess sufficient angular momentum to form a disk (Heger et al. 2003). In fact, observations of mid-infrared emission from a cool disk around an isolated young X-ray pulsar, with an estimated mass of the order of $10M_E$ and an age of more than a million years, were reported by Wang et al. (2006). The disk resembles proto-planetary disks seen around ordinary young stars, suggesting the possibility of planet formation around young neutron stars.

6.4.1.4 Microlensing Events

When an object, generally a galaxy or a galaxy cluster is located between a bright source, like a quasar, and the observer, its gravitational force bends the light emitted from the bright source. In this way, light from the bright source reaches the observer following several different paths, generating identical images of the source, known as gravitational lenses or mirages (Refsdal 1964).

In the past, this effect was used to study light from dim, distant galaxies as their light is bent around closer galaxies. Recently, however, sensitive equipment has been used to observe lensing events from stars in the hub of our own galaxy. The gravitational microlensing effect is similar, and it occurs when the gravitational field of a planet and its parent star act to magnify the light of a distant background object. For the effect to work, the planet and star must pass almost directly between the observer and the distant star. If this is the case, the light from the background object is bent due to the gravity of the foreground object (star + planet).

This light bending from the microlensing event produces a temporary apparent increase in the luminosity of the background object, which is represented in

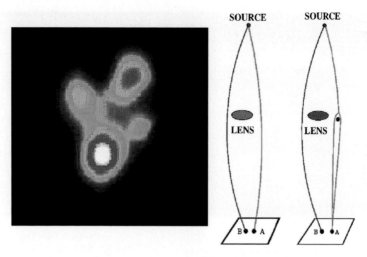

Fig. 6.15 *Left*: Reconstruction of the most famous gravitational lens, the Einstein's Cross or Q2237 + 0305, using bi-dimensional spectroscopy obtained with INTEGRAL at the William Herschel Telescope (Mediavilla et al. 1998). *Right*: Graphical representations of a gravitational lens and a microlensing event. Image credit: Alex Oscoz (IAC)

Fig. 6.15. If the foreground star possesses a planet, a secondary brightness enhancement of the background object occurs due to the planet. In Fig. 6.16, the light curve of a microlensing event clearly betrays the presence of a planet. In fact, the light from the source separates into two components, but separated only by microarcseconds, which cannot be resolved with current instrumentation.

The key advantage of this technique is that it allows low mass (i.e. Earth-like) planets to be detected using available technology. So far, four planets have been discovered by microlensing. Among these, Beaulieu et al. (2006) detected a cool planet with a mass 5.5 times that of the earth, named as OGLE-2005-BLG-290Lb (see Fig. 6.17). More importantly, the detection suggested that such cool, sub-Neptune-mass planets may be more common than gas giant planets, as predicted by the core accretion theory.

Unfortunately, there is also a notable disadvantage to the method: the lensing cannot be repeated because the chance alignment never occurs again. Also, the detected star+planets systems will tend to be several kpc away and follow-up observations of the planet will not be possible. The reason is that one needs to observe very distant stars to improve the chances of alignments.

Despite the repeatability disadvantage, microlensing observation campaigns are not as biased toward large mass exoplanets as radial velocity or transit measurements, as long as the lensing effect is large enough to be detected. Thus, extensive searches via microlensing events are a superb tool to provide useful statistical information about the mass distribution of extrasolar planets and their relative abundance.

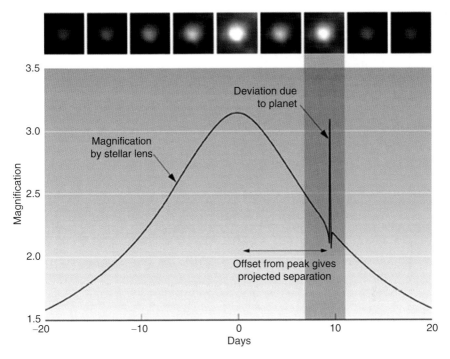

Fig. 6.16 Data from a microlensing event indicate a smooth, symmetric magnification curve as a lens star moves between a source star and an observatory on Earth. The short spike in magnification is caused by a planet orbiting the lens star. Image credit: Lawrence Livermore National Laboratory (PLANET Microlensing Collaboration)

6.4.1.5 Transits

Transit photometry measures the periodic dimming of a star caused by a planet passing in front of it along the line-of-sight of the observer. From Earth, both Mercury and Venus occasionally pass in front of the Sun. When they do, they look like tiny black dots passing across the bright surface. The transit of Mercury in front of the Sun is represented in Fig. 6.18. The figures serve to emphasize how small the rocky planets are compared to their parent star.

Struve (1952) first proposed searching for extrasolar planets using the method of transits. For a transit of a planet across the stellar disk to be observable, the planet must be aligned with the star as seen from Earth with an inclination $i > \theta_T$, where

$$\theta_T = \cos^{-1}[(R_* + R_p)/D]$$

D being the planet-star separation.

The transit of Venus in front of the Sun was observed on 8 June 2004 with the ACRIM 3 instrument on ACRIMSAT (Schneider et al. 2006). During the transit, which lasted ∼5.5 h, the planets angular diameter was approximately 1/32 the

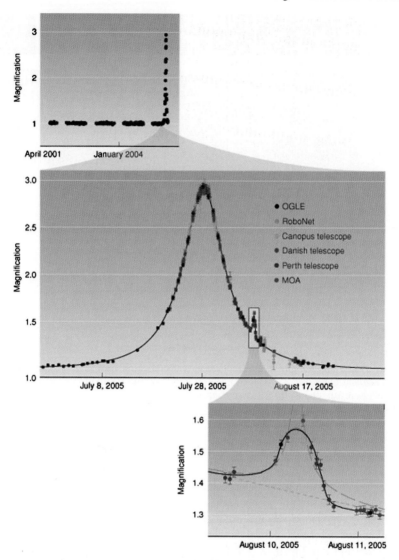

Fig. 6.17 The observed light curve of the OGLE-2005-BLG-290-Lb microlensing event and the best-fit model are plotted as a function of time. The data set consists of 650 data points recorded by the RoboNet, OGLE and MOA collaborations and three of the telescopes used by the PLANET collaboration. The *top panel* shows the OGLE light curve for the previous four years. The *bottom panel* magnifies a 1.5-day interval of the data showing the planetary deviation. Image credit: Lawrence Livermore National Laboratory (PLANET Microlensing Collaboration)

solar diameter, thus covering 0.1% of the stellar surface. A diminution in total solar irradiance of $\sim 1.4\,\mathrm{W\,m^{-2}}$ ($\sim 0.1\%$, closely corresponding to the geometrically occulted area of the photosphere) was measured.

6.4 Methods for the Detection of Exoplanets

Fig. 6.18 Mercury transit observed with Singer H-alpha full disk telescope on 15 November 1999. The image shows a small portion of the full disk image near the solar north pole. The field of view is approximately 470 × 170 arcsec, or 340,000 × 125,000 km, on the Sun. Mercury is clearly discernible as a dark disk with its diameter approximately 10 arcsec. Courtesy: Big Bear Solar Observatory

Several planets, all far larger than the Earth, have already been detected using this technique. The transit light curve of the exoplanet HD209458b, the first to be measured in transit although the planet itself was discovered using radial velocity measurements, is shown in Figs. 6.19 and 6.20.

Giant planets in inner orbits can be detectable, independently of the orbit alignment, based on the periodic modulation of their reflected light. For the 10% of these that have transits, the transit depth can be combined with the mass found from Doppler data to determine the density of the planet as was done, among many others, in the case of HD209458b (Mazeh et al. 2000, and several others). As the orbital inclination must be close to 90° to cause transits, there is very little uncertainty in the mass of any giant planet detected.

Ehrenreich et al. (2005) simulated the transit of different Earth-like planets across G, F and K type stars. From the transit photometry one can derive the ratio of the planet with respect to the parent star (see Fig. 6.21). The shape, depth and duration of the transit also give us important information about the planet (Alonso et al. 2007).

Mercury-size planets can even be detected in the habitable-zone of K and M stars. However, planets with orbital periods greater than 2 years are not readily detectable, since their chances of being properly aligned along the line of sight from the observer to the parent star become very small.

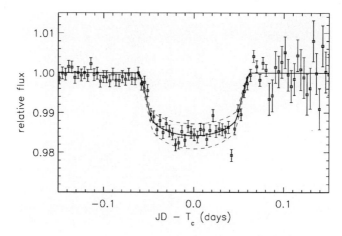

Fig. 6.19 The first detected transit of an extra-solar planet, HD 209458b. The figure shows the measured relative intensity vs. time. Measurement noise increases to the right due to increasing atmospheric air mass. From the detailed shape of the transit, some of the physical characteristics of the planet can be inferred. Adapted from Charbonneau et al. (2000). Reproduced by permission of the American Astronomical Society

Sartoretti and Schneider (1999) computed the detection probability of satellites of extrasolar planets with transit photometry. They conclude that if a satellite is extended enough to produce a detectable drop in the stellar light curve, the probability of detecting it when the planet also transits over the star is nearly one, but the chances decrease if the planet does not transit. The photometric effect of our Moon, observed from an astronomical distance, is only 0.009 mmag (Simon et al. 2007). If a satellite is not extended enough to produce a detectable drop in the stellar light curve, it might still be detected through the time shift of planetary transits resulting from the rotation of the planet around the barycentre of the planet–satellite system. So far, however, no exomoons are known.

At present, extensive exoplanet searches using transits are being undertaken from space. The COROT mission was launched in December 2006, and although data are just becoming available, several new extrasolar planets have already been reported. The KEPLER mission, launched in March 2009, will expand the transit searches from space.

6.4.1.6 Differential Spectro-photometry During Transits

A special observing technique has been developed over the past few years, which is halfway between indirect observations and direct detection. Spectrophotometric observations during primary and secondary eclipses of transiting planets can provide a tool to directly isolate planetary properties. The trick is to make precise observations when the planet passes in front of the star, or before and after the planet is hidden behind the star (secondary eclipse).

6.4 Methods for the Detection of Exoplanets 273

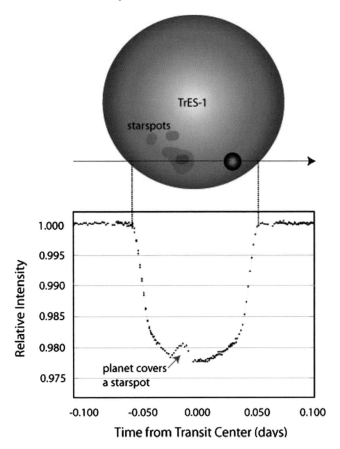

Fig. 6.20 The transit light curve of an exoplanet in front of TrES-1, an 11.8 magnitude star, taken with the STIS spectrograph on the Hubble Space Telescope (Charbonneau et al. 2007). This particular light curve shows a bump during the transit. At that time, the planet covered up a starspot as it traversed the face of the star. Starspots (like sunspots) are cooler, and hence dimmer than their surroundings. When a starspot is occulted by the planet, the fraction of blocked starlight decreases. HST photometry allows a very accurate measurement of the planetary radius, which in turn puts strong constraints on theoretical models of the planetary interior. Image credit: G. Laughlin, http://oklo.org

Based on HST high-precision spectrophotometric observations of four planetary transits of HD 209458b, Charbonneau et al. (2002) detected the absorption from sodium in the planetary atmosphere. Richardson et al. (2007) reported the first extrasolar planet infrared spectra (7.5–13.2 μm) of the transiting HD 209458b. The observations revealed a hot thermal continuum for the planetary spectrum and a broad superposed emission peak centred near 9.65 μm that was attributed to emission by silicate clouds. Harrington et al. (2007) reported the direct detection of thermal emission from the smallest known transiting planet, HD149026b, which indicated a brightness temperature of 2,300 ± 200 K at 8 μm, which make it the

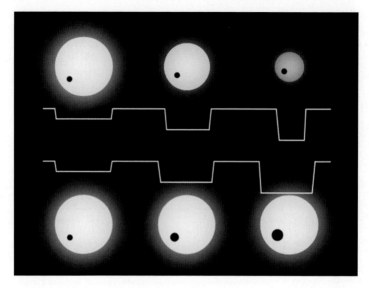

Fig. 6.21 *Top*: The effects of the transit of an exoplanet in front of different star types. *Bottom*: the effects of the transit of different planets above the same star

hottest exoplanet measured so far. Using Spitzer and this same methodology, Tinetti et al. (2007) and Swain et al. (2008) have recently reported the presence of water and methane, respectively, in the atmosphere of the extrasolar planet HD189733b (Fig. 6.22).

It is also possible to make differential measurements of the day and night side of close-in giant planets. The goal is to measure the total light received from the star–planet system as accurately as possible. The light from the star is considered constant with time, and superimposed on top of this constant level is a small fluctuation that is due to the planet. This is because, as the planet orbits the star, first it exposes the bright side (that which faces the host star) and then the night side (that which faces away from the star). These kinds of measurements are possible for every star that harbours a close-in giant planet (Harrington et al. 2006).

6.4.1.7 Miscellaneous Indirect Detection Methods

A number of miscellaneous techniques for detecting the presence of exoplanets have been proposed in the literature, and most of them are included in Fig. 6.11. A planetary mass object, either an exoplanet or a brown dwarf, has recently been detected by timing the eclipses of two eclipsing binaries (Deeg et al. 2008).

Still, the majority of these methods remain unexplored either because of their technical difficulties or because of the low chances of success. Within this category falls the search for radio emissions from exoplanets (Zarka 2007), either natural

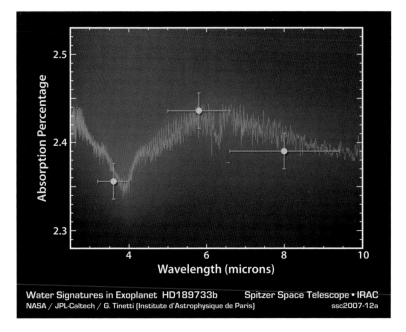

Fig. 6.22 The transmission spectrum of the hot Jupiter HD 189733b. The *solid line* is a synthetic model spectra. The observations are indicated with *yellow dots*

or artificial, or the study of stellar super-flares (Cuntz and Shkolnik 2002). For a detailed summary and review of past, current and planned extrasolar planet detection efforts, visit the Extrasolar Planets Encyclopaedia.

6.4.2 Direct Observations of Exoplanets

The detection of countless extrasolar planets and the determination of some of their large-scale properties, such as mass and radius, is, in itself, a fascinating enterprise. However, these indirect detection methods provide little to no information about other physical characteristics of a planet, such as the composition and structure of its atmosphere or the properties of its surface (if any). To characterize these physical properties of exoplanets, one needs to directly detect planetary radiation.

Eventually, astronomers hope to be able to isolate either the light being reflected by exoplanets or the thermal infrared radiation emanating from the planetary surface/atmosphere itself. These techniques are known as *direct* detection methods, resulting in pictures or spectra from the exoplanet itself (Aime and Vakili 2006; Quirrenbach 2006; Kalas 2007; Beuzit et al. 2007). Direct detections will allow us to analyze the chemical composition and assess the physical state of these distant worlds.

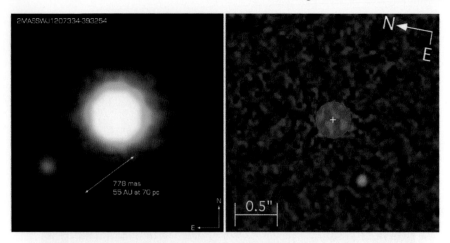

Fig. 6.23 *Left*: False-colour infrared image of the brown dwarf 2MASSWJ1207334-393254 (*blue*) and its planetary companion 2MASSWJ1207334-393254b (*red*), as viewed by the Very Large Telescope. This was the first confirmed extrasolar planet to have been directly imaged. The planet orbits a brown dwarf (*centre-right*) at a distance that is nearly twice as far as Neptune is from the sun. The photo is based on three near-infrared exposures (in the H, K and L' wavebands) with the NACO adaptive-optics facility at the 8.2 m VLT Yepun telescope at the ESO Paranal Observatory. *Right*: Confirmation of the object with infrared HST data by Schneider (2005)

The first extrasolar giant planet detected by direct observations (2M1207-39b), that is resolved with respect to the central object to which it is tied, was announced by Chauvin et al. (2004) and confirmed in 2005 (Chauvin et al. 2005). In this case the primary object is not a star, but a young brown dwarf (Fig. 6.23). So far, according to the Exoplanets Encyclopaedia, four exoplanets have been directly observed, although they are all orbiting far away from their parent star and their masses lie very close to the brown dwarf limit. In the order of (probable) lower mass, these are DHTauB, USCOCTIO108B, GQlupiB, ABPicB and Oph16AB.

In 2008, however, two planetary systems were directly imaged with different instrumentation and almost simultaneously. Kalas et al. (2008) have been studying the star Fomalhaut, located about 25 ly from Earth, for several years. Fomalhaut was known to harbour a belt of cold dust with a structure consistent with gravitational sculpting by an orbiting planet, and Hubble Space Telescope observations separated by 1.73 years revealed counterclockwise orbital motion of this exoplanet candidate, Fomalhaut b. Fomalhaut b lies about 119 astronomical units (AU) from the star and 18 AU from the dust belt, matching predictions of its location, and has an estimated mass of about 3 M_J (Fig. 6.24). High-contrast observations with the Keck and Gemini telescopes have also revealed three planets orbiting the star HR 8799, with projected separations of 24, 38 and 68 AU (Marois et al. 2008). Multi-epoch data show counter clockwise orbital motion for all three imaged planets, with masses between 5 and 13 M_J. It is notable how this system resembles a scaled-up version of the outer portion of our solar system.

6.4 Methods for the Detection of Exoplanets

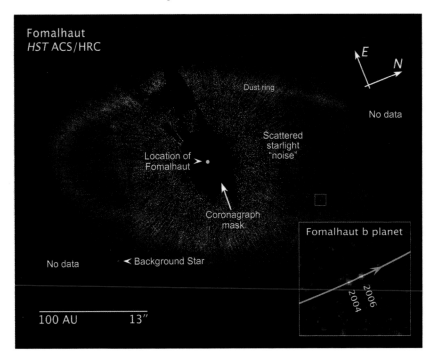

Fig. 6.24 Fomalhaut and orbiting planet. Credit: NASA, ESA and P. Kalas (University of California, Berkeley, USA)

Nevertheless, direct detection is an extremely challenging task, and for the small rocky planets, quite likely it will only be possible from space. At present, the technological development is not sufficient to resolve planets that lie closer to the parent star, but some methodologies have been proposed to reach this objective in a midterm future.

6.4.2.1 Coronagraphy

A coronagraph is an optical system used to see things that are close to a bright object. The name derives from the first such observatories dedicated to solar studies. A solar coronagraph uses a disk to block the Sun's bright surface and thus reveal the faint solar corona, stars, planets and sungrazing comets. In other words, a coronagraph produces an artificial solar eclipse. The French astronomer Bernard Lyot (1897–1952) first introduced the coronagraph in the 1930s (Lyot 1933). In his design, he also introduced a 'Lyot' stop (named after him) in the pupil plane to remove scattered light.

The simplest possible coronagraph is a lens or pinhole camera behind an appropriately aligned occulting disk that blocks direct light from the central object. Coronagraphs operating within the Earth's atmosphere suffer more from scattered light than in outer space, making space telescopes much more effective than the same instruments would be if located on the ground.

Recently, laboratory measurements that reach the contrast limits needed to detect and spectroscopically characterize nearby exoplanetary systems have been reported (Trauger and Traub 2007). By suppressing the diffracted and scattered light near a star-like source, a level of 6×10^{-10} times the star's peak intensity in individual coronagraph images is achieved, which can be further reduced to a level of about 0.1×10^{-10} by combining several images (see Fig. 6.25). This demonstrates that a coronagraphic telescope in space could image an 'Earth-twin' orbiting a nearby star (Guyon et al. 2006; Traub et al. 2006).

A number of exoplanet-hunting coronagraphic space missions have been proposed or are under study (Beichman et al. 1999; Kuchner et al. 2002; Guyon et al. 2006; Cash, 2006), but so far none has been selected by any space agency.

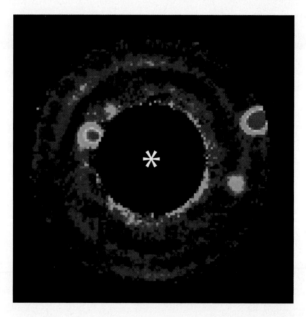

Fig. 6.25 Laboratory images demonstrate contrast at levels required to detect an Earth-twin. Three planet images are shown on the sky. The planets are copies of the measured star but with intensities corresponding to those of Jupiter, half-Jupiter and Earth, respectively. The Earth-twin is at about 4 o' clock, and the Jupiter-twin at 2 o' clock. After some complex processing of the original image sequence, the planets stand out clearly against the residual background noise. Adapted from Trauger and Traub (2007). Reprinted by permission from Macmillan Publishers Ltd., Nature Vol. 446, p. 771. Copyright (2007)

6.4.2.2 Nulling Interferometry

The light from individual telescopes can be combined to simulate collection by a much larger telescope. This technique is called interferometry and was pioneered using radio telescopes (Thompson et al. 1986), but it is now being applied to optical and infrared telescopes.

This method relies on the wave nature of light. A wave has peaks and troughs, and usually, when combining light in an interferometer, the peaks are lined up with one another, boosting the signal. In nulling interferometry, however, the peaks are lined up with the troughs so that they cancel each other out and the star disappears. Planets in orbit around the star show up, however, because they are offset from the central star and their light takes different paths through the telescope system (Angel et al. 1997).

Ground-based nulling interferometers are being developed around the world to perfect the technique. Eventually, sophisticated flotillas of spacecrafts, such as ESA's DARWIN mission or the like, could use nulling interferometry to isolate the light from Earth-like extrasolar planets. As with coronagraphy, several mission proposals and concept designs are under study, but no space agency has yet committed itself to a specific project.

6.4.2.3 Polarimetry

The study of the polarization of a stellar–planet system is a very promising method for extrasolar planet characterization. Using polarization, one can produce an occultation of the starlight, allowing a clear detection of the surrounding planet(s). Moreover, the degree of polarization, P, of the remaining light provides information about the planet's composition. P is determined by

$$P = \frac{I_r - I_l}{I_r + I_l},$$

where I_l and I_r are the intensities polarized parallel and perpendicular to the plane containing the centre of the star and the exoplanet.

The stellar light scattered by the exoplanet is linearly polarized. In the case of an unresolved system, the observed polarization signal is estimated to be about 10^{-5}, depending on the planet's surface and composition.

The flux and state of polarization of the planetary radiation can be described by a Stokes vector, **F**, as

$$\mathbf{F} = [F(\lambda, \alpha), Q(\lambda, \alpha), U(\lambda, \alpha), V(\lambda, \alpha)],$$

with α being the phase angle between the star and the observer as seen from the centre of the planet (Stam et al. 2004). Here, Q and U describe the linearly polarized flux and V the circularly polarized flux. Now, assuming that the light of a solar-like star is not globally polarized (V = 0) and adopting an adequate geometry (U = 0), we have

$$P(\lambda,\alpha) = \frac{Q(\lambda,\alpha)}{F(\lambda,\alpha)} = \frac{Q_{refl}(\lambda,\alpha)}{F_{refl}(\lambda,\alpha) + F_{ther}(\lambda,\alpha)},$$

where $F_{refl}(\lambda,\alpha)$ and $F_{ther}(\lambda,\alpha)$ are the reflected and thermal fluxes from the planet, respectively.

Polarimetric measurements essentially improve the contrast exoplanet/star. Numerical simulations of polarization spectra of giant extrasolar planets have been made by Seager et al. (2000) and Stam et al. (2004). Polarization signatures in planetary transits and microlensing events have also been studied by Carciofi and Magalhães (2005) and Selway and Hendry (2005), respectively. The current instrumental project studies are aiming toward precisions of 10^{-7} given enough photons, see Gisler et al. (2004). Only recently, Berdyugina et al. (2008) have reported the first direct detection of a previously known hot Jupiter, HD189733b, in visible polarized light.

6.5 The Next 20 Years

Several international proposals are already under study for the construction of ground-based Extremely Large Telescopes (ELTs). Current ELT designs range from 30 to 42 m in diameter (Watson et al. 2006). While the technology is not yet readily available for their construction, preliminary studies indicate that the technology to achieve a quantum leap in telescope size is feasible (Hook et al. 2005). For planetary systems, as well as other astrophysical fields, ELTs will offer spectacular advances. In addition to the improved collecting area, they will provide an extremely high spatial resolution (the achievable resolution improves in proportion to the telescope diameter).

Simulations of observations of exoplanets show that a 30 m telescope equipped with adaptive optics should be capable of studying Jupiter-like gas giants at several tens of light years. But only a larger, 100 m class telescope would be able to detect a sample of Earth-size planets (Fig. 6.26).

The number of stars that can be studied is approximately proportional to the spatial resolution to the cube. A 100 m telescope can in principle detect an Earth-like planet around a solar-type star out to a distance of 100 ly (Hainaut et al. 2007). This distance limit means that there are about 1,000 candidate Sun-like stars to be observed. The corresponding numbers are about 200 stars for a 50 m telescope and 30 stars for a 30 m telescope (Hook et al. 2005).

Constructing such large telescopes in space is very costly, but space offers its advantages too. Large space missions capable of directly observing and characterizing

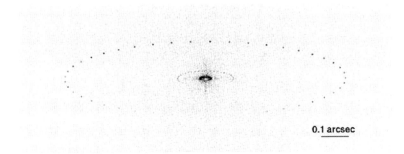

Fig. 6.26 A simulated time-series image of a solar system analogue, containing a Jupiter-like and an Earth-like planet at a distance of 10 pc. The system has been 'observed' at number of epochs as the planets go around in the 15-degree obliquity orbits to illustrate the phase effect. Each epoch is represented by a 100 ksec exposure in the V-band with the OWL 100 m telescope, based on adaptive-optics simulations. The point spread function of the central star has been subtracted from the image. Adapted from Hainaut et al. 2007

Earth-like planets, such as the Terrestrial Planet Finder (TPF)[2] or Darwin, will also need to see improved technologies before they are launched. Technologies such as precise interferometry and flight formation are now being pushed in this pursuit.

The latest report on exoplanets by ESA (Perryman et al. 2005) states that: *Planned space experiments promise a considerable increase in the detections and statistical knowledge arising especially from transit and astrometric measurements over the years 2005–2015, with some hundreds of terrestrial-type planets expected from transit measurements, and many thousands of Jupiter-mass planets expected from astrometric measurements. Beyond 2015, very ambitious space (Darwin/TPF) and ground (E-ELT) experiments are targeting direct detection of nearby Earth-mass planets in the habitable zone and the measurement of their spectral characteristics. Beyond these, 'Life Finder' (aiming to produce confirmatory evidence of the presence of life) and 'Earth Imager' (massive interferometric array providing resolved images of a distant Earth) appear as distant visions.*

Time will tell when and how we will first be able to detect and isolate the light coming from an Earth-like planet.

References

Aime, C., Vakili, F. (eds.): Direct Imaging of Exoplanets: Science & Techniques. Cambridge University Press, Cambridge (2006)

Alonso, R., Brown, T.M., Charbonneau, D., Dunham, E.W., Belmonte, J.A., Deeg, H.J., Fernández, J.M., Latham, D.W., Mandushev, G., O'Donovan, F.T., Rabus, M., Torres, G.: The Transatlantic Exoplanet Survey (TrES): A Review. In: C. Afonso, D. Weldrake, T. Henning (eds.) Transiting Extrapolar Planets Workshop, Astronomical Society of the Pacific Conference Series, vol. 366, pp. 13–22 (2007)

---

[2] At present, this mission has been indefinitely postponed by NASA.

Angel, J.R.P., Woolf, N.J.: An Imaging Nulling Interferometer to Study Extrasolar Planets. Astrophys. J. **475**, 373–379 (1997)

Baraffe, I., Chabrier, G., Allard, F., Hauschildt, P.H.: Evolutionary models for low-mass stars and brown dwarfs: Uncertainties and limits at very young ages. Astron. Astrophys. **382**, 563–572 (2002)

Barrado y Navascués, D., Zapatero Osorio, M.R., Béjar, V.J.S., Rebolo, R., Martín, E.L., Mundt, R., Bailer-Jones, C.A.L.: Optical spectroscopy of isolated planetary mass objects in the σ Orionis cluster. Astron. Astrophys. **377**, L9–L13 (2001)

Basri, G.B.: What Is a "Planet"? Mercury **32**, 27–34 (2003)

Beaulieu, J.P., Bennett, D.P., Fouqué, P., Williams, A., Dominik, M., Jorgensen, U.G., Kubas, D., Cassan, A., Coutures, C., Greenhill, J., Hill, K., Menzies, J., Sackett, P.D., Albrow, M., Brillant, S., Caldwell, J.A.R., Calitz, J.J., Cook, K.H., Corrales, E., Desort, M., Dieters, S., Dominis, D., Donatowicz, J., Hoffman, M., Kane, S., Marquette, J.B., Martin, R., Meintjes, P., Pollard, K., Sahu, K., Vinter, C., Wambsganss, J., Woller, K., Horne, K., Steele, I., Bramich, D.M., Burgdorf, M., Snodgrass, C., Bode, M., Udalski, A., Szymański, M.K., Kubiak, M., Więckowski, T., Pietrzyński, G., Soszyński, I., Szewczyk, O., Wyrzykowski, Ł., Paczyński, B., Abe, F., Bond, I.A., Britton, T.R., Gilmore, A.C., Hearnshaw, J.B., Itow, Y., Kamiya, K., Kilmartin, P.M., Korpela, A.V., Masuda, K., Matsubara, Y., Motomura, M., Muraki, Y., Nakamura, S., Okada, C., Ohnishi, K., Rattenbury, N.J., Sako, T., Sato, S., Sasaki, M., Sekiguchi, T., Sullivan, D.J., Tristram, P.J., Yock, P.C.M., Yoshioka, T.: Discovery of a cool planet of 5.5 Earth masses through gravitational microlensing. Nature **439**, 437–440 (2006)

Beichman, C.A., Woolf, N.J., Lindensmith, C.A.: The Terrestrial Planet Finder (TPF) : a NASA Origins Program to search for habitable planets. National Aeronautics and Space Administration; Pasadena, Calif.: Jet Propulsion Laboratory, California Institute of Technology, (JPL publication; 99–3) (1999)

Béjar, V.J.S., Martín, E.L., Zapatero Osorio, M.R., Rebolo, R., Barrado y Navascués, D., Bailer-Jones, C.A.L., Mundt, R., Baraffe, I., Chabrier, C., Allard, F.: The Substellar Mass Function in σ Orionis. Astrophys. J. **556**, 830–836 (2001)

Benedict, G.F., McArthur, B., Chappell, D.W., Nelan, E., Jefferys, W.H., van Altena, W., Lee, J., Cornell, D., Shelus, P.J., Hemenway, P.D., Franz, O.G., Wasserman, L.H., Duncombe, R.L., Story, D., Whipple, A.L., Fredrick, L.W.: Interferometric Astrometry of Proxima Centauri and Barnard's Star Using HST Fine Guidance Sensor 3: Detection Limits for Substellar Companions. Astrophys. J. **118**, 1086–1100 (1999)

Berdyugina, S.V., Berdyugin, A.V., Fluri, D.M., Piirola, V.: First Detection of Polarized Scattered Light from an Exoplanetary Atmosphere. Astrophys. J. **673**, L83–L86 (2008)

Beuzit, J.L., Mouillet, D., Oppenheimer, B.R., Monnier, J.D.: Direct Detection of Exoplanets. In: B. Reipurth, D. Jewitt, K. Keil (eds.) Protostars and Planets V, pp. 717–732 (2007)

Boss, A.P.: Extrasolar planets: Giant giants or dwarf dwarfs? Nature **409**, 462–463 (2001)

Bouvier, J., Stauffer, J.R., Martin, E.L., Barrado y Navascues, D., Wallace, B., Bejar, V.J.S.: Brown dwarfs and very low-mass stars in the Pleiades cluster: a deep wide-field imaging survey. Astron. Astrophys. **336**, 490–502 (1998)

Burgasser, A.J., Marley, M.S., Ackerman, A.S., Saumon, D., Lodders, K., Dahn, C.C., Harris, H.C., Kirkpatrick, J.D.: Evidence of Cloud Disruption in the L/T Dwarf Transition. Astrophys. J. **571**, L151–L154 (2002)

Burrows, A., Marley, M., Hubbard, W.B., Lunine, J.I., Guillot, T., Saumon, D., Freedman, R., Sudarsky, D., Sharp, C.: A Nongray Theory of Extrasolar Giant Planets and Brown Dwarfs. Astrophys. J. **491**, 856–875 (1997)

Butler, R.P., Marcy, G.W.: A Planet Orbiting 47 Ursae Majoris. Astrophys. J. **464**, L153–L156 (1996)

Butler, R.P., Marcy, G.W., Williams, E., Hauser, H., Shirts, P.: Three New "51 Pegasi–Type" Planets. Astrophys. J. **474**, L115–L118 (1997)

Caballero, J.A., Béjar, V.J.S., Rebolo, R., Eislöffel, J., Zapatero Osorio, M.R., Mundt, R., Barrado Y Navascués, D., Bihain, G., Bailer-Jones, C.A.L., Forveille, T., Martín, E.L.: The

References

substellar mass function in σ Orionis. II. Optical, near-infrared and IRAC/Spitzer photometry of young cluster brown dwarfs and planetary-mass objects. Astron. Astrophys. **470**, 903–918 (2007)

Campbell, B., Walker, G.A.H., Yang, S.: A search for substellar companions to solar-type stars. Astrophys. J. **331**, 902–921 (1988)

Carciofi, A.C., Magalhães, A.M.: The Polarization Signature of Extrasolar Planet Transiting Cool Dwarfs. Astrophys. J. **635**, 570–577 (2005)

Cash, W.: Detection of Earth-like planets around nearby stars using a petal-shaped occulter. Nature **442**, 51–53 (2006)

Chabrier, G., Baraffe, I.: Theory of low mass stars, brown dwarfs and extra-solar giant planets. In: T.R. Bedding, A.J. Booth, J. Davis (eds.) Fundamental stellar properties: the interaction between observation and theory, IAU Symposium, vol. 189, pp. 331–340 (1997)

Chabrier, G., Baraffe, I.: Theory of Low-Mass Stars and Substellar Objects. Annu. Rev. Astron. Astrophys. **38**, 337–377 (2000)

Charbonneau, D., Brown, T.M., Burrows, A., Laughlin, G.: When Extrasolar Planets Transit Their Parent Stars. In: B. Reipurth, D. Jewitt, K. Keil (eds.) Protostars and Planets V, pp. 701–716. University of Arizona Press, USA (2007)

Charbonneau, D., Brown, T.M., Latham, D.W., Mayor, M.: Detection of Planetary Transits Across a Sun-like Star. Astrophys. J. **529**, L45–L48 (2000)

Charbonneau, D., Brown, T.M., Noyes, R.W., Gilliland, R.L.: Detection of an Extrasolar Planet Atmosphere. Astrophys. J. **568**, 377–384 (2002)

Chauvin, G., Lagrange, A.M., Dumas, C., Zuckerman, B., Mouillet, D., Song, I., Beuzit, J.L., Lowrance, P.: A giant planet candidate near a young brown dwarf. Direct VLT/NACO observations using IR wavefront sensing. Astron. Astrophys. **425**, L29–L32 (2004)

Chauvin, G., Lagrange, A.M., Dumas, C., Zuckerman, B., Mouillet, D., Song, I., Beuzit, J.L., Lowrance, P.: Giant planet companion to 2MASSW J1207334-393254. Astron. Astrophys. **438**, L25–L28 (2005)

Close, L.M., Siegler, N., Potter, D., Brandner, W., Liebert, J.: An Adaptive Optics Survey of M8-M9 Stars: Discovery of Four Very Low Mass Binaries with at Least One System Containing a Brown Dwarf Companion. Astrophys. J. **567**, L53–L57 (2002)

Cochran, W.D., Hatzes, A.P., Butler, R.P., Marcy, G.W.: The Discovery of a Planetary Companion to 16 Cygni B. Astrophys. J. **483**, 457–463 (1997)

Cuntz, M., Shkolnik, E.: Chromospheres, flares and exoplanets. Astron. Nachr. **323**, 387–391 (2002)

Davies, J.C., Hunt, G.C., Smith, F.G.: Changing Periodicities in the Pulsars. Nature **221**, 27–29 (1969)

Deeg, H.J., Ocaña, B., Kozhevnikov, V.P., Charbonneau, D., O'Donovan, F.T., Doyle, L.R.: Extrasolar planet detection by binary stellar eclipse timing: evidence for a third body around CM Draconis. Astron. Astrophys. **480**, 563–571 (2008)

Ehrenreich, D., Tinetti, G., Lecavelier des Etangs, A., Vidal-Madjar, A., Selsis, F.: The transmission spectrum of Earth-size transiting planets. In: Bull. Am. Astron. Soc. vol. 37, p. 1565 (2005)

Gaensler, B.M., Frail, D.A.: A large age for the pulsar B1757-24 from an upper limit on its proper motion. Nature **406**, 158–160 (2000)

Gatewood, G.: On the astrometric detection of neighboring planetary systems. Icarus **27**, 1–12 (1976)

Gatewood, G., Breakiron, L., Goebel, R., Kipp, S., Russell, J., Stein, J.: On the astrometric detection of neighboring planetary systems. II. Icarus **41**, 205–231 (1980)

Gatewood, G., Eichhorn, H.: An unsuccessful search for a planetary companion of Barnard's star BD +4 3561. Astrophys. J. **78**, 769–776 (1973)

Geballe, T.R., Knapp, G.R., Leggett, S.K., Fan, X., Golimowski, D.A., Anderson, S., Brinkmann, J., Csabai, I., Gunn, J.E., Hawley, S.L., Hennessy, G., Henry, T.J., Hill, G.J., Hindsley, R.B., Ivezić, Ž., Lupton, R.H., McDaniel, A., Munn, J.A., Narayanan, V.K., Peng, E., Pier, J.R., Rockosi, C.M., Schneider, D.P., Smith, J.A., Strauss, M.A., Tsvetanov, Z.I., Uomoto, A., York, D.G., Zheng, W.: Toward Spectral Classification of L and T Dwarfs: Infrared and Optical Spectroscopy and Analysis. Astrophys. J. **564**, 466–481 (2002)

Gisler, D., Schmid, H.M., Thalmann, C., Povel, H.P., Stenflo, J.O., Joos, F., Feldt, M., Lenzen, R., Tinbergen, J., Gratton, R., Stuik, R., Stam, D.M., Brandner, W., Hippler, S., Turatto, M., Neuhauser, R., Dominik, C., Hatzes, A., Henning, T., Lima, J., Quirrenbach, A., Waters, L.B.F.M., Wuchterl, G., Zinnecker, H.: CHEOPS ZIMPOL. A VLT instrument study for the polarimetric search of scattered light from extrasolar planets. In: A.F.M. Moorwood, M. Iye (eds.) Ground-based Instrumentation for Astronomy, SPIE Conference Series, vol. 5492, pp. 463–474 (2004)

Gold, T.: Rotating Neutron Stars as the Origin of the Pulsating Radio Sources. Nature **218**, 731–732 (1968)

Gray, D.F.: Absence of a planetary signature in the spectra of the star 51 Pegasi. Nature **385**, 795–796 (1997)

Guyon, O., Pluzhnik, E.A., Kuchner, M.J., Collins, B., Ridgway, S.T.: Theoretical Limits on Extrasolar Terrestrial Planet Detection with Coronagraphs. Astrophys. J. Suppl. **167**, 81–99 (2006)

Hainaut, O.R., Rahoui, F., Gilmozzi, R.: Down to Earths, with OWL. In: A.P. Lobanov, J.A. Zensus, C. Cesarsky, P.J. Diamond (eds.) Exploring the Cosmic Frontier: Astrophysical Instruments for the 21st Century, p. 253, Springer (2007)

Halbwachs, J.L., Arenou, F., Mayor, M., Udry, S., Queloz, D.: Exploring the brown dwarf desert with Hipparcos. Astron. Astrophys. **355**, 581–594 (2000)

Harrington, J., Hansen, B.M., Luszcz, S.H., Seager, S., Deming, D., Menou, K., Cho, J.Y.K., Richardson, L.J.: The Phase-Dependent Infrared Brightness of the Extrasolar Planet Andromedae b. Science **314**, 623–626 (2006)

Harrington, J., Luszcz, S., Seager, S., Deming, D., Richardson, L.J.: The hottest planet. Nature **447**, 691–693 (2007)

Hatzes, A.P.: Starspots and exoplanets. Astron. Nachr. **323**, 392–394 (2002)

Hayashi, C., Nakano, T.: Evolution of Stars of Small Masses in the Pre-Main-Sequence Stages. Progr. Theor. Phys. **30**, 460–474 (1963)

Heger, A., Fryer, C.L., Woosley, S.E., Langer, N., Hartmann, D.H.: How Massive Single Stars End Their Life. Astrophys. J. **591**, 288–300 (2003)

Hook, I., OPTICON ELT Science Working Group: Science with extremely large telescopes. Messenger **121**, 2–8 (2005)

Israelian, G., Rebolo, R., Basri, G., Casares, J., Martín, E.L.: Evidence of a supernova origin for the black hole in the system GRO J1655 - 40. Nature **401**, 142–144 (1999)

Jameson, R.F., Sherrington, M.R., Giles, A.B.: A failed search for black dwarfs as companions to nearby stars. Mon. Not. Roy. Astron. Soc. **205**, 39P–41P (1983)

Kalas, P. (ed.): In the Spirit of Bernard Lyot: The Direct Detection of Planets and Circumstellar Disks in the 21st Century. University of California, Berkeley (2007)

Kalas, P., Graham, J.R., Chiang, E., Fitzgerald, M.P., Clampin, M., Kite, E.S., Stapelfeldt, K., Marois, C., Krist, J.: Optical Images of an Exosolar Planet 25 Light-Years from Earth. Science **322**, 1345–1348 (2008)

Kirkpatrick, J.D.: New Spectral Types L and T. Annu. Rev. Astron. Astrophys. **43**, 195–245 (2005)

Kirkpatrick, J.D., Reid, I.N., Liebert, J., Cutri, R.M., Nelson, B., Beichman, C.A., Dahn, C.C., Monet, D.G., Gizis, J.E., Skrutskie, M.F.: Dwarfs Cooler than "M": The Definition of Spectral Type "L" Using Discoveries from the 2 Micron All-Sky Survey (2MASS). Astrophys. J. **519**, 802–833 (1999)

Kuchner, M.J., Traub, W.A.: A Coronagraph with a Band-limited Mask for Finding Terrestrial Planets. Astrophys. J. **570**, 900–908 (2002)

Kumar, S.S.: The Structure of Stars of Very Low Mass. Astrophys. J. **137**, 1121–1125 (1963)

Kürster, M., Endl, M., Rouesnel, F., Els, S., Kaufer, A., Brillant, S., Hatzes, A.P., Saar, S.H., Cochran, W.D.: Terrestrial planets around M dwarfs via precise radial velocities. VLT + UVES observations of Barnard's star = GJ 699. In: M. Fridlund, T. Henning, H. Lacoste (eds.) Earths: DARWIN/TPF and the Search for Extrasolar Terrestrial Planets, ESA Special Publication, vol. 539, pp. 485–489 (2003)

Latham, D.W., Stefanik, R.P., Mazeh, T., Mayor, M., Burki, G.: The unseen companion of HD114762 - A probable brown dwarf. Nature **339**, 38–40 (1989)

References

Lin, D.N.C., Woosley, S.E., Bodenheimer, P.H.: Formation of a planet orbiting pulsar 1829 - 10 from the debris of a supernova explosion. Nature **353**, 827–829 (1991)

Lindegren, L., Perryman, M.A.C.: GAIA: Global astrometric interferometer for astrophysics. Astron. Astrophys. Suppl. **116**, 579–595 (1996)

Lucas, P.W., Roche, P.F.: A population of very young brown dwarfs and free-floating planets in Orion. Mon. Not. Roy. Astron. Soc. **314**, 858–864 (2000)

Luhman, K.L., Rieke, G.H., Young, E.T., Cotera, A.S., Chen, H., Rieke, M.J., Schneider, G., Thompson, R.I.: The Initial Mass Function of Low-Mass Stars and Brown Dwarfs in Young Clusters. Astrophys. J. **540**, 1016–1040 (2000)

Lyot, B., Marshall, R.K.: The Study of the Solar Corona without an Eclipse. J. Roy. Astron. Soc. Canada **27**, 225–234 (1933)

Magazzu, A., Martin, E.L., Rebolo, R.: A spectroscopic test for substellar objects. Astrophys. J. **404**, L17–L20 (1993)

Marois, C., Macintosh, B., Barman, T., Zuckerman, B., Song, I., Patience, J., Lafreniere, D., Doyon, R.: Direct Imaging of Multiple Planets Orbiting the Star HR 8799. Science **322**(5906), 1348–1352 (2008)

Martín, E.L., Delfosse, X., Basri, G., Goldman, B., Forveille, T., Zapatero Osorio, M.R.: Spectroscopic Classification of Late-M and L Field Dwarfs. Astrophys. J. **118**, 2466–2482 (1999)

Mayor, M., Queloz, D.: A Jupiter-Mass Companion to a Solar-Type Star. Nature **378**, 355–359 (1995)

Mayor, M., Udry, S., Lovis, C., Pepe, F., Queloz, D., Benz, W., Bertaux, J.., Bouchy, F., Mordasini, C., Segransan, D.: The HARPS search for southern extra-solar planets. XIII. A planetary system with 3 Super-Earths (4.2, 6.9, and 9.2 Earth masses). Astron. Astrophys. **493**, 639–644 (2009)

Mazeh, T., Latham, D.W., Stefanik, R.P.: Spectroscopic Orbits for Three Binaries with Low-Mass Companions and the Distribution of Secondary Masses near the Substellar Limit. Astrophys. J. **466**, 415–426 (1996)

Mazeh, T., Naef, D., Torres, G., Latham, D.W., Mayor, M., Beuzit, J.L., Brown, T.M., Buchhave, L., Burnet, M., Carney, B.W., Charbonneau, D., Drukier, G.A., Laird, J.B., Pepe, F., Perrier, C., Queloz, D., Santos, N.C., Sivan, J.P., Udry, S., Zucker, S.: The Spectroscopic Orbit of the Planetary Companion Transiting HD 209458. Astrophys. J. **532**, L55–L58 (2000)

Mediavilla, E., Arribas, S., del Burgo, C., Oscoz, A., Serra-Ricart, M., Alcalde, D., Falco, E.E., Goicoechea, L.J., Garcia-Lorenzo, B., Buitrago, J.: Two-dimensional Spectroscopy Reveals an Arc of Extended Emission in the Gravitational Lens System Q2237 + 0305. Astrophys. J. **503**, L27–L30 (1998)

Nakajima, T., Oppenheimer, B.R., Kulkarni, S.R., Golimowski, D.A., Matthews, K., Durrance, S.T.: Discovery of a Cool Brown Dwarf. Nature **378**, 463–465 (1995)

Padoan, P., Nordlund, Å.: The stellar IMF as a property of turbulence. In: E. Corbelli, F. Palla, H. Zinnecker (eds.) The Initial Mass Function 50 Years Later, Astrophysics and Space Science Library, vol. 327, p. 357 (2005)

Perryman, M., Hainaut, O., Dravins, D., Leger, A., Quirrenbach, A., Rauer, H., Kerber, F., Fosbury, R., Bouchy, F., Favata, F., Fridlund, M., Gilmozzi, R., Lagrange, A.M., Mazeh, T., Rouan, D., Udry, S., Wambsganss, J.: ESA-ESO Working Group on "Extra-solar Planets". In: M. Perryman et al., (ed.) ESA-ESO Working Group on "Extra-solar Planets" ESA (2005)

Perryman, M.A.C.: Extra-solar planets. Rep. Progr. Phys. **63**, 1209–1272 (2000)

Pickett, B.K., Durisen, R.H., Cassen, P., Mejia, A.C.: Protostellar Disk Instabilities and the Formation of Substellar Companions. Astrophys. J. **540**, L95–L98 (2000)

Quirrenbach, A.: Direct detection of exoplanets: science and techniques. In: P. Whitelock, M. Dennefeld, B. Leibundgut (eds.) The Scientific Requirements for Extremely Large Telescopes, IAU Symposium, vol. 232, pp. 109–118 (2006)

Rebolo, R., Martin, E.L., Basri, G., Marcy, G.W., Zapatero-Osorio, M.R.: Brown Dwarfs in the Pleiades Cluster Confirmed by the Lithium Test. Astrophys. J. **469**, L53–L56 (1996)

Rebolo, R., Zapatero-Osorio, M.R., Martin, E.L.: Discovery of a Brown Dwarf in the Pleiades Star Cluster. Nature **377**, 129–131 (1995)

Refsdal, S.: The gravitational lens effect. Mon. Not. Roy. Astron. Soc. **128**, 295–306 (1964)

Reipurth, B., Clarke, C.: The Formation of Brown Dwarfs as Ejected Stellar Embryos. Astrophys. J. **122**, 432–439 (2001)

Richardson, L.J., Deming, D., Horning, K., Seager, S., Harrington, J.: A spectrum of an extrasolar planet. Nature **445**, 892–895 (2007)

Saar, S.H., Donahue, R.A.: Activity-related Radial Velocity Variation in Cool Stars. Astrophys. J. **485**, 319–327 (1997)

Sartoretti, P., Schneider, J.: On the detection of satellites of extrasolar planets with the method of transits. Astron. Astrophys. Suppl. **134**, 553–560 (1999)

Saumon, D., Hubbard, W.B., Burrows, A., Guillot, T., Lunine, J.I., Chabrier, G.: A Theory of Extrasolar Giant Planets. Astrophys. J. **460**, 993–1018 (1996)

Schneider, G.: Near-IR Spectrophotometry of 2MASSWJ 1207334-393254B - An Extra-Solar Planetary Mass Companion to a Young Brown Dwarf. In: HST Proposal, p. 6858 (2005)

Schneider, G., Pasachoff, J.M., Willson, R.C.: The Effect of the Transit of Venus on ACRIM's Total Solar Irradiance Measurements: Implications for Transit Studies of Extrasolar Planets. Astrophys. J. **641**, 565–571 (2006)

Seager, S., Whitney, B.A., Sasselov, D.D.: Photometric Light Curves and Polarization of Close-in Extrasolar Giant Planets. Astrophys. J. **540**, 504–520 (2000)

Seiradakis, J.: Pulsar astronomy: Older than they look. Nature **406**, 139–140 (2000)

Selway, K.L., Hendry, M.A.: Modeling the Polarimetric Signatures of Extra-Solar Planets During Microlensing Events. In: A. Adamson, C. Aspin, C. Davis, T. Fujiyoshi (eds.) Astronomical Polarimetry: Current Status and Future Directions, Astronomical Society of the Pacific Conference Series, vol. 343, pp. 176–180 (2005)

Simon, A., Szatmáry, K., Szabó, G.M.: Determination of the size, mass, and density of "exomoons" from photometric transit timing variations. Astron. Astrophys. **470**, 727–731 (2007)

Sozzetti, A.: Astrometric Methods and Instrumentation to Identify and Characterize Extrasolar Planets: A Review. Publ. Astron.Soc. Pac. **117**, 1021–1048 (2005)

Stam, D.M., Hovenier, J.W., Waters, L.B.: Using polarimetry to detect and characterize Jupiter-like extrasolar planets. Astron. Astrophys. **428**, 663–672 (2004)

Struve, O.: Proposal for a project of high-precision stellar radial velocity work. The Observatory **72**, 199–200 (1952)

Swain, M.R., Vasisht, G., Tinetti, G.: The presence of methane present in the atmosphere of an extrasolar planet. Nature **452**, 329–331 (2008)

Tarter, J.C.: Brown Dwarfs, Lilliputian Stars, Giant Planets and Missing Mass Problems. In: Bulletin of the American Astronomical Society, vol. 8, p. 517 (1976)

Thompson, A.R., Moran, J.M., Swenson, G.W.: Interferometry and synthesis in radio astronomy. Wiley, New York (1986)

Tinetti, G., Vidal-Madjar, A., Liang, M.C., Beaulieu, J.P., Yung, Y., Carey, S., Barber, R.J., Tennyson, J., Ribas, I., Allard, N., Ballester, G.E., Sing, D.K., Selsis, F.: Water vapour in the atmosphere of a transiting extrasolar planet. Nature **448**, 169–171 (2007)

Traub, W.A., Levine, M., Shaklan, S., Kasting, J., Angel, J.R., Brown, M.E., Brown, R.A., Burrows, C., Clampin, M., Dressler, A., Ferguson, H.C., Hammel, H.B., Heap, S.R., Horner, S.D., Illingworth, G.D., Kasdin, N.J., Kuchner, M.J., Lin, D., Marley, M.S., Meadows, V., Noecker, C., Oppenheimer, B.R., Seager, S., Shao, M., Stapelfeldt, K.R., Trauger, J.T.: TPF-C: status and recent progress. In: Advances in Stellar Interferometry. Edited by Monnier, John D.; Schöller, Markus; Danchi, William C. Proceedings of the SPIE, vol. 6268 (2006)

Trauger, J.T., Traub, W.A.: A laboratory demonstration of the capability to image an Earth-like extrasolar planet. Nature **446**, 771–773 (2007)

Van de Kamp, P.: Astrometric study of Barnard's star from plates taken with the 24-inch Sproul refractor. Astrophys. J. **68**, 515 (1963)

Van de Kamp, P.: The Nearby Stars. Annu. Rev. Astron. Astrophys. **9**, 103–126 (1971)

Van de Kamp, P.: Unseen astrometric companions of stars. Annu. Rev. Astron. Astrophys. **13**, 295–333 (1975)

Van de Kamp, P.: The planetary system of Barnard's star. Vistas Astron. **26**, 141–157 (1982)

References

Walker, G.A.H., Bohlender, D.A., Walker, A.R., Irwin, A.W., Yang, S.L.S., Larson, A.: Gamma Cephei - Rotation or planetary companion? Astrophys. J. **396**, L91–L94 (1992)

Walker, G.A.H., Walker, A.R., Irwin, A.W., Larson, A.M., Yang, S.L.S., Richardson, D.C.: A search for Jupiter-mass companions to nearby stars. Icarus **116**, 359–375 (1995)

Wang, Z., Chakrabarty, D., Kaplan, D.L.: A debris disk around an isolated young neutron star. Nature **440**, 772–775 (2006)

Watson, F.G., Hook, I.M., Colless, M.M.: Astronomy with Extremely Large Telescopes, p. 363. Astrophysics Update 2 (2006)

Whitworth, A., Bate, M.R., Nordlund, Å., Reipurth, B., Zinnecker, H.: The Formation of Brown Dwarfs: Theory. In: B. Reipurth, D. Jewitt, K. Keil (eds.) Protostars and Planets V, pp. 459–476 (2007)

Wolszczan, A.: Confirmation of Earth Mass Planets Orbiting the Millisecond Pulsar PSR:B1257 + 12. Science **264**, 538–542 (1994)

Wolszczan, A., Frail, D.A.: A planetary system around the millisecond pulsar PSR1257 + 12. Nature **355**, 145–147 (1992)

Zapatero Osorio, M.R., Béjar, V.J.S., Martín, E.L., Rebolo, R., Barrado y Navascués, D., Bailer-Jones, C.A.L., Mundt, R.: Discovery of Young, Isolated Planetary Mass Objects in the σ Orionis Star Cluster. Science **290**, 103–107 (2000)

Zapatero Osorio, M.R., Caballero, J.A., Béjar, V.J.S., Rebolo, R., Barrado Y Navascués, D., Bihain, G., Eislöffel, J., Martín, E.L., Bailer-Jones, C.A.L., Mundt, R., Forveille, T., Bouy, H.: Discs of planetary-mass objects in σ Orionis. Astron. Astrophys. **472**, L9–L12 (2007)

Zarka, P.: Plasma interactions of exoplanets with their parent star and associated radio emissions. Plant. Space Sci. **55**, 598–617 (2007)

Chapter 7
The Worlds Out There

As we described in Chap. 2, an extraterrestrial might observe different types of Earths along the history of our blue planet. This evolutionary aspect must be taken into account in the future when a demographic study of exoplanets becomes available. From the beginning, the Earth's evolution was well constrained by certain fixed parameters: mass, distance to the Sun, chemical composition, etc. We can easily imagine that by changing some of these parameters, other types of 'Earth-like' planets could be obtained. In this chapter, we are interested in describing this potentially rich variety of planets. This field of research is rapidly evolving and some of the statements included in these last three chapters may well become outdated in the next few years.

The mass of an astronomical body is the main parameter used to characterize its physical properties. From a certain value, we have stars undergoing some kind of thermonuclear fusion and below this value we have planets. Over the past few decades it has become evident that all planetary systems are born as a consequence of a unique process: the contraction of an interstellar cloud. The mass, rotation and chemical composition of this cloud will determine the mass distribution of the different bodies in the system. At the end of the dynamical interactions during the first eon of evolution of the planetary system, only the survivors in the battle for available space will remain. In the case of our solar system, this resulted in one star and eight planets.

Let us start by describing the physical parameters of what we can call a planet, and the two broad types, giants and terrestrial (rocky), that we can expect to find in other systems. The recent discussion on the planetary character of Pluto will be the starting point (see Weintraub 2007 for a historical background).

7.1 Definition of a Planet

During its General Assembly, held in Prague in August 2006, the International Astronomical Union (IAU) adopted the following resolution:

1. A classical planet is a celestial body that (a) is in orbit around the Sun, (b) has sufficient mass for its self-gravity to overcame rigid body forces so that it

assumes a hydrostatic equilibrium (nearly round) shape, and (c) has cleared the neighbourhood around its orbit.
2. A dwarf planet is a celestial body that (a) is in orbit around the Sun, (b) has sufficient mass for its self-gravity to overcame rigid body forces so that it assumes a hydrostatic equilibrium (nearly round) shape, (c) has not cleared the neighbourhood around its orbit, and (d) is not a satellite.
3. All other objects orbiting the Sun shall be referred to collectively as 'Small Solar System bodies'. These currently include most of the asteroids, most Trans-Neptunian Objects (TNOs), comets and other small bodies.

This decision was founded in clear objective facts, which we describe now in some detail (see Soter 2006 and Basri and Brown 2006).

Stern and Levison (2002) derived a parameter Λ to quantify the extent to which a body of mass M and orbital period P scatters smaller masses out of its orbital zone in a Hubble time,[1] dynamically dominating the zone. Figure 7.1 clearly delimits the planets from the dwarf planets. In the solar system, the objects with $\Lambda > 1$ are effectively solitary:

$$\Lambda = \frac{kM^2}{P},$$

where k is a constant.

A related parameter, μ, proposed by Stoter (2006), describes whether a body of mass M is an end product of a disk accretion

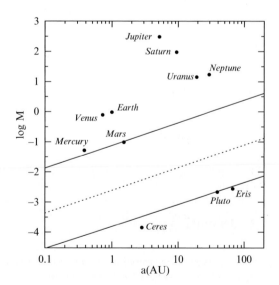

Fig. 7.1 Plot of mass M (in Earth masses) vs. the semi-major axis a for heliocentric bodies. *The dashed line* corresponds to $\Lambda = 1$. Based on data of Soter (2006)

---

[1] Inverse of the Hubble constant, H.

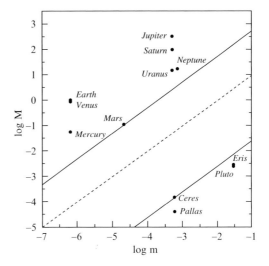

Fig. 7.2 Logarithmic plot of mass M of a body vs. the aggregated mass m in its orbital zone. Both masses are expressed in Earth masses. The *dashed line* correspond to $\mu = 100$. Based on data of Soter (2006)

$$\mu = \frac{M}{m},$$

where m is the aggregate mass of all other bodies that currently share its orbital zone (Fig. 7.2).

The upper mass limit for a planet is often taken to be about 13 Jupiter masses (4,100 Earth masses), above which deuterium fusion occurs and the body is called a brown dwarf (see Chap. 6). For planets a basic distinction is made between terrestrial (Earth-like) and giants, based on the mass and size and also on the chemical composition.

7.2 Our Solar System

7.2.1 General Facts

In our solar system:

- Planets encircle the Sun moving along direct and almost coplanar orbits.
- Most of the angular momentum of the solar system is concentrated in the planets and especially in the gas giants. Although the Sun contains more than 99% of the mass of the solar system, it contains less than 0.5% of the angular momentum due to a very slow spinning, which still needs to be satisfactorily explained.
- The planets were formed from cold matter. The presence on Earth and perhaps on the other planets of high abundances of light elements such as lithium, beryllium and boron that are unstable under the nuclear reactions inside the stars is consistent with the idea that planet-forming matter came from regions outside the star.

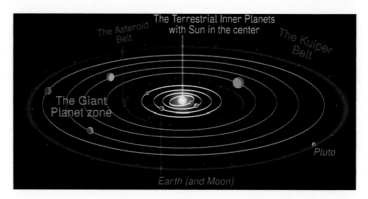

Fig. 7.3 The regions of the solar system

- The planets are divided into terrestrial and gas giants. It seems clear that planets formed close to the Sun would be less capable of retaining a significant atmosphere than those formed further away. However, the existence of outer icy giant planets as well as the intermediate gas giants must be explained in terms of their evolution within a global context, which also explains the observed features in other planetary systems.
- All the planets exhibit some tilt on their spin axes. In the case of Venus and Uranus, the tilt is so pronounced that they have retrograde rotation.
- The apparent regularity in the distribution of planetary distances (Nieto, 1972) was first published by J.B. Titius (1729–1796) in 1766 and popularized later by J.E. Bode (1747–1826) and is known as the Titius–Bode law.

The solar system is believed to have formed according to the nebular hypothesis, which holds that it emerged from a giant molecular cloud 4.6 Ga ago. As the nebula collapsed, conservation of angular momentum made it rotate faster. The material within the nebula condensed and the atoms within it began to collide with increasing frequency. The centre, where most of the mass collected, became increasingly hotter than the surrounding disc. As gravity, gas pressure, magnetic fields and rotation acted on the contracting nebula, it began to flatten into a spinning protoplanetary disc with a hot, dense protostar at the centre (see next chapter for more details).

Figure 7.3 shows a scheme of the main regions of the solar system, which we will describe later in some detail.

7.2.2 Chemical Abundances in the Solar System

The Sun contains 99.85% of all the matter in the solar system and it is composed primarily by hydrogen (74.9%), helium (23.8%) and the rest by heavier elements that receive the generic name of metals. When compared to the values obtained

7.2 Our Solar System

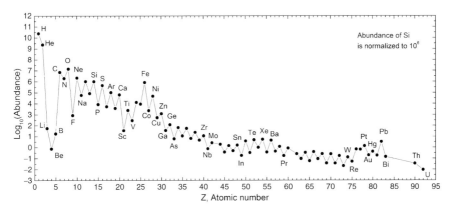

Fig. 7.4 Chemical abundances in the solar system. Source: Lodders (2003). Permission granted under the terms of the GNU Free Documentation License

from the spectra of the solar photosphere, meteorites are depleted in noble gases and CHON (carbon, hydrogen, oxygen and nitrogen), which readily form gaseous components.

The abundances of chemical elements can only be differentially determined with respect to a standard. Because of the large differences in the abundances, these are expressed on a logarithmic scale. For a determined element, *el*, we have

$$\log \frac{N_{el}}{N_{st}}.$$

In many astrophysical contexts, hydrogen is usually taken as the standard reference, but for solid bodies silicon is used.

The primordial distribution of abundances of chemical elements constitutes an essential characteristic of the solar system. Figure 7.4 represents their abundance measured in number of atoms found per 1,000,000 atoms of silicon.

7.2.3 Giant Planets

The Gas Giants, Jupiter and Saturn in our system, have primary atmospheres, captured directly from the original solar nebula, rich in primordial hydrogen and helium.

Uranus and Neptune are depleted in H and He and show significant amounts of ice and rocks. They also receive the name of Neptunian or Icy planets.

Table 7.1 includes the main data of these planets.

Table 7.1 Gas and ice giants of the solar system

Planet	a (AU)	Eq. Radius (R_E)	Mass (M_E)	Orbital Period (year)	Eccentricity
Jupiter	5.204	11.209	317.8	11.859	0.04877
Saturn	9.582	9.449	95.152	29.657	0.05572
Uranus	19.229	4.007	14.536	84.323	0.04441
Neptune	30.104	3.883	17.147	164.790	0.01121

1 solar mass = 1047.7 M_J. Data Source: Wikipedia

Table 7.2 Terrestrial planets of the solar system

Planet	a (AU)	Eq. Radius (R_E)	Mass (M_E)	Orbital Period (year)	Eccentricity
Mercury	0.387	0.383	0.055	0.241	0.20563
Venus	0.723	0.950	0.815	0.615	0.00680
Earth	1.000	1.000	1.000	1.000	0.01671
Mars	1.524	0.533	0.107	1.881	0.09331

1 Earth Mass (M_E) = 5.9736×10^{24} kg. Earth Equatorial Radius = 6378.1 km. Data Source: Wikipedia

7.2.4 Terrestrial Planets

In Chap. 2 we studied in some detail the characteristics of the Earth's structure. Earth-like planets receive in the literature different names such as Terrestrial, Rocky or Inner planets. They have a central metallic core, mostly iron, with a surrounding silicate mantle. Some of them also possess an atmosphere. Table 7.2 summarizes the main data for the terrestrial planets of the solar system.

During the process of formation of a planetary system, a temperature gradient is established across the protoplanetary disk (see Chap. 8 for more details). Refractory elements[2] are condensed close to the star, where also less materials are available, forming the terrestrial or rocky planets with an upper mass limit estimated in 13 Earth masses. In the outer parts, volatile elements[3] condense (Fig. 7.5). The condensation sequence provides a simple explanation of the two classes of planets: terrestrial and giants. Condensation temperatures relevant for the solar system are given by Lodders (2003).

In view of its cosmic abundance, water was the dominant ice (see Encrenaz 2008 for a review on the water on the solar system). In this context, an apparent paradox arises. The building blocks of a terrestrial planet are essentially dry due to the high temperatures close to the star. Its water is delivered by collisions with embryos

---

[2] These are any chemical element that condenses from a gas at high temperatures. For example, iron–nickel alloys and silicate minerals.

[3] Volatile elements vapourize at relatively low temperatures, such as carbon dioxide (CO_2), methane (CH_4) and water (H_2O).

7.2 Our Solar System

Fig. 7.5 Condensation sequence of substances in the solar system

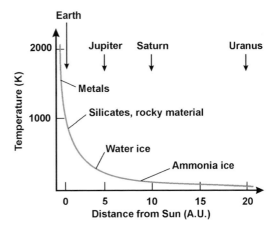

formed beyond the snow line, a_{snow}, a minimum distance from the star with luminosity L_{star} at which ice could have condensed out of the gas at $T_{snow} \sim 50$–200 K, corresponding approximately to a distance of 2.7 AU from a solar-like star.[4] In other words, it is the line that separates the rocky from the ice planets:

$$a_{snow} \sim \frac{L_{star}}{T^4_{snow}}.$$

7.2.5 Dwarf Planets and Other Minor Bodies

In the solar system dwarf planets (Table 7.3) and other minor bodies are concentrated in three main belts: the asteroid, the Kuiper belts and the Oort cloud

7.2.5.1 Asteroid Belt

The asteroid belt is the region of the solar system located roughly between the orbits of the planets Mars and Jupiter. It is occupied by numerous irregularly shaped bodies called asteroids or minor planets.

It has been estimated that the total mass of the Main Asteroid Belt may total less than 1/1,000th the mass of the Earth (Krasinsky 2002). This belt is only a small remnant of the material that once resided in the region between Mars and Jupiter,

---

[4] This is the distance at which water sublimation becomes a significant fraction of the cometary activity, as the comets approach the Sun.

Table 7.3 Main data of the dwarf planets of the solar system

Planet	a (AU)	Eq. Radius (R_E)	Mass (M_E)	Orbital Period (year)	Eccentricity
Ceres	2.77	0.074	0.0002	4.60	0.07934
Pluto	39.48	0.0021	0.0021	248.09	0.24881
Eris	67.67	0.0025	0.0025	557	0.44177

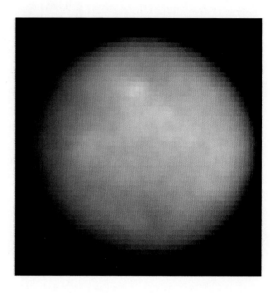

Fig. 7.6 Image of Ceres obtained by the Hubble Space Telescope. The spatial resolution of the image is about 18 km per pixel, enhancing the contrast in these images to bring out features on Ceres' surface, which are both brighter and darker than the average which absorbs 91% of sunlight falling on it. Courtesy: NASA, ESA, J. Parker (Southwest Research Institute), P. Thomas (Cornell University) and L. McFadden (University of Maryland, College Park)

but once may have contained between 2 and 10 Earth masses of material, most of them lost by a strong wind from the Early Sun and by dynamical interactions with other bodies.

The largest body is Ceres,[5] which belongs to the category of dwarf planet (see Fig. 7.6). Its surface is probably made of a mixture of water ice and various hydrated minerals like carbonates and clays (Rivkin et al. 2006). Ceres appears to be differentiated into a rocky core and ice mantle (Thomas et al. 2005), sharing some of the basic properties of the terrestrial planets. It has been suggested that it may harbour an ocean of liquid water, a critical element for life, capped by an upper crust of refrozen ice (McCord and Sotin 2005).

The current belt consists primarily of three categories of asteroids: C-type or carbonaceous asteroids, S-type or silicate asteroids and M-type or metallic asteroids (Chapman et al. 1975; Tholen 1989).

---

[5] With an equatorial radius of 483.3 km, Ceres is located between 2.54 and 2.76 AU from the Sun. It was discovered on 1 January 1801 by the monk Giuseppe Piazzi (1746–1826). In 2015, it will be visited by the DAWN spacecraft.

(1.) Carbonaceous asteroids: As their name suggests are carbon-rich and dominate the belt's outer regions. Together they comprise over 75% of the visible asteroids. Chemically, their spectra match the primordial composition of the early solar system, with only the lighter elements and volatiles removed. They are extremely dark (albedos in the 0.03–0.1 range).

(2.) S-type or silicate-rich asteroids (C-type): They are more common toward the inner region of the belt, within 2.5 AU of the Sun. The spectra of their surfaces reveal the presence of silicates and some metal, but no significant carbonaceous compounds. This indicates that their materials have been significantly modified from their primordial composition, probably via melting and reformation. They have a relatively high albedo and form about 17% of the total asteroid population.

(3.) M-type (metal-rich) asteroids: Form about 10% of the total population; their spectra resemble that of iron–nickel. Some are believed to have formed from the metallic cores of differentiated progenitor bodies that were disrupted through collision. However, there are also some silicate compounds that can produce a similar appearance. Within the main belt, the number distribution of M-type asteroids peaks at a semi-major axis of about 2.7 AU.

The asteroids are the source of dust in the inner solar system evidenced by the Zodiacal Light (Fig. 7.7), first observed by G.D. Cassini (1625–1712) in 1683, and correctly interpreted by N. Fatio de Duiliers (1664–1753) in 1684 as produced by sunlight reflected from small particles orbiting the Sun.[6]

7.2.5.2 Kuiper Belt

The Kuiper belt is a relatively thick torus (or 'doughnut') of space, extending from the orbit of Neptune (at 30 AU) to approximately 50 AU from the Sun (Edgeworth 1943). The objects in this region (KBOs) consist of two main populations: (a) the classical Kuiper belt objects (or 'cubewanos'), which lie in orbits untouched by Neptune, and (b) the resonant Kuiper belt objects, those that Neptune has locked into a precise orbital ratio such as 3:2 (the plutinos) and 2:1 (the twotinos). For a description of the resonances see Chap. 8.

We also have the scattered disc, the place where bodies with extreme eccentricity[7] and high inclination are the norm and circular orbits are exceptional. However, the difference between the Kuiper belt and the scattered disc is not clear-cut, and many astronomers see the scattered disc not as a separate population but as an outward region of the Kuiper belt (Morbidelli and Brown 2004).

Pluto, discovered in 1930, for many years was considered a planet. Now it is a part of a growing number of dwarf planets located in the Kuiper belt. It has three

---

[6] The amount of material needed to produce the zodiacal light is very small, about a particle every 8 km.
[7] Their orbits are the result of gravitational scattering by the gas giants.

Fig. 7.7 Two fundamental planes of planet Earth's sky compete for attention in this remarkable wide-angle vista, recorded on 23rd January. Arcing above the horizon and into the night at the left is a beautiful band of Zodiacal Light – sunlight scattered by dust in the solar system's ecliptic plane. Its opponent on the right is composed of the faint stars, dust clouds and nebulae along the plane of our Milky Way Galaxy. Both celestial bands stand above the domes and towers of the Teide Observatory on the island of Tenerife. Courtesy: Daniel López (IAC)

known moons (Charon, Nix and Hydra) and it will be explored in detail by the spacecraft 'New Horizons' from 2015 to 2020. Like other members of the belt, it is composed primarily of rock and ice. Its surface seems to have more than 98% nitrogen ice, with traces of methane and carbon monoxide (Owen et al. 1993).

The change of Pluto's status was no doubt motivated by the discovery of similar bodies in this region (Fig. 7.8).[8] Many other candidates have been also imaged in this region and probably they will join the list of dwarf planets. At present, more than 1,200 objects (KBOs) have been discovered. Some irregular satellites of giant planets, such as Phoebe, might be a captured KBO (Di Sisto and Brunini 2007). The collective mass of this belt is roughly a tenth the mass of the Earth (Iorio 2007).

7.2.5.3 Oort Cloud

The orbits of long period comets are highly elliptical and isotropically distributed, which suggests that they originated from a spheroidal source of $\sim$100,000 AU in

---

[8] There are at least 70,000 bodies with diameters larger than 100 km in the 30–50 AU region.

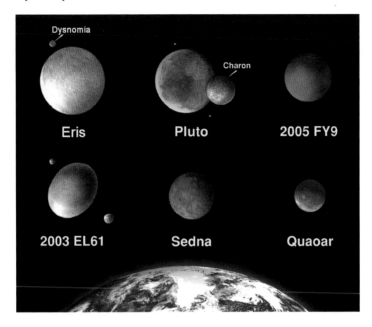

Fig. 7.8 Largest known Kuiper belt objects. At the *bottom* the Earth curvature is shown for comparison

extent, the Oort cloud (Oort 1950). So far, only two objects with orbits indicating that they may belong to the Oort Cloud have been discovered: 90377 Sedna and 2000 CR105. Figure 7.9 shows the location of Sedna and its spatial relation with other minor bodies of the solar system. Sedna shows a strong red colour and its spectrum suggests the existence of different types of ices on its surface, particularly nitrogen and methane (Barucci et al. 2005) and probably water ice (Emery et al. 2007) .

7.3 Planetary Atmospheres

The existence of an atmosphere is a characteristic factor of an astronomical body of planetary mass. Two parameters constrain the possibilities of a given body to have this gaseous envelope: the escape velocity (V_{esc})[9] and the average thermal speed (V_{th}), given by

$$V_{esc} = \left(\frac{2GM_p}{R_p}\right)^{0.5},$$

---

[9] Minimum speed the particles of an atmosphere need to 'escape' from the gravitational field of the planet.

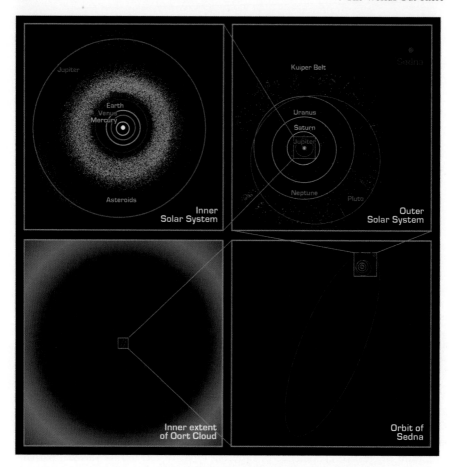

Fig. 7.9 These *four panels* show the location of 'Sedna', which lies in the farthest reaches of our solar system. Each panel, moving counterclockwise from the upper left, successively zooms out to place Sedna in context. The *first panel* shows the orbits of the inner planets, including Earth, and the asteroid belt that lies between Mars and Jupiter. In the *second panel*, Sedna is shown well outside the orbits of the outer planets and the more distant Kuiper Belt objects. Sedna's full orbit is illustrated in the third panel along with the object's current location. Sedna is nearing its closest approach to the Sun; its 10,000 year orbit typically takes it to far greater distances. The final panel zooms out much farther, showing that even this large elliptical orbit falls inside what was previously thought to be the inner edge of the Oort Cloud. Credit: Planetary Photojournal, NASA/CALTECH

$$V_{th} = \frac{3kT}{Mm_H},$$

where a body can retain an atmosphere of average temperature T composed of a particular element of molecular weight, M, if $V_{esc} \gg V_{th}$. Figure 7.10 illustrates this dichotomy, plotting the surface temperatures against their escape velocities for

7.3 Planetary Atmospheres

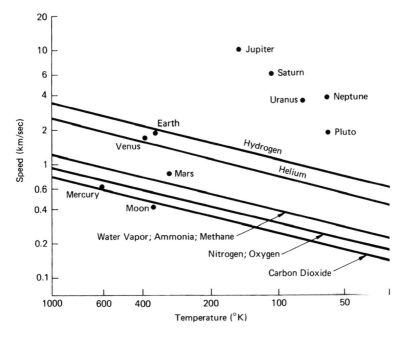

Fig. 7.10 Surface temperatures vs. escape velocities. Lines are drawn for different gases and represent six times the average thermal speed

different bodies of the solar system. It also explains clearly the differences between gas giants and terrestrial planets.

Atmospheres can gain material through the following processes:

- Bombardment by other small-mass bodies of the planetary system. A good example was the delivery of water and organics to the Early Earth
- Evapouration/sublimation of substances in the surface
- Outgassing through volcanic eruptions.

External injection of energy can also produce the removal of an existing atmosphere, often temporary. Among these mechanisms we can mention the following:

- Bombardment by other small-mass bodies of the planetary system. Only impactors with a diameter greater than the scale height of the atmosphere impart a substantial part of energy to the atmosphere (cf. Ahrens 1993)
- Condensation, a gas turns into liquid or ices on the surface when cooled
- Chemical reactions leading to a gas becoming bound into surface rocks or liquids.

In the following section, we describe the main characteristics of the extrasolar gas giants already detected and the expected properties of the exoearths.

7.4 Statistical Properties of the Extrasolar Giant Planets

Quoted values of the frequency, f_p, of giant planets around solar-type stars range between 4% and 9% for planetary masses in the range $1M_J \leq M_p < 10M_J$ and orbital radii a ≤ 3 AU, and $0.5M_J \leq M_p < 10M_J$ and a ≤ 4 AU, respectively (Sozzetti, 2005).

The details of the demographics of substellar-mass objects will be worked out in the next decade, providing basic constraints to our understanding of star and planet formation.

For reviews on the statistical properties of exoplanets see Marcy et al. (2005), Udry and Santos (2007) and Udry (2008). For online catalogues of exoplanets see http://exoplanets.org (Jones et al. 2008) and http://exoplanets.eu (The Extrasolar Planets Encyclopaedia). Different monographs (Deming and Seager 2003; Arnold 2006; Deeg et al. 2008; Mason 2008) and popular books (Lemonick 1998; Mayor and Frei 2003; Casoli and Encrenaz 2007) offer an ample review on this topic.

7.4.1 Mass Distribution

Any theory that attempts to explain the process of stellar formation in clusters should reproduce the initial census of stars, brown dwarfs and planets, the so-called *initial mass function*. The brown dwarf desert is defined as the lack of these objects close to stars. Grether and Lineweaver (2006) found that with decreasing mass the number of objects decreases two orders of magnitude in the range [1 solar mass – brown dwarf] and then rises rapidly again toward the giant Jupiter-like planets.

Figure 7.11 shows the mass distribution of the detected exoplanets, indicating a dramatic increase toward small masses. Marcy et al. (2005) calculated that this decrease in mass is roughly characterized by a power law, $dN/dM \sim M^{-1.16}$. It is still premature to speak about gaps between different intervals of planetary mass. For the physical meaning of a power-law and its application to a biological context see Pérez–Mercader (2002). The low-mass tail of the planet distribution is still under construction due to the observational constraints explained in the previous chapter.

A good deal of observational work remains to be done. This is especially relevant for the lack of terrestrial planets at small orbital distances (Fig. 7.12).

7.4.2 Hot Jupiters

The discovery of *Hot Jupiters*[10] (see Fig. 7.13) was one of the first surprises following the discovery of the first exoplanets. As these bodies could not have been

---

[10] Planets with $M_P > M_J$, temperatures close to 1,500 K and located at orbital distances of 0.05 AU.

7.4 Statistical Properties of the Extrasolar Giant Planets

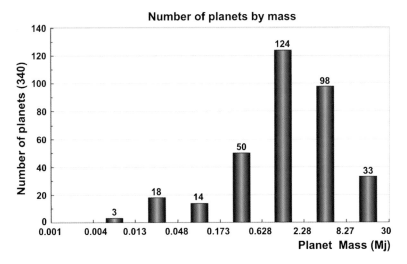

Fig. 7.11 Minimum mass distribution of the 305 known nearby exoplanets. The planetary mass, expressed in M_J, units is represented in a logarithmic scale. The mass distribution shows a dramatic decrease in the number of planets at high masses. This distribution represents results from many surveys, and so is drawn from an inhomogeneous sample. Source of data: J. Schneider (The Extrasolar Planets Encyclopaedia)

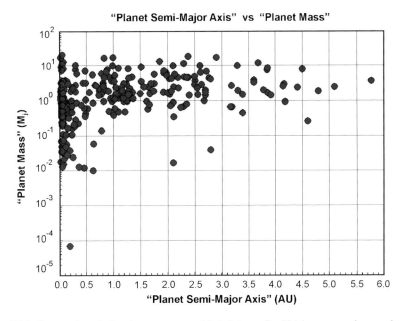

Fig. 7.12 Scatter plot of planetary mass vs. orbital distance for 294 known nearby exoplanets. The planetary mass, expressed in M_J units, is represented in a logarithmic scale. This distribution represents results from many surveys, and so is drawn from an inhomogeneous sample. Data and plot: J. Schneider (The Extrasolar Planets Encyclopaedia)

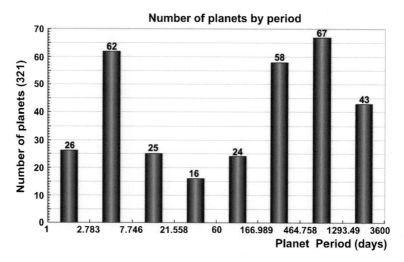

Fig. 7.13 Distribution of orbital periods among 321 exoplanets. There is a 'pile-up' of planets with orbital periods near three days (Hot Jupiters). Doppler surveys generally have uniform sensitivity to hot Jupiters, and so for massive planets, there is no important selection effect contributing to the three-day pile-up. This distribution represents results from many surveys, and so is drawn from an inhomogeneous sample. Data and plot: J. Schneider (The Extrasolar Planets Encyclopaedia)

formed so close to the star, clearly we need (1) a migration mechanism and (2) a stopping mechanism to avoid the planet from being engulfed by the parent star. Other common characteristics of Hot Jupiters are the following:

- They have a much greater chance of transiting their star as seen from a further outlying point than planets of the same mass in larger orbits
- Because of high levels of insolation, they have a lower density than otherwise would be the case. This has implications for radius determination, because due to limb darkening of the planet against its background star during a transit, the planet's ingress and egress boundaries are harder to determine
- They all have low eccentricities. This is because their orbits have been circularized or are being circularized. This also causes the planet to synchronize its rotation and orbital periods, and so it always presents the same face to its parent star, that is the planet becomes tidally locked. The atmospheric circulation in such conditions has been studied by Goodman (2009).

We can also expect to find *Hot Neptunes* ($T \sim 700$ K), exoplanets with masses resembling the core and envelope mass of Uranus and Neptune. In fact, the planet orbiting the M2 dwarf Gliese 436 at 0.02 AU seems to correspond to this type (Butler et al. 2004). The planet's atmosphere is probably composed of hydrogen and helium that scatters blue light preferentially by Rayleigh scattering, as in the Earth's atmosphere, making the atmosphere blue (see Chap. 3).

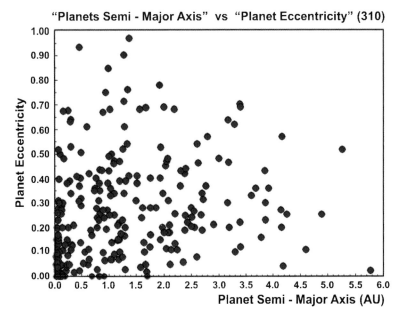

Fig. 7.14 Scatter plot of the eccentricity of the planet vs. its orbital distance (expressed in AU). This distribution represents results from many surveys, and so is drawn from an inhomogeneous sample. Data and plot: J. Schneider (The Extrasolar Planets Encyclopaedia)

7.4.3 Eccentric Planets

Figure 7.14 shows the range of the eccentricity of planetary orbits at different orbital distances. This plot shows a remarkable circularisation for planets placed close to the star.

The overall distribution of eccentricities vs. orbital periods is similar to that of stellar binaries, although their process of formation is different (Halbwachs et al. 2005).

7.4.4 Role of the Metallicity

In astronomical convention the metallicity, Z, is defined for the mass fraction of all the elements heavier than hydrogen, X, and helium, Y. In total we have $X + Y + Z = 1.0$. Usually the metallicity is represented by the abundance of iron, [Fe/H], an element showing numerous absorption and emission lines in the stellar spectra. Its value is normalized to that of the Sun:

$$[\text{Fe/H}] = \log \frac{n(\text{Fe})}{n(\text{H})_{star}} - \log \frac{n(\text{Fe})}{n(\text{H})_{sun}}.$$

A clear correlation has been established between the presence of giant planets and the metallicity of the parent star (Santos et al. 2005; Ida and Lin 2005; Fischer and Valenti 2005; Israelian 2006). Hot Jupiters are found almost exclusively around highly metallic stars. However, the relationship exoplanets–metallicity does not seem to exist for lower mass objects, although the number of planets is still low to have a statistical relevance. The Neptune-mass planets found so far have a rather flat metallicity distribution. This conclusion is strengthened by the fact that a significant portion of them have been found, thanks to follow-up studies of (metal-rich) stars already known to harbour giant planets, a fact that should have biased the sample toward higher metallicities. Moreover, considering systems with only hot Neptunes (without any other Jupiter mass analog), though the number is still small, the metallicity distribution becomes slightly metal-poor (see Udry and Santos 2007).

Zinnecker (2004) studied the formation of terrestrial planets in metal-poor environments, finding that these will be smaller and less massive than those of our solar system. Such conditions are to be found in regions with dust-to-gas ratios of 1/1000 or less. Examples of such regions are the Magellanic Clouds or the Galactic Halo.

7.4.5 Stellar Masses

The range of stellar masses explored up to date is quite narrow, concentrated on solar-like stars (Fig. 7.15).

The search for exoplanets around massive stars is hindered by their hot and rapid rotating atmospheres. This can be partially avoided by observing these stars in their

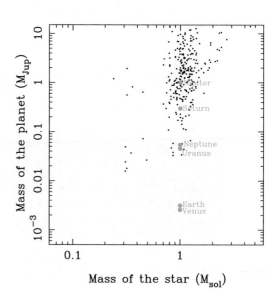

Fig. 7.15 Mass of the exoplanet vs. mass of the parent star. Courtesy: M.R. Zapatero-Osorio (IAC)

post-MS phase (G sub-giant and giant stars). In the past few years, there has been growing interest in the exploration of M dwarfs, mainly due to their habitability potential (Tarter et al. 2007).

7.5 Types of Terrestrial Planets

While awaiting progress in observing techniques that will allow the detection of terrestrial planets, we can still speculate on the possible Earths we may find. We can easily imagine that by changing some of the parameters of the planetary structure, while keeping the character of a terrestrial planet, we can envisage different types of planets (Fig. 7.16). The diverse properties of the terrestrial planets of the solar system already illustrate the possible variety of these kinds of bodies in extrasolar systems. For this purpose we recommend the reader to review the basic concepts presented in Chaps. 2 and 4.

Sudarsky et al. (2000, 2003) proposed a five-stage classification of giant planets based on their atmospheric composition and emergent spectrum. This idea has yet to be developed for the forthcoming terrestrial planets. The following compounds will be changed in the numerical simulations to form different terrestrial planets: water, silicates/carbon and iron.

Fig. 7.16 Artistic view illustrating the idea that rocky, terrestrial planets may be plentiful and diverse in the Universe. Credit: NASA/JPL-Caltech

7.5.1 Rocky Planets

The three large terrestrial planets of the solar system were formed by similar processes and probably had similar primordial atmospheres composed predominantly by carbon dioxide (Fig. 7.17).

Venus was originally in the habitable zone of a fainter Sun and probably had some water on its surface. On the other hand, Mars soon lost its greenhouse atmosphere because it was too small to have prevented its thick original atmosphere from escaping or to maintain the tectonic activity. Along its history, a planet like the Earth has undergone major changes in the chemical composition of its atmosphere, something that we can also expect in other Earth-like planets to be detected in the future.

An important parameter for terrestrial planets is the ratio I/R between the mass of ice, I, and the quantity of refractory elements, R (metals and rocks) (Sotin et al. 2007; Ehrenreich and Cassan 2007).

For I/R $\sim 10^{-4}$ we have a rocky planet with a water content similar to that of the Earth. For I/R in the range 0.33–0.5, the planet should be an intermediate between the Jupiter satellites Europa and Ganymede, with a liquid ocean under an ice-shell (Greenberg, 2005).

Finally, for I/R ~ 1, we have a complete ocean planet. Kuchner (2003) figured that planets formed beyond the snow line with masses $M_p < 10 M_E$ could retain their atmospheres, rich in volatiles like H_2O and NH_3, when migrating toward the inner parts of the planetary system. Figure 7.18 shows how the atmosphere of these bodies can survive against the EUV-driven escape.

These volatile-rich planets, also known as Hot-Earths, could be common in the habitable zones of young stars and around M-stars. At a given mass they can be confused with rocky planets, being identifiable by a slightly larger radius. They can be so small as 0.1 M_E.

Fig. 7.17 Venus, Earth and Mars: three rocky planets with atmosphere

7.5 Types of Terrestrial Planets

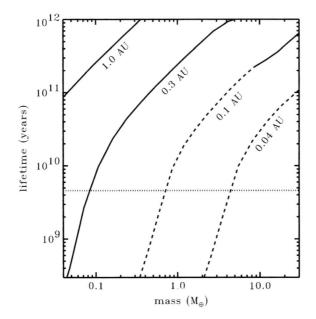

Fig. 7.18 Survival time for EUV-driven atmospheric escape as a function of mass for volatile planets at different AU. The curves become dashed where the atmosphere becomes 10% of the mass of the planet. The dotted line corresponds to the age of the solar system. Adapted from Kuchner (2003) Fig. 1. Reprinted with permission from the American Astronomical Society

7.5.2 Super-Earths

Super-Earths are defined as planets ranging from 1–13 Earth masses (Fig. 7.19) and not dominated by an atmosphere.[11] They can either have a rocky surface or be ocean worlds. To discriminate between the two cases, future observations will need to observe the mass of the planet with a 10% or better accuracy and the planet radius with a 5% accuracy, well within the technical reach of CoRoT and Kepler.

Boss (2006) describes a mechanism whereby UV radiation from a nearby massive star strips off the gaseous envelope of a gas giant, exposing a super-Earth. M dwarfs that form in regions of future high-mass star formation would then be expected to have super-Earths orbiting at distances of several AU and beyond, while those that form in regions of low-mass star formation would be expected to have gas giants at the same distances.

7.5.2.1 Internal Structure

Valencia et al. (2006, 2007a) have studied the internal structure of these planets, with special emphasis on the tectonic activity. They demonstrate that as planetary

---

[11] $H \ll R_p$, where H is the height of the atmosphere.

Fig. 7.19 Artistic view of a super-Earth orbiting a red dwarf star and with a hypothetical moon. Credit: MicroFun Collaboration, Cfa, NSF

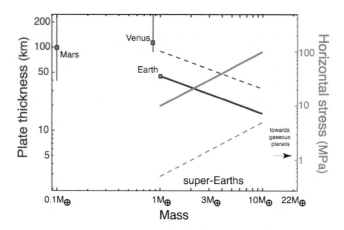

Fig. 7.20 Plate thickness (*blue and left axes*) and horizontal stresses (*green and right axes*) for super-Earths. *Dashed lines* represent the case of a reduced radioactive flux. Adapted from Valencia et al. (2007a) Fig. 1. Reprinted by permission from the American Astronomical Society

mass increases, the stress available to overcome resistance to plate motion increases while the plate thickness decreases, because a more vigorous convective interior can transport heat more efficiently to the surface (Fig. 7.20). These effects contribute favourably to the subduction of the lithosphere, an essential component of plate tectonics.

7.5 Types of Terrestrial Planets

Table 7.4 Discovered Super-Earths (M < 13 (M_E)) and their relevant data

Name	Star spectral type	Mass planet (M_E)	Distance (AU)	Reference
CoRoT-7b	G9V	11.13 < 21	0.017	Rouan et al. (2009)
				Leger et al. (2009)
55 Cnc e	G8V	10.81	0.038	McArthur et al. (2004)
Gliese 876d	M3.5V	8.41	0.021	Rivera et al. (2005)
HD 181433	K5V	7.5	0.080	Bouchy et al. (2009)
HD 40307d	K2V	9.2	0.134	Mayor et al. (2009a)
HD 40307c		6.9	0.081	
HD 40307b		4.2	0.047	
OGLE-2005-BLG-390Lb		5.5	2.0–4.1	Beaulieu et al. (2006)
GJ 436b	M2.5V	5		Ribas et al. (2008)
MOA-2007-BLG-192Lb	BD	3.3	0.080	Bennett et al. (2008)
GJ 581 d	M3	7.7	0.22	Udry et al. (2007)
GJ 581 c		5.36	0.07	
GJ 581 e		1.9	0.03	Mayor et al. (2009b)

BD Brown Dwarf. See 'The Extrasolar Planets Encyclopaedia' for more details

A different view has been presented by O'Neill and Lenardic (2007), who claim that simply by increasing the planetary radius the ratio of driving forces to the resistive strength of the lithosphere decreases. Thus super-Earth tectonics would exist as an episodic or stagnant lid regime, similar to Venus.

The different behaviours of plate tectonics between planets of similar mass, like Venus and Earth, indicate that parameters other than mass may be important for the dynamics of the lithosphere. The lower level of hydration of Venus's mantle probably plays a significant role.

On a super-Earth, continents and mountains would be much lower than on Earth, because the temperatures in the crust would increase faster with depth, until the fluid point would be reached in the crust instead of the mantle, as on 'our' Earth. Mountains can only pile up until the underlying pressures are about 3,000–3,500 atmospheres and that pressure would be reached at shallower and shallower depths on a bigger Earth.

Around ten super-Earths have been discovered to date (Table 7.4), a number that will increase dramatically in the coming years. Valencia et al. (2007b) have modelled structures and properties of Gliese 876d, the first Super-Earth ever discovered.

Mayor et al. (2009a) announced the discovery of three Super-Earths orbiting a K2V star with marked sub-solar metallicity. This opens the possibility that small planets could present a different relation with the metallicity of the host star. These results agree with those of Sousa et al. (2008), who found that, in contrast to their Jovian counterparts, Neptune-like planets do not form preferentially around metal-rich stars.

Elkins-Tanton and Seager (2008b) discuss the different ways that Super-Earth exoplanets could obtain their atmospheres, leading to a wide range of atmospheric mass and composition.

7.5.2.2 Surface Appearance and Habitability

Two main varieties of Super-Earths have been proposed: rocky and ocean (Fig. 7.21). Both types originate in a well-mixed structure of silicates and volatiles. The formation is predictable: iron and siderophile elements precipitate into a core with the volatiles above it. If water is present in adequate quantities, an ocean world can be formed, with a thick layer of water kept in the form of ice under high pressures. Kuchner (2003) and Leger et al. (2004) have proposed that these so-called *ocean planets* may form from icy Neptunian-type planets that migrate inward and melt. The expected primary volatiles are H_2O, NH_3 and CO_2, and the composition of ice should be similar to that of comets (90% H_2O, 5% NH_3 and 5% CO_2).

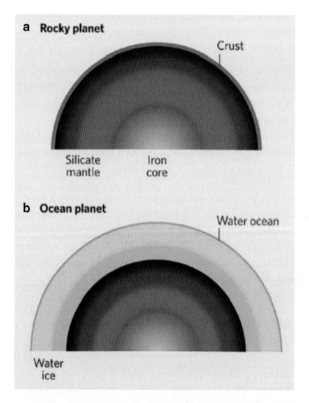

Fig. 7.21 Varieties of Super-Earths: Rocky and Ocean Worlds. Reprinted with permission from Macmillan Publishers Ltd: (D.D. Sasselov, Nature Vol. 451, p. 29, Fig. 2), copyright 2008

7.5 Types of Terrestrial Planets

Located in the habitable zone, Super-Earths would be able to develop stable oceans of water. The surfaces of these planets would be liquid water exclusively, that is no continents. The oceans would be around 100 km deep covering thick mantles of high pressure phases of water ice. If the Earth was larger, the volume of water would increase faster than the increase in surface water, so the ocean would be deeper. Therefore, the pressures at the bottoms of these oceans would be much higher.

We have already stressed, in Chap. 2, the importance of the CO_2 cycle for the stability of the climate and habitability of the Earth. The silicate crust of Super-Earths would be recycled very rapidly through local vulcanism and 'hotspots'. However, the only weathering that would be possible would be chemical, because all the volatiles would be released into the oceans rather than the atmosphere. In summary, the cycle of important gases for life would be enhanced.

For a fixed amount of greenhouse gases in the atmosphere, the planet temperature would depend on the distance to the star and we could speculate on the evolutionary changes in its climate with the increase of stellar luminosity. However, at close distances to the star, the formation of a kind of Super-Venus would be difficult because of the larger escape velocities.

Although Kuchner (2003) expects the liquid surface to be entirely obscured by a thick steam atmosphere, Leger et al. (2004) model a liquid ocean under either an obscuring cloud or a clear atmosphere. In the former case, the specular reflection from the ocean would not be visible, and the evidence of liquid water would be indirect: the mean density could be inferred from transits and radial velocities, and the surface temperature could be inferred from the star's incident flux and the planet albedo, or from the planet infrared emissivity.

Super-Earths could be identified by precise measurements with the transits technique, allowing the simultaneous determination of mass and radius. Adams et al. (2008) warns that such low-density planets cannot be distinguished from rocky planets with a thick hydrogen-rich atmosphere. In this context, Miller-Ricci and Seager (2008) proposed a method to discriminate between hydrogen-rich and hydrogen-poor atmospheres based only on the transmission spectrum of the planet.

Von Bloh et al. (2007) studied the limits of habitability for different planetary masses and fractions of surface occupied by continents, r_c. The maximum lifetime of the biosphere, t_{max}, is approached by the analytical function (see also the thesis by C. Bounama 2008)

$$t_{max}(r_c, M_P) = t_{max,E}(r_c) \left(\frac{M_P}{M_E}\right)^c,$$

where $t_{max,E}(r_c)$ is the maximum lifetime of a planet with one Earth mass and c is a fit-parameter variable with r_c. More recently, the same group (Von Bloh et al. 2009) has extended the calculations to the Red Giant Branch finding that only Ocean Worlds are able to remain habitable beyond the stellar main-sequence.

7.5.3 Carbon–Oxygen Ratio: The Carbon Planets

The Sun shows a C/O ratio of about 0.5 and therefore this oxidative characteristic has led to the formation of rocky planets with CO_2 atmospheres and surfaces composed mainly of silicates.

For C/O ~ 1, the condensation sequence changes dramatically and the substances condensed at the highest temperatures are carbon-rich compounds, such as silicon carbide, known as an industrial abrasive (Lodders 2004). This material could have accumulated in a ring called the tar-line around the star.[12] The uppermost crust of the planets formed in that ring could consist of graphite (Fig. 7.22), but deeper down into the planet high pressure would transform graphite into diamonds.

Such a planet would probably have an iron-rich core like the terrestrial planets. Silicon carbide and titanium carbide layers would form the mantle. Above that would lie a layer of carbon in the form of graphite, possibly with a few kilometres of a thick substratum of diamond, if there were sufficient pressure. Figure 7.23 illustrates the differences between carbon and silicate planets. During volcanic eruptions diamonds from the interior would come up to the surface, creating mountains of SiC and diamonds.

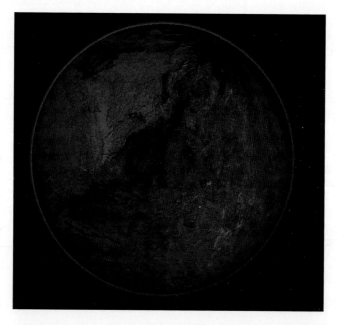

Fig. 7.22 An artistic view of a carbon planet. The surface is dark and reddish from hydrocarbon deposits. Credit: Wikimedia Commons. Author: Luyten

---

[12] In our Solar System, the carbonaceous chondrites would be the bodies that illustrate this process locally.

7.5 Types of Terrestrial Planets

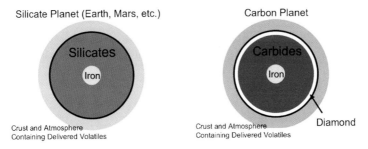

Fig. 7.23 The internal structure of silicates and carbon planets. Credit: M. Kuchner and S. Seager

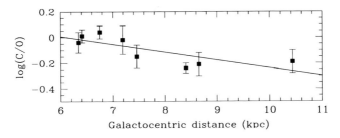

Fig. 7.24 C/O radial abundance gradients of the Galactic disk from HII region abundances. The lines indicate the least-squares line fit to the data. The Sun is located at 8 kpc. Courtesy: C. Esteban (IAC). Reprinted with permission from the American Institute of Physics (Esteban et al. 2005)

The atmospheres of carbon planets could be smoggy and composed mainly of methane and carbon monoxide. The surfaces would be covered with oceans and lakes of hydrocarbons, similar to those recently imaged on Titan by the Cassini–Huygens mission.

In general, C/O ratios in stars and HII regions increase with metallicity and towards the galactic centre (Esteban et al. 2005 and Fig. 7.24). Therefore, we should expect a larger proportion of carbon planets in this direction. Moreover, the entire Galaxy is growing richer in carbon and in the future the number of such planets could increase.

Environments suitable for carbon stars may exist around pulsars and in the disk around β Pictoris.

7.5.4 Super-Mercuries

Table 7.5 indicates the internal composition of the terrestrial planets of the solar system. In this table, Mercury can clearly be seen as exceptional, showing a comparatively very large core. Three major theories have been put forward to explain why Mercury is so much denser and more metal-rich than Earth, Venus and Mars. Each theory predicts a different composition for the rocks on Mercury's surface (Benz et al. 2007).

Fig. 7.25 Artistic view of a pulsar planet in a globular cluster. Courtesy: NASA and G. Bacon (STsI)

Table 7.5 Percentage of the mass contained in the mantle and core of the terrestrial planets and parameters related with the dynamo (Magnetic field strength and rotational period)

Planet	Mantle	Core	Rotation Period	Magnetic Field (Gauss)
Mercury	35	65	59 Earth days	0.003
Venus	68	32	117 Earth days	<0.00003
Earth	70	30	24 hours	0.305
Mars	88	12	24.6 hours	<0.0003

According to one theory, before Mercury formed, drag by solar nebular gas near the Sun mechanically sorted silicate and metal grains, with the lighter silicate particles preferentially slowed and lost to the Sun; Mercury later formed from material in this region and is consequently enriched in metal. This process does not predict any change in the composition of the silicate minerals making up the rocky portion of the planet, only the relative amounts of metal and rock.

Another theory holds that tremendous heat in the early nebula vapourized part of the outer rock layer of proto-Mercury and left the planet strongly depleted in volatile elements (Cameron 1985). This model predicts a rock composition poor in easily evaporated elements such as sodium and potassium.

The third theory is that a giant impact, after proto-Mercury had formed and differentiated, stripped off the primordial crust and upper mantle (Benz et al. 1988). This idea predicts that the present-day surface is made of rocks highly depleted in the elements that would have been concentrated in the crust, such as aluminium and calcium.

Another puzzle to be solved is the relatively large magnetic field of Mercury (see Table 7.5). Three requisites are needed for a dynamo to operate: (a) an electrically conductive fluid medium, (b) kinetic energy provided by planetary rotation and (c) an internal energy source to drive convective motions within the fluid. The detection of magnetic fields on Earth and Mercury are to be explained in this context. The Messenger spacecraft, now in orbit around Mercury, and the future Beppi-Colombo mission will help to understand better the internal structure of Mercury.

A hypothetical planet different from Mercury, a coreless silicate exoplanet, has been suggested by Elkins-Tanton and Seager (2008). They discuss two accretionary paths to such a planet, where the oxidation state of the material would play the most important role, leading to a trapping of the iron in the mantle.

7.5.5 Planets Around Pulsars in Metal-Poor Environments

Planets around pulsars were the first ones discovered in history and probably belong to the carbon planets category (Roberge et al. 2006).

Different kinds of planets can also be expected in metal-poor environments such as globular clusters. In fact, a planet, 12.7 Ga old and with 2.5 M_J, was detected in a binary system composed of a white dwarf and a pulsar, in the M4 globular cluster (Thorsett et al. 1993; Ford et al. 2000; Sigurdsson et al. 2003).

7.5.6 Terrestrial Planets Around Giant Planets: The Rocky Moons

Although they do not meet all the IAU conditions for being called a planet, several objects have been proposed that share many of the properties of the terrestrial planets, namely the mass range and the astrobiological interest. They correspond to the Classes III and IV habitats proposed by Lammer et al. (2009).

In our own solar system we can first mention Titan, which is being observed in detail by the Cassini–Huygens mission (Lorenz and Mitton 2008). It is in some aspects considered as an analogue of the Early Earth, although at a much lower temperature. Both atmospheres are composed mainly of nitrogen, are rich in organic compounds, and the cycle of methane in Titan plays a similar role to that of water in our present planet (Coustenis and Taylor 1999, 2008; Clarke and Ferris 1997; Trainer 2006; Waite 2007; Hirtzig 2009).

The jovian satellite Europa has probably an ocean of liquid water under the surface and it is therefore also a possible candidate to host some kind of life (Greenberg 2005, 2008). More recently, the interest has been focussed in Enceladus, the sixth largest satellite of Saturn, where surface structures indicate the existence of an important source of internal energy. It also shows a water-rich plume venting from the South Pole, where organic material has been detected (McKay 2008; Parkinson 2008; Waite et al. 2009). .

Orbiting extrasolar giant planets, rocky moons can be detected by small perturbations in the motion of the parent body and detected in its spectrum (Williams

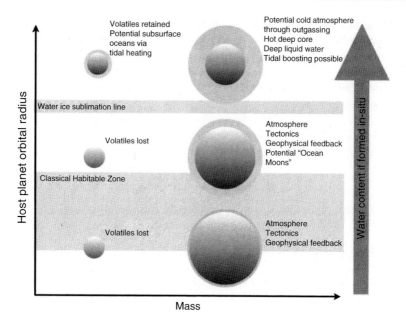

Fig. 7.26 Schema of the possible Moons existing around giant planets. Source: Fig. 11.7 of Scharf (2008). Copyright: Springer

and Knacke 2004). The main source of energy would be tidal heating (cf. Reynolds et al. 1987). Williams et al. (1997) were the first to study the habitability of such objects, estimating that a minimum of 0.23 M_E is necessary for the planet to sustain its own plate tectonics, a critical parameter for habitability. They also suggest that sputtering by charged particles trapped within the magnetosphere of the giant is an efficient mechanism for volatile loss, but the moon could be shielded with its own magnetic field. Scharf (2006) demonstrate that large moons (M > $0.1 M_E$), at orbital radii similar to the Galilean satellites, could maintain habitable conditions during episodes of heat dissipation in the order of 1–100 times that currently seen on Io.

Figure 7.26 represent the possible classes of moons we can find as a function of mass and distance from the parent star (Scharf 2008).

7.5.7 Free-Floating Planets

Zinnecker (2001) speculates on the possibility that a good fraction of planets born in binary systems will in the long run be subject to ejection due to gravitational perturbations. Therefore, he expects that there should exist a free-floating population of Jupiter-like or even Earth-like planets in interstellar space. During the process of stellar formation bodies of planetary mass can be ejected. Zapatero Osorio et al. (2000) first discovered such bodies inside a star cluster in the Orion nebula.

They also receive the name of rogue or orphan planets. Stevenson (1999) hypothesized that they could possibly sustain a thick atmosphere that would not freeze due to radiative heat loss. He proposed that atmospheres are preserved by the pressure-induced far infrared radiation opacity of a thick hydrogen-containing atmosphere. Such bodies may therefore have water oceans whose surface pressure and temperature are like those found at the base of the Earth's oceans, but these potential homes for life will be difficult to detect.

7.6 Characterization of Exoplanets

7.6.1 Mass–Radius Relationships

The internal structure of a planet can be determined empirically by measuring its free oscillation using planetary seismology. So far only the Moon has been scanned this way (Lognonné 2005) with data supplied by the Apollo Seismic Network. Future projects are now planned for other bodies of the Solar System. However, exoplanets are far away and only primitive diagnostics can be used to probe their internal structure such as the ratio between their mass and radius.

For a hypothetical homogeneous body this ratio is simple, but real planets have multiple layers of different chemical composition and therefore detailed theoretical models must be elaborated to be compared with the observations. Zapolsky and Salpeter (1969) established the basic background for these calculations.

Figure 7.27 shows the relationship for the terrestrial planets, dwarf planets and some satellites of the solar system, showing the existence of a common internal structure.

The transit technique allows the determination of the planetary radius. Together with the mass from radial velocity measurements, the R/M ratio for the known transiting exoplanets discovered so far can be calculated (Fig. 7.28). These observations are limited to giant planets and are clearly different from those of the terrestrial planets due to the distinct chemical composition and layering distribution. Five of the planets have densities between those of Jupiter and Saturn, despite their much shorter orbital periods, and one of them has a lower density. These results suggest an unforeseen variety of extrasolar worlds

Baraffe et al. (2008) have modelled the R_p/M_p ratio for planets in the range [1 M_J – 10 M_E] at different ages and levels of heavy metals enrichment, providing grids of planetary evolution to be compared with future transit observations. Valencia et al. (2007c) obtained the following relationship (values expressed in Earth units) as a function of Ice Mass Fraction (IMF), which represents the fraction of water for planets in the mass range 1–10 M_E.

$$R = (1 + 0,56\,\text{IMF})(M)^{0,262(1-0,138\cdot\text{IMF})}.$$

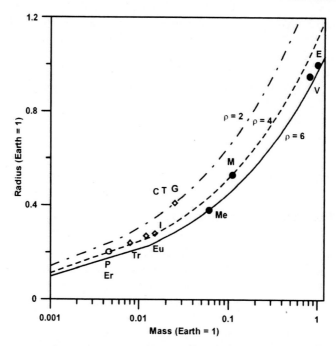

Fig. 7.27 Mass–Radius relationship for the terrestrial planets and large satellites of the solar system. E (Earth), V (Venus), Me (Mercury), M (Mars), Galilean satellites (I, G, Eu,C), T (Titan), P (Pluto) and Eris (Er). Three iso-density curves are also plotted. Adapted from Fig. 7.5 of Sánchez Lavega (2008). Reprinted with permission from Cambridge University Press

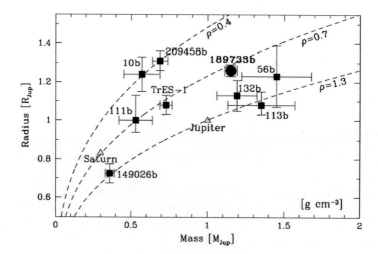

Fig. 7.28 Mass–radius relation of extrasolar giant planets from the OGLE and TrES. Three iso-density curves and values for Jupiter and Saturn are also plotted. Data: Udry (2008) and http://www.exoplanets.eu/transits/TRANSITS.htm

7.6 Characterization of Exoplanets

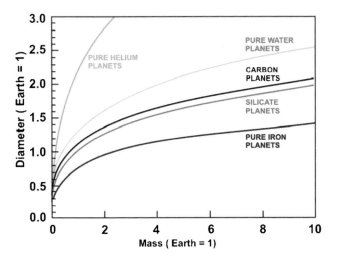

Fig. 7.29 Mass–radius relation for solid planets. Adapted from Seager et al. (2007)

Seager et al. (2007) have calculated M–R relationships for solid exoplanets. They follow simple laws, probably as a result of the fact that the building blocks of solid planets have equations of state well approximated by a polytrope of the form $\rho = \rho_0 + cP^n$. They find that in such plots (Fig. 7.29) carbon planets' curves overlap with those of water and silicate planets. This is because the zero-pressure density of graphite (2.25 g cm^{-3}) is similar to that of water–ice VII [13](1.46 g cm^{-3}), and that of SiC to that of MgSiO$_3$. If such planets are ever discovered, the spectroscopic signatures of their atmospheres will be necessary for their correct identification.

Grasset et al. (2009) have proposed M–R relationships for solid planets and solid cores ranging from 1 to 100 Earth–mass planets. The Super-Earth family includes four classes of planets: iron-rich, silicate-rich, water-rich or with a thick atmosphere. For a given mass, the planetary radius increases significantly from the iron-rich to the atmospheric-rich planet. Even if some overlaps are likely, M–R measurements could be accurate enough to ascertain the discovery of an Earth-like planet.

Fortney et al. (2007)[14] have computed radii of pure H–He, water, rock and iron planets, along with various mixtures. Water planets are 40–50% larger than rocky planets, independently of mass. They obtained the following analytical function for planets with different fractions of rocky and iron material given by the rock mass fraction (RMF)[15]

$$R = (0.0592\,\text{RMF} + 0.0975)(\log M)^2 + (0.2337\,\text{RMF} + 0.4938)\log M \\ + (0.3102\,\text{RMF} + 0.7932)$$

---

[13] Subjected to varying pressures and temperatures, ice can form in roughly a dozen different phases differentiated by their crystalline structure, ordering and density.

[14] See a posterior Erratum. The file in astro-ph is already corrected.

[15] RMF = 1 corresponds to pure rocky material and RMF = 0 to pure iron planets.

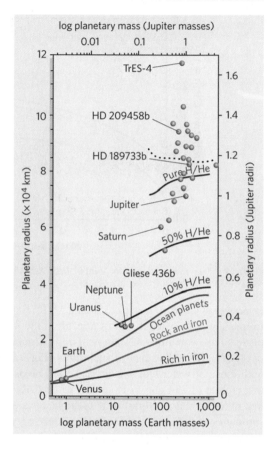

Fig. 7.30 Various planets within (*red circles*) and beyond (*grey circles*) the solar system are compared with models of different interior composition: from an interior made of rock and iron alone (Venus and Earth approximate this state) to a gaseous composition of just hydrogen and helium (of which Jupiter is the closest example in our Solar System). Reprinted by permission from Macmillan Publishers Ltd: (D.D. Sasselov, Nature Vol. 451, p. 29, Fig. 1), copyright 2008

where R and M are expressed in Earth normalized units. Table 7.5 gave the values of RMF for the terrestrial planets of the solar system.

As a summary, Fig. 7.30 shows the M–R relationship for planets of different compositions. A big challenge to the theoretical models is given by TrES-4, the largest known exoplanet, with an apparent mean density of 0.24 g cm^{-3} (Mandushev et al. 2007).

7.6.2 Atmospheres of Exoplanets

Following the detection and measurements of their basic parameters, the next step in the characterization of exoplanets would be the determination of the chemical composition and physical parameters of the atmosphere. Some atmospheres of exoplanets have already been studied spectroscopically, but the sample is limited to giant planets. Familiar attributes of solar system giant planet atmospheres (see Sánchez Lavega 2008), including hot stratospheres, clouds and redistribution of heat

7.6 Characterization of Exoplanets

Fig. 7.31 Temperature map of the exoplanet HD189733b based on Spitzer observations. Credit: NASA/JPL-Caltech/ H. Knutson (Harvard-Smithsonian CfA)

by winds, have been recognized and modelled in some of these exotic atmospheres (Marley 2007). We have a fairly complete atmospheric characterization of two Hot Jupiters observed with the transit technique.

7.6.2.1 HD 189733b

It is a planet of 1.15 M_J orbiting a K dwarf at 0.0313 AU, a distance that makes the transits occur frequently (Bouchy et al. 2005). No evidence of the presence of Earth-sized satellites and/or Saturn-type debris rings have been detected (Pont et al. 2007). Tinetti et al. (2007) detected water vapour, verified by Swain et al. (2008), who also found methane.

Knutson et al. (2007) have produced the temperature map of this exoplanet (Fig. 7.31), which reveals a warm spot on its 'sunlit' side (the planet is tidally locked). Probably strong winds exist in its atmosphere. Recently, Lecavelier des Etangs et al. (2008) have analysed a transit spectrum in the range of 550–1050 nm. They show that the slope of the absorption is typical of an atmosphere dominated by Rayleigh scattering, identifying $MgSiO_3$ as a possible abundant condensate with particles of submicron size (see also Pont et al. 2008).

7.6.2.2 HD 209458b

A giant planet of about 0.7 M_J orbiting very close (0.047 AU) to a yellow star, located 150 light-years from the Earth. Its volume is some 145% greater than that of Jupiter. Charbonneau et al. (2002) first observed sodium in its atmosphere when the planet transited across the stellar disk. Subsequent observations indicated that hydrogen, oxygen and carbon are evapourating at such a rate that relatively soon the planet may become a dead rocky core, a so-called Chthonian planet (Vidal-Madjar et al. 2003, 2004; Hebrard et al. 2004). This escape was later verified by Ballester et al. (2007). It is thought that the atmosphere of this exoplanet will not evapourate

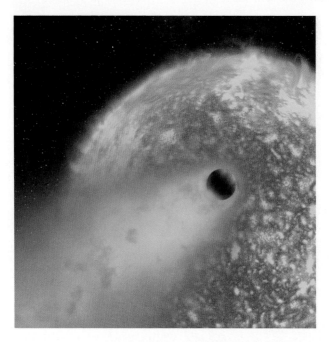

Fig. 7.32 Artistic view of the photoevaporation in HD209458b. Credit: ESA, NASA, Hubble Space Telescope and Alfred Vidal-Madjar (IAP/CNRS, Paris)

entirely, although it may have lost up to 7% of its mass over its estimated lifetime of 5 Ga (See Fig. 7.32 for an artistic view of the process). Recently, Holmström et al. (2008) interpreted the transit observations in the light of the Lyman-α line as produced by the interaction between the exosphere of the planet and a hot and slow stellar wind.

Deming et al. (2005) detected the presence of clouds composed probably of carbon monoxide. According to the fits of theoretical models to observations, the presence of a stratosphere and an inversion layer are required (Burrows et al. 2007). Richardson et al. (2007) recorded an IR spectrum, finding a broad emission peak centred near 9.65 microns, that they attribute to emission by silicate clouds. Barman (2007) combined former HST observations and theoretical modelling to detect the presence of water vapour. Spitzer observations by Knutson et al. (2008) detected the presence of an inversion layer high in the atmosphere, leading to significant water emission.

Lecavelier des Etangs et al. (2008) show that the rise in absorption depth at short wavelengths can be interpreted as Rayleigh scattering within the atmosphere. They also determine a temperature of 2200 ± 260 K at 33 mbar pressure. García Muñoz (2007) has studied the physical and chemical aeronomy of the planet. He showed that tidal forces may enhance the escape rate, thus shortening the lifetime of the planet to a few Ga.

Barnes and O'Brien (2002) estimate that no primordial satellites with masses greater than $7 \times 10^{-7} M_E$ ($\sim$70 km radius for a density of $3\,\mathrm{gm\,cm^{-3}}$) could have

7.6.2.3 Terrestrial Planets

In the preceding chapters we described the chemical composition of the Earth's atmosphere and the major spectral features which, in principle, we expect to detect in alien Earth-like atmospheres. The exception could be the features of biological origin such as oxygen. Spectroscopic observations during the Mercury and Venus transits are an adequate reference (Schleicher et al. 2004).

Modelling of the spectral appearance of terrestrial exoplanets has been carried out by Ehrenreich et al. (2005) and Tinetti (2006). Tinetti et al. (2006) developed a spatially and spectrally resolved model of the Earth, based on observations of our planet from different spacecrafts. Smith et al. (2004) studies the effect of ionizing radiation on these kinds of atmospheres.

Assuming that we know the atmospheric pressure and temperature at the planetary surface, we can derive the state of the water (vapour, liquid or ice) reading on the corresponding phase diagram (Fig. 7.33).

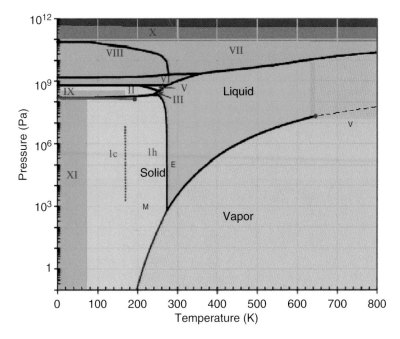

Fig. 7.33 Water phase diagram. The T–P regimes of vapour, liquid and ice are shown. The average values of terrestrial planets of the solar system are indicated (Venus, V; Earth, E; and Mars, M)

7.6.3 Radio Emission of Exoplanets

We described briefly in Chap. 4 how a magnetized planet can emit non-thermal radioemission (see Fig. 4.27). There is a source of energetic (keV) electrons in their magnetospheres, from auroral processes or as a result of magnetic coupling between the planet and a satellite. Therefore, we can detect exoplanets also in this wavelength range, although the number of possible targets is not very high (Farrell et al. 2004; Griessmeier et al. 2007).

So far, various attempts with several Hot-Jupiters have given negative results (Bastian et al. 2000; Shiratori et al. 2006; Lazio and Farrell 2007). Radio emission is most likely to be detected from planets around stars with high-density coronae, which are therefore likely to be bright X-ray sources (Jardine and Comeron 2008).

7.7 Terraformed Planets

In the last section of this chapter, we provide room for speculation. Future exoplanet observations could reveal worlds with atmospheric properties that lie outside our physical models. These artificial states could have been triggered by an extraterrestrial civilization with the aim of rendering these worlds more suitable for its purposes, whatever they may be. The atmosphere would be the easiest feature to modify through changes in its chemical composition. Sagan (1961), MacElroy and Averner (1976), and Berman et al. (1976) were probably the first to popularize this concept.

Terraforming is the hypothetical process by which the climate of a planet would be changed into something similar to that if the Earth, with the goal of making it habitable. Mars has been the target of most of these studies (see Allaby and Lovelock 1985) and also of science-fiction novels.[16] Fogg (1995) and Zubrin and Mc Kay (1997) discusses the technological requirements for this work. Figure 7.34 shows a simulated sequence of this transformation of Mars.

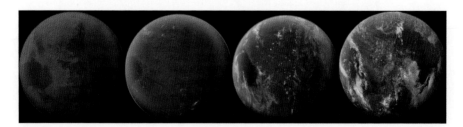

Fig. 7.34 Sequence of the terraforming process of Mars

---

[16] See the trilogy of Kim Stanley Robinson: Red Mars, Green Mars and Blue Mars.

Table 7.6 Column amounts necessary for the terraforming of Mars and lifetimes against photolysis in the present terrestrial atmosphere

Gas	Column per m^2	Lifetime (yr)
$CF_3CF_2CF_3$	1.1×10^{22}	$>10^8$
$CF_3SCF_2CF_3$	1.4×10^{22}	8950
SF_6	5.2×10^{21}	3200
SF_5CF_3	1.0×10^{22}	4050
$SF_4(CF_3)_2$	8.0×10^{21}	3070

Table 3 of Gerstell et al. (2001)

An increase in the current temperatures to levels comparable to the Earth's could be possible by adding powerful greenhouse gases to the atmosphere (Gerstell et al. 2001). Table 7.6 summarizes the most relevant properties of these gases. Other proposals contemplate the installation of large mirrors (125 km) in orbit that would reflect sunlight to heat the Martian surface (Birch 1992; Zubrin and Mc Kay 1997).

The solar flux in the Mars orbit is sufficient for photosynthesis, but probably the level of UV solar radiation would be lethal for the microorganisms growing after the first phase of terraforming. Oxygen producing bacteria could mimic the process occurring on Earth. A previous step for producing oxygen might begin with the production of hydrochloric acid from common salt, NaCl, believed to be common in Martian soil (Fogg 1992). The overall reaction yielding 2.5 moles of oxygen for every mole of nitrogen is given by

$$4NaNO_3 + 2H_2O \longrightarrow 4NaOH + 2N_2 + 5O_2.$$

An important problem would be the lack of a magnetosphere. The technological requirements for the creation of an artificial global field are more in the domain of dreams. Any extraterrestrial civilization will no doubt be constrained by the availability of usable energy.

A recent monograph by Beech (2009) summarizes the current knowledge in this field. This author deems that this may turn out to be a long-term solution to the energy crisis and problems of population growth that civilization on Earth is now experiencing.

7.8 Expect the Unexpected

In this chapter we have summarized the current knowledge on exoplanets and the different kinds of exoplanets that can be detected. However, the window for observing other Earths still needs to be opened. When dealing with extrasolar planets, we will need to expect the unexpected and all kinds of situations are possible. A classification of different types of planets certainly implies the existence of hybrids or intermediate stages.

The principle of plenitude, first suggested by Lovejoy (1976) tracing back to Aristotle, established that what can be done must be done. This can also be applied to the variety of planetary worlds. We will be constrained only by the laws of physics and the distribution and evolution of the chemical elements in our Galaxy. Our final tribute goes to the artists who imagined these worlds some decades in advance of the future astronomical observations (e.g. Carroll 2007).

References

Adams, E.R., Seager, S., Elkins-Tanton, L.: Ocean planet or thick atmosphere: on the mass-radius relationship for solid exoplanets with massive atmospheres. Astrophys. J. **673**, 1160–1164 (2008)

Ahrens, T.J.: Impact erosion of terrestrial planetary atmospheres. Annu. Rev. Earth Planet. Sci. **21**, 525–555 (1993)

Allaby, M., Lovelock, J.: Greening of Mars. Warner Books (1985)

Arnold, L., Bouchy, F., Moutou, C. (eds.): Tenth Anniversary of 51 Peg-b: Status of and prospects for hot Jupiter studies (2006)

Ballester, G.E., Sing, D.K., Herbert, F.: The signature of hot hydrogen in the atmosphere of the extrasolar planet HD 209458b. Nature **445**, 511–514 (2007)

Baraffe, I., Chabrier, G., Barman, T.: Structure and evolution of super-Earth to super-Jupiter exoplanets: I. Heavy element enrichment in the interior. Astron. Astrophys. **482**, 315–332 (2008)

Barman, T.: Identification of absorption features in an extrasolar planet atmosphere. Astrophys. J. **661**, L191–L194 (2007)

Barnes, J.W., O'Brien, D.P.: Stability of satellites around close-in extrasolar giant planets. Astrophys. J. **575**, 1087–1093 (2002)

Barucci, M.A., Cruikshank, D.P., Dotto, E., Merlin, F., Poulet, F., Dalle Ore, C., Fornasier, S., de Bergh, C.: Is sedna another triton? Astron. Astrophys. **439**, L1–L4 (2005)

Basri, G., Brown, M.E.: Planetesimals to brown dwarfs: What is a planet? Annu. Rev. Earth Planet. Sci. **34**, 193–216 (2006)

Bastian, T.S., Dulk, G.A., Leblanc, Y.: A Search for Radio Emission from Extrasolar Planets. Astrophys. J. **545**, 1058–1063 (2000)

Beaulieu et al., J.P.: Discovery of a cool planet of 5.5 Earth masses through gravitational microlensing. Nature **439**, 437–440 (2006)

Beech, M.: Terraforming: The Creation of Habitable Worlds. Springer, Heidelberg (2009)

Bennett, D.P., Bond, I.A., Udalski, A., Sumi, T., Abe, F., Fukui, A., Furusawa, K., Hearnshaw, J.B., Holderness, S., Itow, Y., Kamiya, K., Korpela, A.V., Kilmartin, P.M., Lin, W., Ling, C.H., Masuda, K., Matsubara, Y., Miyake, N., Muraki, Y., Nagaya, M., Okumura, T., Ohnishi, K., Perrott, Y.C., Rattenbury, N.J., Sako, T., Saito, T., Sato, S., Skuljan, L., Sullivan, D.J., Sweatman, W.L., Tristram, P.J., Yock, P.C.M., Kubiak, M., Szymański, M.K., Pietrzyński, G., Soszyński, I., Szewczyk, O., Wyrzykowski, Ł., Ulaczyk, K., Batista, V., Beaulieu, J.P., Brillant, S., Cassan, A., Fouqué, P., Kervella, P., Kubas, D., Marquette, J.B.: A low-mass planet with a possible sub-stellar-mass host in microlensing event MOA-2007-BLG-192. Astrophys. J. **684**, 663–683 (2008)

Benz, W., Anic, A., Horner, J., Whitby, J.A.: The origin of mercury. Space Sci. Rev. **132**, 189–202 (2007)

Benz, W., Slattery, W.L., Cameron, A.G.W.: Collisional stripping of mercury's mantle. Icarus **74**, 516–528 (1988)

Berman, S., Kuhn, W.R., Langhoff, P.W., Rogers, S.R., Thomas, J.W., MacElroy, R.D., Averner, M.M.: On the habitability of Mars. An approach to planetary ecosynthesis. NASA SP-414. National Technical Information Service, Springfield (1976)

References

Birch, P.: Terraforming Mars quickly. J. Br. Interplanet. Soc. **45**, 331–340 (1992)

Boss, A.P.: Rapid formation of super-earths around M dwarf stars. Astrophys. J. **644**, L79–L82 (2006)

Bouchy, F., Mayor, M., Lovis, C., Udry, S., Benz, W., Bertaux, J.L., Delfosse, X., Mordasini, C., Pepe, F., Queloz, D., Segransan, D.: The HARPS search for southern extra-solar planets. XVII. Super-Earth and Neptune-mass planets in multiple planet systems HD 47 186 and HD 181 433. Astron. Astrophys. **496**, 527–531 (2009)

Bouchy, F., Udry, S., Mayor, M., Moutou, C., Pont, F., Iribarne, N., da Silva, R., Ilovaisky, S., Queloz, D., Santos, N.C., Ségransan, D., Zucker, S.: ELODIE metallicity-biased search for transiting Hot Jupiters. II. A very hot Jupiter transiting the bright K star HD 189733. Astron. Astrophys. **444**, L15–L19 (2005)

Burrows, A., Hubeny, I., Budaj, J., Knutson, H.A., Charbonneau, D.: Theoretical spectral models of the planet HD 209458b with a thermal inversion and water emission bands. Astrophys. J. **668**, L171–L174 (2007)

Butler, R.P., Vogt, S.S., Marcy, G.W., Fischer, D.A., Wright, J.T., Henry, G.W., Laughlin, G., Lissauer, J.J.: A Neptune-Mass planet orbiting the nearby M dwarf GJ 436. Astrophys. J. **617**, 580–588 (2004)

Cameron, A.G.W.: The partial volatilization of Mercury. Icarus **64**, 285–294 (1985)

Carroll, M.: Space Art: How to draw and paint planets, moons, and landscapes of alien worlds. Watson-Guptill (2007)

Casoli, F., Encrenaz, T.: The new worlds: Extrasolar planets. Springer (2007)

Chapman, C.R., Morrison, D., Zellner, B.: Surface properties of asteroids - A synthesis of polarimetry, radiometry, and spectrophotometry. Icarus **25**, 104–130 (1975)

Charbonneau, D., Brown, T.M., Noyes, R.W., Gilliland, R.L.: Detection of an extrasolar planet atmosphere. Astrophys. J. **568**, 377–384 (2002)

Clarke, D.W., Ferris, J.P.: Chemical Evolution on Titan: Comparisons to the Prebiotic Earth. Orig. Life Evol. Biosph. **27**, 225–248 (1997)

Coustenis, A., Taylor, F.: Titan: The Earth-like Moon. World Scientific (1999)

Coustenis, A., Taylor, F.: Titan: Exploring an Earth-like World. World Scientific (2008)

Deeg, H., Belmonte, J.A., Aparicio, A. (eds.): Extrasolar Planets, IAC Winter School. Cambridge University Press, Cambridge (2008)

Deming, D., Brown, T.M., Charbonneau, D., Harrington, J., Richardson, L.J.: A New Search for Carbon Monoxide Absorption in the Transmission Spectrum of the Extrasolar Planet HD 209458b. Astrophys. J. **622**, 1149–1159 (2005)

Deming, D., Seager, S. (eds.): Scientific Frontiers in Research on Extrasolar Planets, Astronomical Society of the Pacific Conference Series, vol. 294 (2003)

Di Sisto, R.P., Brunini, A.: Possible origin of the Saturn satellite, Phoebe. Boletin de la Asociacion Argentina de Astronomia La Plata Argentina **50**, 27–30 (2007)

Edgeworth, K.E.: The evolution of our planetary system. J. Br. Astron. Assoc. **53**, 181–188 (1943)

Ehrenreich, D., Cassan, A.: Are extrasolar oceans common throughout the Galaxy? Astron. Nachr. **328**, 789–792 (2007)

Ehrenreich, D., Tinetti, G., Lecavelier Des Etangs, A., Vidal-Madjar, A., Selsis, F.: The transmission spectrum of Earth-size transiting planets. Astron. Astrophys. **448**, 379–393 (2006)

Elkins-Tanton, L., Seager, S.: Coreless Terrestrial Exoplanets. Astrophys. J. **808** (2008)

Elkins-Tanton, L., Seager, S.: Ranges of Atmospheric Mass and Composition of Super Earth Exoplanets. Astrophys. J. **808** (2008)

Emery, J.P., Dalle Ore, C.M., Cruikshank, D.P., Fernández, Y.R., Trilling, D.E., Stansberry, J.A.: Ices on (90377) Sedna: confirmation and compositional constraints. Astron. Astrophys. **466**, 395–398 (2007)

Encrenaz, T.: Water in the Solar System. Annu. Rev. Astron. Astrophys. **46**, 57–87 (2008)

Esteban, C., García-Rojas, J., Peimbert, M., Peimbert, A., Ruiz, M.T., Rodríguez, M., Carigi, L.: Carbon and Oxygen Galactic Gradients: Observational Values from H II Region Recombination Lines. Astrophys. J. **618**, L95–L98 (2005)

Farrell, W.M., Lazio, T.J.W., Zarka, P., Bastian, T.J., Desch, M.D., Ryabov, B.P.: The radio search for extrasolar planets with LOFAR. Planet. Space Sci. **52**, 1469–1478 (2004)

Fischer, D.A., Valenti, J.: The Planet-Metallicity Correlation. Astrophys. J. **622**, 1102–1117 (2005)

Fogg, M.J.: A synergic approach to terraforming Mars. J. Br. Interplanet. Soc. **45**, 315–329 (1992)

Fogg, M.J.: Terraforming: Engineering Planetary Environments. SAE International (1995)

Ford, E.B., Joshi, K.J., Rasio, F.A., Zbarsky, B.: Theoretical Implications of the PSR B1620-26 Triple System and Its Planet. Astrophys. J. **528**, 336–350 (2000)

Fortney, J.J., Marley, M.S., Barnes, J.W.: Planetary Radii across Five Orders of Magnitude in Mass and Stellar Insolation: Application to Transits. Astrophys. J. **659**, 1661–1672 (2007)

García Muñoz, A.: Physical and chemical aeronomy of HD 209458b. Planet. Space Sci. **55**, 1426–1455 (2007)

Gerstell, M.F., Francisco, J.S., Yung, Y.L., Boxe, C., Aaltonee, E.T.: Keeping Mars warm with new super greenhouse gases. Proc. Natl. Acad. Sci. **98**, 2154–2157 (2001)

Goodman, J.: Thermodynamics of Atmospheric Circulation on Hot Jupiters. Astrophys. J. **693**, 1645–1649 (2009)

Grasset, O., Schneider, J., Sotin, C.: A Study of the Accuracy of Mass-Radius Relationships for Silicate-Rich and Ice-Rich Planets up to 100 Earth Masses. Astrophys. J. **693**, 722–733 (2009)

Greenberg, R.: Europa: The Ocean Moon: Search for and alien biosphere. Springer Praxis (2005)

Greenberg, R.: Unmasking Europa: The search for Life on Jupiter's Ocean Moon. Springer (2008)

Grether, D., Lineweaver, C.H.: How Dry is the Brown Dwarf Desert? Quantifying the Relative Number of Planets, Brown Dwarfs, and Stellar Companions around Nearby Sun-like Stars. Astrophys. J. **640**, 1051–1062 (2006)

Grießmeier, J.M., Zarka, P., Spreeuw, H.: Predicting low-frequency radio fluxes of known extrasolar planets. Astron. Astrophys. **475**, 359–368 (2007)

Halbwachs, J.L., Mayor, M., Udry, S.: Statistical properties of exoplanets. IV. The period-eccentricity relations of exoplanets and of binary stars. Astron. Astrophys. **431**, 1129–1137 (2005)

Hébrard, G., Lecavelier Des Étangs, A., Vidal-Madjar, A., Désert, J.M., Ferlet, R.: Evaporation Rate of Hot Jupiters and Formation of Chthonian Planets. In: J. Beaulieu, A. Lecavelier Des Etangs, C. Terquem (eds.) Extrasolar Planets: Today and Tomorrow, Astronomical Society of the Pacific Conference Series, vol. 321, pp. 203–204 (2004)

Hirtzig, M., Tokano, T., Rodriguez, S., Le Mouélic, S., Sotin, C.: A review of Titan's atmospheric phenomena. Astron. Astrophys. Rev. **17**, 105–147 (2009)

Holmström, M., Ekenbäck, A., Selsis, F., Penz, T., Lammer, H., Wurz, P.: Energetic neutral atoms as the explanation for the high velocity hydrogen around HD 209458b. Nature **802** (2008)

Ida, S., Lin, D.N.C.: Dependence of Exoplanets on Host Stars' Metallicity and Mass. Progr. Theor. Phys. Suppl. **158**, 68–85 (2005)

Iorio, L.: Dynamical determination of the mass of the Kuiper Belt from motions of the inner planets of the Solar system. Mon. Not. Roy. Astron. Soc. **375**, 1311–1314 (2007)

Israelian, G.: Chemical abundances in stars with extrasolar planetary systems. In: L. Arnold, F. Bouchy, C. Moutou (eds.) 10th Anniversary of 51 Peg-b: Status of and prospects for hot Jupiter studies, pp. 35–45 (2006)

Jardine, M., Cameron, A.C.: Radio emission from exoplanets: the role of the stellar coronal density and magnetic field strength. Astron. Astrophys. **490**, 843–851 (2008)

Jones, H.R.A., Butler, R.P., Wright, J.T., Marcy, G.W., Fischer, D.A., Vogt, S.S., Tinney, C.G., Carter, B.D., Johnson, J.A., McCarthy, C., Penny, A.J.: A Catalogue of Nearby Exoplanets. In: N.C. Santos, L. Pasquini, A.C.M. Correia, M. Romaniello (eds.) Princeton Series in Astrophysics, pp. 205–206 (2008)

Knutson, H.A., Charbonneau, D., Allen, L.E., Burrows, A., Megeath, S.T.: The 3.6-8.0 μm Broadband Emission Spectrum of HD 209458b: Evidence for an Atmospheric Temperature Inversion. Astrophys. J. **673**, 526–531 (2008)

Knutson, H.A., Charbonneau, D., Allen, L.E., Fortney, J.J., Agol, E., Cowan, N.B., Showman, A.P., Cooper, C.S., Megeath, S.T.: A map of the day-night contrast of the extrasolar planet HD 189733b. Nature **447**, 183–186 (2007)

Krasinsky, G.A., Pitjeva, E.V., Vasilyev, M.V., Yagudina, E.I.: Hidden Mass in the Asteroid Belt. Icarus **158**, 98–105 (2002)

Kuchner, M.J.: Volatile-rich Earth-Mass Planets in the Habitable Zone. Astrophys. J. **596**, L105–L108 (2003)

Lammer, H., Bredehöft, J.H., Coustenis, A., Khodachenko, M.L., Kaltenegger, L., Grasset, O., Prieur, D., Raulin, F., Ehrenfreund, P., Yamauchi, M., Wahlund, J.E., Griessmeier, J.M., Stangl, G., Cockell, C.S., Kulikov, Y.N., Grenfell, J.L., Rauer, H.: What makes a planet habitable? Astron. Astrophys. Rev. **17**, 181–249 (2009)

Lazio, T.J.W., Farrell, W.M.: Magnetospheric Emissions from the Planet Orbiting τ Bootis: A Multiepoch Search. Astrophys. J. **668**, 1182–1188 (2007)

Lecavelier Des Etangs, A., Vidal-Madjar, A., Désert, J.M., Sing, D.: Rayleigh scattering by H_2 in the extrasolar planet HD 209458b. Astron. Astrophys. **485**, 865–869 (2008)

Leger, A., Rouan, D., Schneider, J., Barge, P., Fridlund, M., Samuel, B., Ollivier, M., Guenther, E., Deleuil, M., Deeg, H.J., Auvergne, M., Alonso, R., Aigrain, S., Alapini, A., Almenara, J.M., Baglin, A., Barbieri, M., Bruntt, H., Borde, P., Bouchy, F., Cabrera, J., Catala, C., Carone, L., Carpano, S., Csizmadia, S., Dvorak, R., Erikson, A., Ferraz-Mello, S., Foing, B., Fressin, F., Gandolfi, D., Gillon, M., Gondoin, P., Grasset, O., Guillot, T., Hatzes, A., Hebrard, G., Jorda, L., Lammer, H., Llebaria, A., Loeillet, B., Mayor, M., Mazeh, T., Moutou, C., Paetzold, M., Pont, F., Queloz, D., Rauer, H., Renner, S., Samadi, R., Shporer, A., Sotin, C., Tingley, B., Wuchterl, G.: Transiting exoplanets from the CoRoT space mission VIII. CoRoT-7b: the first Super-Earth with measured radius. Astron. Astrophys. (2009)

Léger, A., Selsis, F., Sotin, C., Guillot, T., Despois, D., Mawet, D., Ollivier, M., Labèque, A., Valette, C., Brachet, F., Chazelas, B., Lammer, H.: A new family of planets? "Ocean-Planets". Icarus **169**, 499–504 (2004)

Lemonick, M.D.: Other Worlds: The Search for Life in the Universe. Simon and Schuster (1998)

Lodders, K.: Solar System Abundances and Condensation Temperatures of the Elements. Astrophys. J. **591**, 1220–1247 (2003)

Lodders, K.: Jupiter Formed with More Tar than Ice. Astrophys. J. **611**, 587–597 (2004)

Lognonné, P.: Planetary Seismology. Annu. Rev. Earth Planet. Sci. **33**, 571–604 (2005)

Lorenz, R.D., Mitton, J.: Titan Unveiled: Saturn Mysterious Moon Explored. Princeton University Press (2008)

Lovejoy, A.: The Great Chain of Being. Harvard University Press, London (1976)

MacElroy, R.D., Averner, M.M.: Atmospheric engineering of Mars. Adv. Eng. Sci. **3**, 1203–1214 (1976)

Mandushev, G., O'Donovan, F.T., Charbonneau, D., Torres, G., Latham, D.W., Bakos, G.Á., Dunham, E.W., Sozzetti, A., Fernández, J.M., Esquerdo, G.A., Everett, M.E., Brown, T.M., Rabus, M., Belmonte, J.A., Hillenbrand, L.A.: TrES-4: A Transiting Hot Jupiter of Very Low Density. Astrophys. J. **667**, L195–L198 (2007)

Marcy, G., Butler, R.P., Fischer, D., Vogt, S., Wright, J.T., Tinney, C.G., Jones, H.R.A.: Observed Properties of Exoplanets: Masses, Orbits, and Metallicities. Progr. Theor. Phys. Suppl. **158**, 24–42 (2005)

Marley, M.: Characterization of Extrasolar Planets: Lessons From Atmospheres Modeling. In: P. Kalas (ed.) In the Spirit of Bernard Lyot: The Direct Detection of Planets and Circumstellar Disks in the 21st Century (2007)

Mason, J. (ed.): Exoplanets: Detection, Formation, Properties, Habitability. Springer/Praxis (2008)

Mayor, M., Bonfils, X., Forveille, T., Delfosse, X., Udry, S., Bertaux, J., Beust, H., Bouchy, F., Lovis, C., Pepe, F., Perrier, C., Queloz, D., Santos, N.C.: The HARPS search for southern extrasolar planets XVIII. An Earth-mass planet in the GJ 581 planetary system. Astron. Astrophys. (2009)

Mayor, M., Frei, P.Y.: New Worlds in the Cosmos: The discovery of Exoplanets. Cambridge University Press, Cambridge (2003)

Mayor, M., Udry, S., Lovis, C., Pepe, F., Queloz, D., Benz, W., Bertaux, J.., Bouchy, F., Mordasini, C., Segransan, D.: The HARPS search for southern extra-solar planets. XIII. A planetary system with 3 Super-Earths (4.2, 6.9, 9.2 Earth masses). Astron. Astrophys. **493** (2008)

McArthur, B.E., Endl, M., Cochran, W.D., Benedict, G.F., Fischer, D.A., Marcy, G.W., Butler, R.P., Naef, D., Mayor, M., Queloz, D., Udry, S., Harrison, T.E.: Detection of a Neptune-Mass Planet in the ρ Cancri System Using the Hobby-Eberly Telescope. Astrophys. J. **614**, L81–L84 (2004)

McCord, T.B., Sotin, C.: Ceres: Evolution and current state. J. Geophys. Res. **110**(9), 5009 (2005)

McKay, C.P., Porco Carolyn C., Altheide, T., Davis, W.L., Kral, T.A.: The Possible Origin and Persistence of Life on Enceladus and Detection of Biomarkers in the Plume. Astrobiology **8**, 909–919 (2008)

Miller-Ricci, E., Rowe, J.F., Sasselov, D., Matthews, J.M., Guenther, D.B., Kuschnig, R., Moffat, A.F.J., Rucinski, S.M., Walker, G.A.H., Weiss, W.W.: MOST Space-based Photometry of the Transiting Exoplanet System HD 209458: Transit Timing to Search for Additional Planets. Astrophys. J. **682**, 586–592 (2008)

Miller-Ricci, E., Sasselov, D., Seager, S.: The Atmospheric Signatures of Super-Earths: How to Distinguish Between Hydrogen-Rich and Hydrogen-Poor Atmospheres. Astrophys. J. **808** (2008)

Morbidelli, A., Brown, M.E.: The Kuiper belt and the primordial evolution of the solar system. In: M.C. Festou, H.U. Keller, H.A. Weaver (eds.) Comets II, pp. 175–191. University of Arizona Press (2004)

Nieto, M.M.: The Titius-Bode Law of Planetary Distances: Its History and Theory. Pergamon Press (1972)

O'Neill, C., Lenardic, A.: Geological consequences of super-sized Earths. Geophys. Res. Lett. **34**, 19,204 (2007)

Oort, J.H.: The structure of the cloud of comets surrounding the Solar System and a hypothesis concerning its origin. Bull. Astron. Inst. Neth. **11**, 91–110 (1950)

Owen, T.C., Roush, T.L., Cruikshank, D.P., Elliot, J.L., Young, L.A., de Bergh, C., Schmitt, B., Geballe, T.R., Brown, R.H., Bartholomew, M.J.: Surface ices and the atmospheric composition of Pluto. Science **261**, 745–748 (1993)

Parkinson, C.D., Liang, M.C., Yung, Y.L., Kirschivnk, J.L.: Habitability of Enceladus: Planetary Conditions for Life. Orig. Life Evol. Biosph. **38**, 355–369 (2008)

Pérez-Mercader, J.: Scaling phenomena and the emergence of complexity in astrobiology, pp. 337–360. Astrobiology. The quest for the conditions of life. Gerda Horneck, Christà Baumstark-Khan (eds.). Physics and astronomy online library. Berlin: Springer (2002)

Pont, F., Gilliland, R.L., Moutou, C., Charbonneau, D., Bouchy, F., Brown, T.M., Mayor, M., Queloz, D., Santos, N., Udry, S.: Hubble Space Telescope time-series photometry of the planetary transit of HD 189733: no moon, no rings, starspots. Astron. Astrophys. **476**, 1347–1355 (2007)

Pont, F., Knutson, H., Gilliland, R.L., Moutou, C., Charbonneau, D.: Detection of atmospheric haze on an extrasolar planet: The 0.55 - 1.05 micron transmission spectrum of HD189733b with the Hubble Space Telescope. Mon. Not. Roy. Astron. Soc. **712** (2008)

Reynolds, R.T., McKay, C.P., Kasting, J.F.: Europa, tidally heated oceans, and habitable zones around giant planets. Adv. Space Res. **7**, 125–132 (1987)

Ribas, I., Font-Ribera, A., Beaulieu, J.P.: A $\sim$ 5 ME super-earth orbiting GJ436? The power of near-grazing transits. Astrophys. J. **677**, L59–L62 (2008)

Richardson, L.J., Deming, D., Horning, K., Seager, S., Harrington, J.: A spectrum of an extrasolar planet. Nature **445**, 892–895 (2007)

Rivera, E.J., Lissauer, J.J., Butler, R.P., Marcy, G.W., Vogt, S.S., Fischer, D.A., Brown, T.M., Laughlin, G., Henry, G.W.: Ã 7.5 M_E Planet Orbiting the Nearby Star, GJ 876. Astrophys. J. **634**, 625–640 (2005)

Rivkin, A.S., Volquardsen, E.L., Clark, B.E.: The surface composition of Ceres: Discovery of carbonates and iron-rich clays. Icarus **185**, 563–567 (2006)

Roberge, A., Feldman, P.D., Weinberger, A.J., Deleuil, M., Bouret, J.C.: Stabilization of the disk around β Pictoris by extremely carbon-rich gas. Nature **441**, 724–726 (2006)

Sagan, C.: The planet Venus. Science **133**, 849–858 (1961)

Sánchez-Lavega, A.: The perspective: a panorama of the Solar System. In: H. Deeg, J.A. Belmonte, A. Aparicio (eds.) Extrasolar Planets; XVI Canary Islands Winter School of Astrophysics, pp. 178–216. Cambridge University Press, Cambridge (2008)

Santos, N.C., Israelian, G., Mayor, M., Bento, J.P., Almeida, P.C., Sousa, S.G., Ecuvillon, A.: Spectroscopic metallicities for planet-host stars: Extending the samples. Astron. Astrophys. **437**, 1127–1133 (2005)

Sasselov, D.D.: Astronomy: Extrasolar planets. Nature **451**, 29–31 (2008)

Scharf, C.A.: The Potential for Tidally Heated Icy and Temperate Moons around Exoplanets. Astrophys. J. **648**, 1196–1205 (2006)

Scharf, C.A.: Moons of Exoplanets: Habitats for Life? In: J.W. Masson (ed.) Exoplanets: Detection, Formation, Properties, Habitability, pp. 285–303. Praxis/Springer (2008)

Schleicher, H., Wiedemann, G., Wöhl, H., Berkefeld, T., Soltau, D.: Detection of neutral sodium above Mercury during the transit on 2003 May 7. Astron. Astrophys. **425**, 1119–1124 (2004)

Seager, S., Kuchner, M., Hier-Majumder, C.A., Militzer, B.: Mass-Radius Relationships for Solid Exoplanets. Astrophys. J. **669**, 1279–1297 (2007)

Shiratori, Y., Yokoo, H., Saso, T., Kameya, O., Iwadate, K., Asari, K.: Ten years of quests for radio bursts from extrasolar planets. In: L. Arnold, F. Bouchy, C. Moutou (eds.) Tenth Anniversary of 51 Peg-b: Status of and prospects for hot Jupiter studies, pp. 290–292 (2006)

Sigurdsson, S., Richer, H.B., Hansen, B.M., Stairs, I.H., Thorsett, S.E.: A Young White Dwarf Companion to Pulsar B1620-26: Evidence for Early Planet Formation. Science **301**, 193–196 (2003)

Smith, D.S., Scalo, J., Wheeler, J.C.: Transport of ionizing radiation in terrestrial-like exoplanet atmospheres. Icarus **171**, 229–253 (2004)

Soter, S.: What Is a Planet? Astron. J. **132**, 2513–2519 (2006)

Sotin, C., Grasset, O., Mocquet, A.: Mass radius curve for extrasolar Earth-like planets and ocean planets. Icarus **191**, 337–351 (2007)

Sousa, S.G., Santos, N.C., Mayor, M., Udry, S., Casagrande, L., Israelian, G., Pepe, F., Queloz, D., Monteiro, M.J.P.F.G.: Spectroscopic parameters for 451 stars in the HARPS GTO planet search program. Stellar [Fe/H] and the frequency of exo-Neptunes. Astron. Astrophys. **487**, 373–381 (2008)

Sozzetti, A.: Astrometric Methods and Instrumentation to Identify and Characterize Extrasolar Planets: A Review. Publ. Astron. Soc. Pac. **117**, 1021–1048 (2005)

Stern, S.A., Levison, H.F.: Regarding the criteria for planethood and proposed planetary classification schemes. Highlights of Astronomy **12**, 205–213 (2002)

Stevenson, D.J.: Life-sustaining planets in interstellar space? Nature **400**, 32 (1999)

Sudarsky, D., Burrows, A., Hubeny, I.: Theoretical Spectra and Atmospheres of Extrasolar Giant Planets. Astrophys. J. **588**, 1121–1148 (2003)

Sudarsky, D., Burrows, A., Pinto, P.: Albedo and Reflection Spectra of Extrasolar Giant Planets. Astrophys. J. **538**, 885–903 (2000)

Swain, M.R., Vasisht, G., Tinetti, G.: Methane present in an extrasolar planet atmosphere. Nature **802** (2008)

Tarter, J.C., Backus, P.R., Mancinelli, R.L., Aurnou, J.M., Backman, D.E., Basri, G.S., Boss, A.P., Clarke, A., Deming, D., Doyle, L.R., Feigelson, E.D., Freund, F., Grinspoon, D.H., Haberle, R.M., Hauck II, S.A., Heath, M.J., Henry, T.J., Hollingsworth, J.L., Joshi, M.M., Kilston, S., Liu, M.C., Meikle, E., Reid, I.N., Rothschild, L.J., Scalo, J., Segura, A., Tang, C.M., Tiedje, J.M., Turnbull, M.C., Walkowicz, L.M., Weber, A.L., Young, R.E.: A Reappraisal of The Habitability of Planets around M Dwarf Stars. Astrobiology **7**, 30–65 (2007)

Tholen, D.J.: Asteroid taxonomic classifications. In: R.P. Binzel, T. Gehrels, M.S. Matthews (eds.) Asteroids II, pp. 1139–1150. University of Arizona Press, USA (1989)

Thomas, P.C., Parker, J.W., McFadden, L.A., Russell, C.T., Stern, S.A., Sykes, M.V., Young, E.F.: Differentiation of the asteroid Ceres as revealed by its shape. Nature **437**, 224–226 (2005)

Thorsett, S.E., Arzoumanian, Z., Taylor, J.H.: PSR B1620-26 - A binary radio pulsar with a planetary companion? Astrophys. J. **412**, L33–L36 (1993)

Tinetti, G.: Characterizing Extrasolar Terrestrial Planets with Reflected, Emitted and Transmitted Spectra. Orig. Life Evol. Biosph. **36**, 541–547 (2006)

Tinetti, G., Meadows, V.S., Crisp, D., Fong, W., Fishbein, E., Turnbull, M., Bibring, J.P.: Detectability of Planetary Characteristics in Disk-Averaged Spectra. I: The Earth Model. Astrobiology **6**, 34–47 (2006)

Tinetti, G., Vidal-Madjar, A., Liang, M.C., Beaulieu, J.P., Yung, Y., Carey, S., Barber, R.J., Tennyson, J., Ribas, I., Allard, N., Ballester, G.E., Sing, D.K., Selsis, F.: Water vapour in the atmosphere of a transiting extrasolar planet. Nature **448**, 169–171 (2007)

Trainer, M.G., Pavlov, A.A., Dewitt, H.L., Jimenez, J.L., McKay, C.P., Toon, O.B., Tolbert, M.A.: Inaugural Article: Organic haze on Titan and the early Earth. Proc. Natl. Acad. Sci. **103**, 18,035–18,042 (2006)

Udry, S., Bonfils, X., Delfosse, X., Forveille, T., Mayor, M., Perrier, C., Bouchy, F., Lovis, C., Pepe, F., Queloz, D., Bertaux, J.L.: The HARPS search for southern extra-solar planets. XI. Super-Earths (5 and 8 M_E) in a 3-planet system. Astron. Astrophys. **469**, L43–L47 (2007)

Udry, S., Santos, N.C.: Statistical Properties of Exoplanets. Annu. Rev. Astron. Astrophys. **45**, 397–439 (2007)

Valencia, D., O'Connell, R.J., Sasselov, D.: Internal structure of massive terrestrial planets. Icarus **181**, 545–554 (2006)

Valencia, D., O'Connell, R.J., Sasselov, D.D.: Inevitability of Plate Tectonics on Super-Earths. Astrophys. J. **670**, L45–L48 (2007)

Valencia, D., Sasselov, D.D., O'Connell, R.J.: Detailed Models of Super-Earths: How Well Can We Infer Bulk Properties? Astrophys. J. **665**, 1413–1420 (2007)

Valencia, D., Sasselov, D.D., O'Connell, R.J.: Radius and Structure Models of the First Super-Earth Planet. Astrophys. J. **656**, 545–551 (2007)

Vidal-Madjar, A., Désert, J.M., Lecavelier des Etangs, A., Hébrard, G., Ballester, G.E., Ehrenreich, D., Ferlet, R., McConnell, J.C., Mayor, M., Parkinson, C.D.: Detection of Oxygen and Carbon in the Hydrodynamically Escaping Atmosphere of the Extrasolar Planet HD 209458b. Astrophys. J. **604**, L69–L72 (2004)

Vidal-Madjar, A., Lecavelier des Etangs, A., Désert, J.M., Ballester, G.E., Ferlet, R., Hébrard, G., Mayor, M.: An extended upper atmosphere around the extrasolar planet HD209458b. Nature **422**, 143–146 (2003)

von Bloh, W., Bounama, C., Cuntz, M., Franck, S.: The habitability of super-Earths in Gliese 581. Astron. Astrophys. **476**, 1365–1371 (2007)

von Bloh, W., Cuntz, M., Schroeder, K.P., Bounama, C., Franck, S.: Habitability of Super-Earth Planets around Other Suns: Models including Red Giant Branch Evolution. Astrobiology (2009)

Waite, J.H., Young, D.T., Cravens, T.E., Coates, A.J., Crary, F.J., Magee, B., Westlake, J.: The Process of Tholin Formation in Titans Upper Atmosphere. Science **316**, 870–875 (2007)

Waite Jr., J.H., Lewis, W.S., Magee, B.A., Lunine, J.I., McKinnon, W.B., Glein, C.R., Mousis, O., Young, D.T., Brockwell, T., Westlake, J., Nguyen, M.J., Teolis, B.D., Niemann, H.B., McNutt, R.L., Perry, M., Ip, W.H.: Liquid water on Enceladus from observations of ammonia and ^{40}Ar in the plume. Nature **460**, 487–490 (2009)

Weintraub, D.A.: Is Pluto a Planet? Princeton University Press, Princeton (2007)

Williams, D.M., Kasting, J.F., Wade, R.A.: Habitable moons around extrasolar giant planets. Nature **385**, 234–236 (1997)

Williams, D.M., Knacke, R.F.: Looking for Planetary Moons in the Spectra of Distant Jupiters. Astrobiology **4**, 400–403 (2004)

Zapatero Osorio, M.R., Béjar, V.J.S., Martín, E.L., Rebolo, R., Barrado y Navascués, D., Bailer-Jones, C.A.L., Mundt, R.: Discovery of Young, Isolated Planetary Mass Objects in the σ Orionis Star Cluster. Science **290**, 103–107 (2000)

Zapolsky, H.S., Salpeter, E.E.: The Mass-Radius Relation for Cold Spheres of Low Mass. Astrophys. J. **158**, 809–813 (1969)

Zinnecker, H.: A Free-Floating Planet Population in the Galaxy? In: J.W. Menzies, P.D. Sackett (eds.) Microlensing 2000: A New Era of Microlensing Astrophysics, Astronomical Society of the Pacific Conference Series, vol. 239, pp. 223–227 (2001)

Zinnecker, H.: Chances for Earth-Like Planets and Life Around Metal-Poor Stars. In: R. Norris, F. Stootman (eds.) Bioastronomy 2002: Life Among the Stars, IAU Symposium, vol. 213, pp. 45–50 (2004)

Zubrin, R.M., McKay, C.P.: Technological requirements for terraforming Mars. J. Br. Interplanet. Soc. **50**, 83–92 (1997)

Chapter 8
Extrasolar Planetary Systems

Human being is conditioned by the environment where they are born. This is also the case for a planetary system. The destiny of an Earth-like planet is not independent of the surroundings where it is embedded. This applies perfectly to the type of planetary system to which our terrestrial planet belongs. In the first chapter, we outlined the main characteristics of our planetary system, now we explore this subject further. The detection of other exoplanets, described previously, has enabled other possible configurations of the planetary systems to be described.

We begin this chapter by presenting the current knowledge on the process of formation of planetary systems, starting with our own.

8.1 The Origin of the Solar System: Early Attempts

Several topics must be taken into account to develop a theory for the origin of the Solar System. A good theory must solve and explain all features of the Solar System from its origin to its present state.

8.1.1 Nebular Theory

The first physical model of the Solar System was proposed by R. Descartes (1596–1650) in his *Principles of Philosophy*, published in 1644 (see Aiton 1972). He suggested the existence of vortexes formed in the primordial gas that filled the system. The Sun condensed out of a large vortex while the planets and satellites formed from smaller vortexes. This idea was developed by I. Kant (1724–1804) in his book *Universal Natural History and Theory of the Heavens: An essay on the constitution and mechanical origin of the whole Universe according to Newton's principles*. He envisioned the solar system as having formed from a clotting mass of gas and dust. The mutual gravitational attractions of the particles caused them to start moving and colliding, at which point chemical forces kept them bonded together.

Pierre Simon Laplace (1749–1827) established the basis of its model, known as the 'Laplace nebular theory', in his *Exposition du Système du Monde* published in

1796. It is based on the collapse of an initial slowly spinning spherical cloud of gas and dust that accelerate its rotation and flattens along the spin axis as it contracts. Several annular rings left behind during contraction evolved into condensed spherical bodies leading to a planetary system. The conservation of angular momentum explains this process.

This theory was severely criticized by J.C. Maxwell (1831–1879), who argued that progressive condensation of annular rings to form the planets could not be possible under self gravitation as differential rotation in the outer and inner parts of the rings would destroy any initial condensation. Another major difficulty for the theory was based on the distribution of angular momentum in the Solar System as there was no known mechanism to explain why the planets would have most of the angular momentum while the Sun has most of the mass.

In 1943, C. Von Weizsäcker (1912–2007) elaborated his vortex theory, which suggests that the protoplanetary disk is formed by a pattern of clockwise rotating eddies within an anticlockwise rotating system in pentagonal symmetry (Fig. 8.1). Material would collide at the boundary between vortexes. In these regions material would coalesce to give condensations that would eventually form planets. The orbital radii match well with Bode's law.

Ter Haar (1950) showed that the formation and dissolution of primary vortexes would lead to innumerable small condensations throughout the solar nebula instead of a few large ones; a possible obstacle to overcome only if we take into account the gravitational forces (Kuiper 1951). The vortexes must have a critical density to protect them from tidal disruption from the Sun. The spacing of the planets from the Sun is determined by the ratio of the density in the solar nebula to the critical density. Systems lighter than the Solar System by a factor of 2–3 would have only billions of comet-like bodies instead of planets. For larger densities we would have only a multiple stellar system.

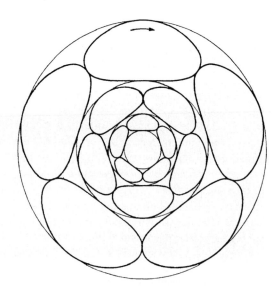

Fig. 8.1 The vortex model of Carl Von Weizsäcker

8.1.2 Catastrophic Theories

G.L. Leclerc, Comte de Buffon (1707–1788), suggested in his *Les époques de la nature*, published in 1778, an origin of the Solar System by a catastrophic collision of a comet against the Sun's surface. He remarked that various regularities observed in the system strongly suggest that it is the product of a single process of formation.

T. Chamberlain (1843–1928) and F. Moulton (1872–1952) proposed that the Solar System derived from huge filaments of gas ejected from an active early Sun under the tidal influence of a massive passing star that prevented the erupted matter from returning to the parent star (Fig. 8.2). The matter ejected by these eruptions would later condense into planets (Moulton 1905).

H. Jeffreys (1891–1989) argued that collisions between filaments would heat up the matter rather than allowing it to cool down and condensate. Together with J. Jeans (1877–1946) he suggested an alternative scenario in which a grazing collision drew out a massive filament of matter (Fig. 8.3). Since the cloud was flatter in the centre than in the ends, the theory explained that the most massive planets are located at an intermediate distance from the Sun. However, it soon became clear that not enough angular momentum could be transferred to the ejected material.

Since the 1960s, new ideas, or developments of the old ones, have arisen in the context of the Solar System's origin and planetary formation; one of these is the Protoplanet Theory. A proponent of this theory, McCrea (1960, 1988), proposed that initially a dense interstellar cloud existed, which eventually produced a cluster of stars. Dense regions in the cloud formed and coalesced. The planets are smaller blobs captured by the star.

Along other lines, Russell (1935) suggested that the Sun could be a binary system with the second star being smaller than the Sun. Both collided in the past and planets were formed. Later Schmidt (1944) and Lyttleton (1961) developed the idea that a star passing through an interstellar cloud of gas and dust would be partially captured

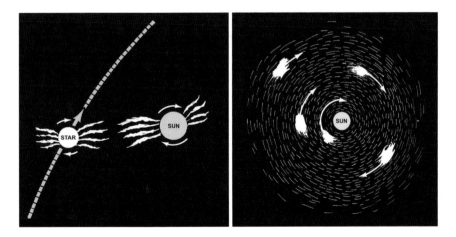

Fig. 8.2 The cosmogony of Chamberlain–Moulton

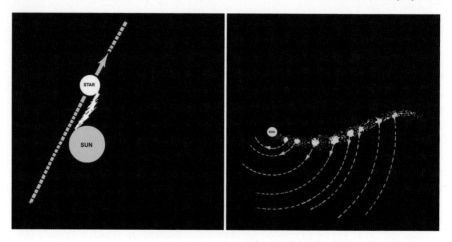

Fig. 8.3 The cosmogony of Jeffreys–Jeans

forming an envelope of gas and dust that eventually would form the planets. Finally, we should mention the Capture Theory (Woolfson 1964, 1978) that considers the idea of planets formed from captured and fissioned filaments (blobs) by an almost condensed Sun from other less massive diffuse proto-stars in a primitive stellar cluster scenery.

In the framework of these theories, the number of possible planetary systems would be rather small, given the reduced density of stars in the Galaxy. The main problem with these proposals is that hot gas expands, instead of contracting. So bubbles of hot gas would not form planets. The current explanation for the fact that most of the angular momentum is in the planets is that the Sun has lost angular momentum by magnetic braking (Ivanova and Taam 2003). For monographs on the historical development of these ideas see Jaki (1977), Brahic (1982) and Brush (1996, 2006).

8.2 Formation of Planetary Systems

8.2.1 Stellar Formation

The starting point of star formation is the collapse of a molecular cloud[1] (Cameron 1973). Hayashi (1961) was the first to describe the early phases of this gravitational collapse. See Larson (2003) for a recent review on the subject.

The angular momentum of the primitive solar nebula was predominantly of random turbulent origin, and perhaps it is plausible that the primitive solar nebula

---

[1] A type of interstellar cloud, cool and dense enough to allow for the formation of molecules.

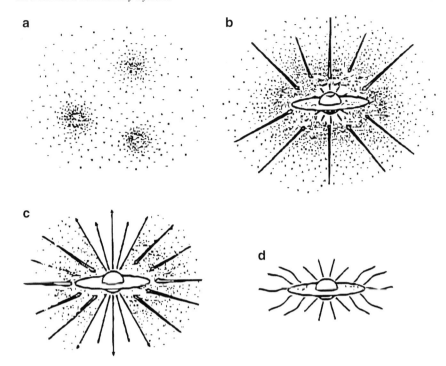

Fig. 8.4 The four stages of star formation. (**a**) Cores form within molecular clouds. (**b**) A protostar with a surrounding nebular disk forms at the center of a cloud core collapsing from inside-out. (**c**) A stellar wind breaks out along the rotational axis of the system. (**d**) The infall terminates with a newly formed star and a circumstellar disk. Adapted from Shu et al. (1987) Fig. 7. Reprinted with permission from Annual Reviews

should have possessed satellite nebulae in highly elliptical orbits, where the comets were formed.

Shu et al. (1987) summarize the knowledge on the formation of solar-like stars. This formation process has four main phases: (1) Formation of concentrations in the molecular cloud, (2) Formation of a protostar with a nebular disk, (3) Formation of a T Tauri star, with a strong wind, which dissipates part of the material of the nebula and (4) Formation of a protoplanetary disk around the star (Fig. 8.4).

Observational evidence of this process in other stellar systems was essential to corroborate the different theories. Since the material around the star is relatively cool, observations in the infrared range of the spectrum are the most adequate tool for this purpose. The projects IRAS (Infrared Astronomical Satellite 1983–1985), ISO (Infrared Space Observatory 1995–1999) and Spitzer Space Telescope marked successive highlights in the infrared observations of star-forming regions (Meyer et al. 2006) .

The Orion nebula is one of the closest ($\sim$1,350 light years away), and therefore, one of the best stellar nurseries to observe this process (Fig. 8.5). Within this region the Sigma Orionis cluster deserves special attention (see Caballero (2008) for a catalogue of its members).

Fig. 8.5 Image of the Orion nebula taken by the Advanced Camera for Surveys (ACS) aboard the NASA's Hubble Space Telescope. The bright central region is the home of the four heftiest stars in the nebula, called the Trapezium. Credit: NASA, ESA, M. Robberto (Space Telescope Institute, ESA) and the Hubble Space Telescope Orion Treasury Project Team

8.2.2 The Early Accretion Phase

Originally, the protoplanetary disk is composed of gas and dust. Some dust grains travel with the gas when the cloud collapses, while more dust condenses from the gas phase within the central plane of the disks. The dynamics of the dust is dominated by the gravity of the star and the aerodynamic forces within the gas, including turbulence. In the beginning, gravitational interactions between dust grains are small, becoming dominant when they reach sizes of 1–100 km, the planetesimals (Goldreich and Ward 1973). These planetesimals will accrete to form cores of the giant planets, and rocky and ice planets.

Early in the process, the inner portions of the disk reach higher temperatures than the outer ones, establishing a gradient across the disk. The 'equilibrium condensation model' explains the condensation from gas to solids for the refractory materials contained in the inner parts, followed by the volatiles in the outer zones (see Fig. 7.5).

The temperature profile across a typical disk is usually given by

$$T \simeq 2.8 \times 10^2 \left(\frac{a}{1\,\mathrm{AU}}\right)^{-1/2} \left(\frac{L_{\mathrm{star}}}{L_{\mathrm{sun}}}\right)^{1/4}.$$

8.2.3 The Protoplanetary and Debris Disks

Planets are known to have formed around stars and brown dwarfs. To understand how common planetary systems are, the first step is to determine what is the most adequate environment for planetary formation. Here we concentrate on stars with masses lower than two solar masses. The ancestors of the Sun and the solar-like stars are the 'T Tauri stars' (Bertout 1989). The mass of the disk around a classical T Tauri star is about 1–3% of the stellar mass, mostly made of gas with only 1% of the disk in the form of dust (Natta et al. 2007). T Tauri stars are the link between deeply embedded protostars and low-mass main-sequence stars. They show an excess of infrared emission resulting from the presence of a circumstellar disk of dust. The disk eventually disappears due to accretion onto the central star, planet formation and photoevaporation by UV-radiation from the central and nearby stars (Adams et al. 2004). As a result, the young star becomes a weakly lined T Tauri star, which slowly evolves into a solar-like star.

Debris disks are observed around main-sequence stars and are usually defined as the non-planetary component of a planetary system. In the Solar system, the debris are concentrated in two belts, which have evolved with time (Wyatt et al. 2003; Wyatt 2008). Aumann et al. (1984) discovered the first of such disks around Vega (α Lyrae). Another interesting object is Beta Pictoris, a bluish white main sequence dwarf star of spectral type A5 V. A circumstellar disk of dust and gas was first detected around Beta Pictoris by Smith and Terrile (1984). Golimowski et al. (2006) reported the presence of a second disk inclined by 5° from the main disk. Freistetter et al. (2007) suggested that a giant planet at 12 AU could account for the main features observed in the disk of β Pictoris. Figure 8.6 show a recent picture of this disk.

Far-IR observations at 70 μm suggest that 10–15% of solar-like stars possess cool outer dust disks that are massive analogs of the Solar Systems's Kuiper Belt (Bryden et al. 2006). So far, debris disks have been detected around 63 solar-like stars placed within 65 parsecs (Hillebrand et al. 2008; Trilling et al. 2008).

Further observations of A stars conclude that the IR excess decays with the stellar age (Rieke et al. 2005). Sub-millimetre and far-infrared observations have been able to resolve disks around other stars such as Vega (Su et al. 2005), Epsilon Eridani and Formalhaut[2] (Fig. 8.7). Clear regions observed in the disks probably indicate

---

[2] From the Arabic 'Fun al Hut' (the fish mouth), it is also known as HD216956 and alpha-Piscis Austrini, a 200-Ma-old A3 star located at a distance of 22 light years.

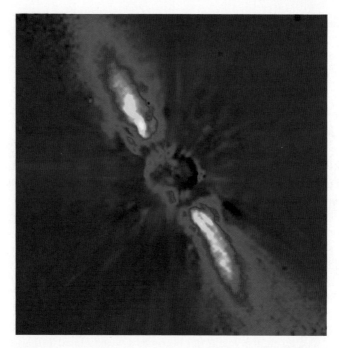

Fig. 8.6 Detection of a young disk of dust around the young star Beta Pictoris. Observation at 1.25 μm with the ESO 3.6 m telescope. Courtesy of European Southern Observatory

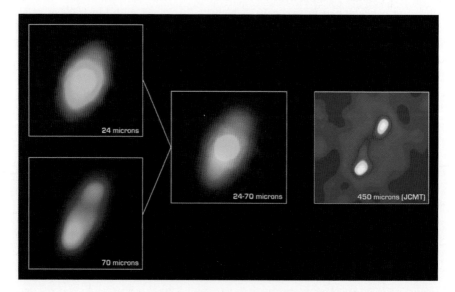

Fig. 8.7 Observation of a protoplanetary disk around the star Formalhaut with the Spitzer Space Telescope at three different wavelengths. Credit: NASA/JPL-Caltech/K. Stapelfeldt (JPL)

8.2 Formation of Planetary Systems

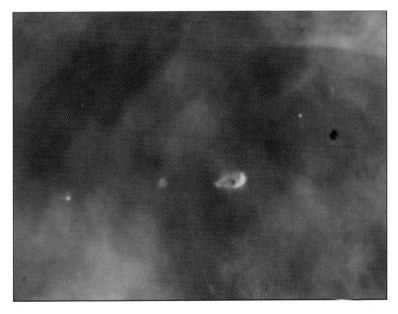

Fig. 8.8 Portion of the Orion nebula with five young stars. The field of view is 0.14 light years across. Credit: C.R. O'Dell, Rice University. Courtesy of Hubble Space Telescope ESA/NASA

the presence of planets (Forrest et al. 2004; Jura et al. 2004; Pontoppidan et al. 2008; Kalas et al. 2008).

Stellar nebula or protoplanetary disks have now been observed in the Orion nebula, and other star-forming regions, by astronomers using the Hubble Space Telescope. Some of these are as large as 1,000 AU in diameter. Our Solar System probably formed inside a cosmic Orion-like nebular maelstrom (Boss 1998, 2002) and not in a calm nebula. Figure 8.8 shows four young stars in Orion surrounded by gas and dust.

The accretion disks disappear on time scales of 1–10 million years (Pascucci et al. 2006), suggesting that planet formation may have already started. A model to account for planetary growth at a time scale similar or less than the lifetime of the circumstellar discs is still needed. It is also remarkable that the fraction of stars with disks increases with decreasing stellar masses, which may be due to a mass-dependent timescale for the dissipation of the internal discs (Zapatero et al. 2007).

Most of the dust in discs around newborn stars is made of silicates. In the outer parts, the dust grains are small and amorphous, becoming larger and crystalline in the vicinity of the star via stellar irradiation. This process has been observed in two stars by Van Boekel et al. (2004).

Beichman et al. (2005) found a clear similarity between the spectrum of the disk around the star HD 69830 and that of the comet Hale-Bopp (Fig. 8.9). Between 8 and 35 μm, the spectrum is dominated by strong features attributable to crystalline silicates, called forsterite, with an emitting surface area more than 1,000 times that

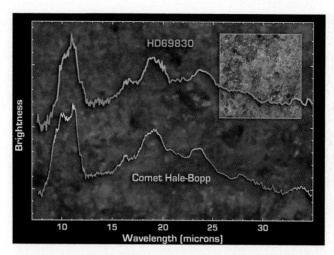

Fig. 8.9 The spectrum of the disk around the star HD69830, obtained by Spitzer Space Telescope, and the one corresponding to the comet Hale-Bopp (ISO observations). The inset shows a picture of forsterite, courtesy of Dr. George Rossman, California Institute of Technology, Pasadena. Credit: NASA/JPL-Caltech, C. Beichman (JPL)

of our zodiacal cloud. Since no excess is detected at 70 μm, the emitting material must be quite warm and be confined to within a few Angstrom unit of the star.

Figure 8.10 shows an infrared spectrum of two stars, young enough to still possess deep dusty disks. The broad depression signifies the presence of silicates. Other dips in the spectra are attributed to water ice, methanol ice (red) and carbon dioxide ice. The presence of these substances in solid form clearly shows that the protoplanetary disks have low temperatures (Lisse et al. 2007).

8.2.4 Formation of Giant Planets

The traditional view, 'core accretion', is that giant planets first form a rocky core, several times the Earth's size, and then accrete material for the outer gaseous envelope. At large orbital radii, beyond the snow line, the temperature is low enough that ices as well as rocky materials can condense. The main problem was the amount of time necessary to form a Jupiter-like planet (see Fig. 8.11); however, the initial estimates in the range 10–1,000 Ma have been reduced to 2–3 Ma at 5 AU (Inaba and Ikoma, 2003; Inaba et al. 2003). In principle, this mechanism explains the formation of ice giants and the mentioned correlation (see Chap. 7) between stellar metallicities and the presence of exoplanets.

Working with this model, Laughlin et al. (2004) found that the formation of Jupiter-mass planets orbiting M dwarf stars is seriously inhibited at all radial locations, in sharp contrast to solar-type stars. On the contrary, Kennedy and Kenyon

8.2 Formation of Planetary Systems

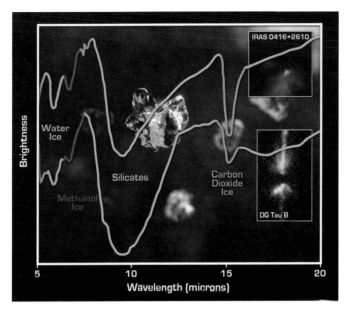

Fig. 8.10 Ice in protoplanetary disks. Spitzer spectra of two young stars indicating the presence of water ice (*blue*), methanol ice (*red*) and carbon dioxide ice (*green*). Credit: NASA/JPL-Caltech/ D. Watson (University of Rochester)

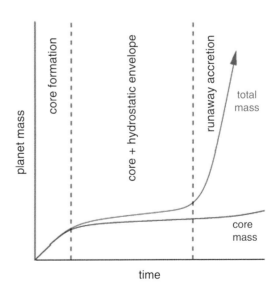

Fig. 8.11 Schematic illustration of the growth of giant planets via the core accretion mechanism. The *blue line* shows the time evolution of the core mass, the *red line* shows the growth of the total (core plus envelope) mass. Credit: P. Armitage (2008) Scholarpedia 3(3)4479

(2008) have found that giant planets form over a wide range of spectral types. The probability that a given star has at least one gas giant increases linearly with stellar mass from 0.4 to 3 solar masses.

A paradox existed in the Solar System that apparently would contradict this model: Jupiter seemed to have a rocky core of about 3 M_E, or no core at all, while Saturn possesses one in the range 10–20 M_E (Saumon and Guillot 2004). However, recent simulations by Militzer et al. (2008) increase the size of Jupiter's core to 17 M_E, also suggesting that the core is made of layers of metals, rocks and ices of methane, ammonia and water. The question may finally be settled with the arrival of the Juno mission, in 2016, which will measure the planet's magnetic field and gravity.

An alternative theory, called 'disk instability', is based on a gravitational process where the action takes place in the protoplanetary disk (Boss 1997, 2007). A gas disk with surface density, Σ, sound speed, c_S, and angular velocity, Ω, is unstable if the Toomre parameter, Q, is smaller than 1

$$Q = \frac{c_S \Omega}{\pi G \Sigma}.$$

If, additionally, the disk is able to cool on an orbital time scale, then the instability produces fragmentation of the disk into giant planets.

This model offers an explanation to the question of how Jupiter and Saturn formed fast enough to explain the great thickness of their atmospheres, some thousands of kilometres deep. In this scenario, gas giant planets form rapidly (around 1,000 years), which is fast enough to prevent the gas loss of the protoplanetary disks. Hence, the thick atmospheres of the gas giant planets no longer seem so mysterious.

However, an additional mechanism is necessary to explain the thin atmospheres of the icy giant planets (Uranus and Neptune) with the disk instability model, a process that we have discussed briefly in the previous chapter (Boss 2006): the photoevaporation of material produced by the EUV emission of a massive nearby star, which is intense enough to heat the nebular gas throughout the Solar System's protoplanetary disk of gas and dust (Boss 2003). The proto-Uranus and proto-Neptune planets were located far from the Sun, at a distance where gravity is comparatively weak. Heated gas in their vicinity escapes more quickly into deep space, and as a result, the icy giants have comparatively thin atmospheres. Terrestrial planet formation is still plausible in this scenario even if gas giants formed faster.

8.2.5 Formation of Terrestrial Planets

In the absence of direct observations or suitable laboratory experiments, much of what we know about terrestrial planet formation comes from numerical simulations. It is generally accepted that terrestrial planets (TP) grew by the collisional accumulation of planetesimals (Chambers 2004).

8.2 Formation of Planetary Systems

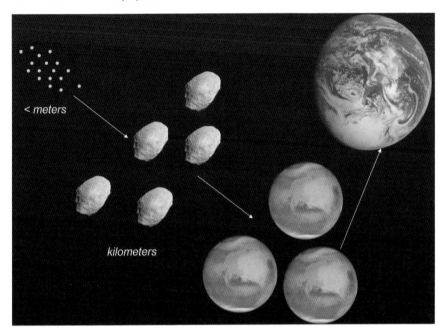

Fig. 8.12 Phases in the formation of terrestrial planets

Our current understanding of the formation of terrestrial planets can be described through the following steps (Fig. 8.12):

- Runaway accretion in a disk of small bodies ($<10^6$ years). Large bodies typically grow quicker than small ones due to differences in their orbital eccentricities and inclinations.
- Oligarchic growth period in which planetary embryos grow at the expense of smaller bodies (10^5-10^6 years), until they have swept up most of the smaller planetesimals.
- The former process generates planetary embryos, which have lunar to martian masses. The system becomes unstable and the orbits begin to intersect (10^8 years). In the inner region of the planetary system the embryos collide and coalesce into Earth-size bodies.

A crucial point is the separation between silicates and heavy metals and the formation of the planetary cores. Radioactive dating of meteorites suggests that the Earth and Mars cores were formed only 29 and 13 million years after the birth of the Solar System (Yin et al. 2002; Kleine et al. 2002).

Simulations by Levison et al. (1998) indicate that terrestrial planets will outnumber giant planets. This is supported by the steep rise of the mass distribution of planets with decreasing mass. Around two-thirds of the stars have enough solids to form Earth-like planets (Greaves et al. 2007). Boss (2006) have studied the mechanisms of formation of Super-Earths around M dwarfs, concluding that these stars should form significantly more Super-Earths than giant planets.

Raymond et al. (2005) investigated the formation of TPs in disks with varying density profiles, Σ, given by a law $\Sigma(1\mathrm{AU})r^{-\alpha}$, where r is the radial distance to the parent star. They find that for steep profiles (high α), TPs are more numerous, massive and form closer to the star, showing higher iron and lower water contents. Raymond et al. (2007) suggest that the fraction of systems with sufficient disk mass to form planets with $M > 0.3\ M_E$ decreases for low-mass stars. Water delivery in such systems (e.g. around M stars) is inefficient and therefore TPs should be small and dry.

The number and masses of terrestrial planets may vary from one planetary system to another due to differences in the amount of solid material available and the presence or absence of giant planets, as well as the highly stochastic nature of planet formation. Moreover, the initial configuration of the planets in a given system can undergo drastic changes produced by mutual gravitational interactions. It is therefore relevant to investigate the dynamical stability of planetary systems.

8.3 Planetary Orbits

Most of the Solar System planets have almost circular orbits. However, in exoplanetary systems, large and chaotic variations in the eccentricity of giant planets could produce severe perturbations in the orbit of Earth-like planets, affecting their habitability conditions. Here we review the basic concepts on orbital configurations, migrations and dynamical stability, and how they will affect the Earth-like planets. For monographs on celestial mechanics see Moulton (1970), Gónzalez Martínez-Pais (2003) and Celletti and Perozzi (2007).

8.3.1 Basic Orbital Elements

Orbital elements are the parameters needed to specify the orbit of a celestial body (e.g. a planet) uniquely. In Fig. 8.13, the basic elements are represented. The reference plane is the ecliptic. The two points where the ecliptic intersects the orbit of the planet are called nodes. The reference direction is given by the ascending node, the vernal point. Based on this reference frame, the orbit of a body is defined by the following elements, referred at a determined epoch, the time origin $t = 0$.

- Semi-major axis, a. Its length is the distance between the geometric centre of the orbital ellipse with the periapsis (point of closest approach to the central body), passing through the focal point where the centre of mass resides.
- Eccentricity ($0 < e < 1$), determining the shape of the orbit, describing how flattened it is compared to a circle.
- The inclination, i, orients the orbital plane with respect to the plane of the ecliptic. Imagine the angle being formed by pivoting the orbital plane through an axis of rotation coinciding with the line of nodes.

8.3 Planetary Orbits

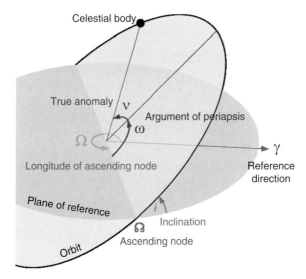

Fig. 8.13 Orbital elements. In this diagram, the orbital plane (*yellow*) intersects a reference plane, the ecliptic (*gray*). See text for a detailed description

- The longitude of the ascending node, $0 < \Omega < 360°$, orients the ascending node with respect to the vernal point. Imagine the angle being formed by pivoting the orbital plane through an axis of rotation perpendicular to the plane of the ecliptic and passing through the centre of mass.
- The argument of periapsis (perihelion) ($0 < \omega < 360°$) orients the semi-major axis with respect to the ascending node. Imagine the angle being formed by pivoting the orbital plane through an axis of rotation perpendicular to itself and passing through the centre of mass.
- The true anomaly, v also written T, orients the celestial body in space. It is the angle between the direction of periapsis and the current position of the object on its orbit, measured at the focus of the ellipse.
- The time of passage for the perihelion, T_0, usually expressed as a Julian date.
- Longitude of perihelion: $\bar{\omega} = \Omega + \omega$.
- Mean anomaly of the epoch, M, the fraction of the orbital period that has elapsed since the last passage at periapsis, expressed as an angle: $M - M_0 = n(t - t_0)$, where M_0 is the mean anomaly at time t_0 and n the mean motion, a measure of how fast a body progresses around its orbit.

Table 8.1 lists the main orbital parameters of the planets of the Solar System.[3]

---

[3] For a complete list of the orbital elements of the planets of the Solar System, see http://ssd.jpl.nasa.gov

Table 8.1 List of the main orbital parameters of the planets of the Solar System

	Semimajor axis (AU)	Eccentricity	Orbital period [years (days)]	Ω	Mass [M_E(M_J)]
Mercury	0.387	0.206	0.241	48.332	0.055
Venus	0.723	0.007	0.615	76.681	0.86 (0.0027)
Earth	1.000	0.017	1 (365.25)	−11.261	1.00 (0.0031)
Mars	1.524	0.093	1.881	49.759	0.11 (0.0003)
Jupiter	5.203	0.048	11.862	100.556	318 (1.00)
Saturn	9.537	0.054	29.458	113.715	95 (0.299)
Uranus	19.191	0.047	84.01	74.230	14.5 (0.046)
Neptune	30.069	0.009	164.79	131.722	17.2 (0.054)

8.3.2 Keplerian Orbits

J. Kepler (1571–1630) established his famous three laws of planetary motion by studying the observations made by Tycho Brahe (1546–1601). These are the following:

1. The orbit of every planet is an ellipse with the sun at one of the foci.
2. A line joining a planet and the sun sweeps out equal areas during equal intervals of time.
3. The squares of the orbital periods of planets are directly proportional to the cubes of the semi-major axis of the orbits.

I. Newton (1643–1727) exposed in his *Philosophiae Naturalis Principia Mathematica*[4] (1687) his three laws of motion together with the law of universal gravitation, formulated as follows: Two bodies, with masses M_1 and M_2, and located at a distance d will attract each other with a gravitational force given by

$$F = G \frac{M_1 M_2}{d^2},$$

where G is the gravitational constant, first accurately determined by H. Cavendish (1731–1810) in 1798. See Fixler et al. (2007) for a modern measurement.

Newton found that two bodies with gravitational attraction will follow trajectories described by conical sections. In his model, Geometry and Physics were connected perfectly. If both bodies are bound gravitationally, the orbit it will describe is an ellipse. In the case that the second body manages to win the gravitational attraction of the first, it will describe a hyperbola. In a case of equality it will describe a parabola (see Fig. 8.14).

The two bodies will move around the common centre of mass or barycentre (Fig. 8.15). The distance from the centre of the first body to the barycentre is given by

---

[4] It was preceded, in November 1684, by a short manuscript entitled *De motu corporum in gyrum* (On the motion of bodies in an orbit), sent by Newton to E. Halley, who later edited the Principia.

8.3 Planetary Orbits

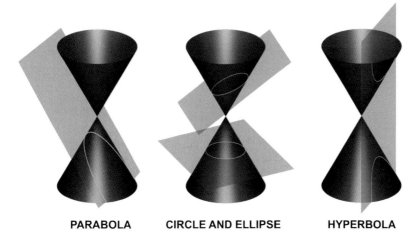

PARABOLA   **CIRCLE AND ELLIPSE**   **HYPERBOLA**

Fig. 8.14 Conic sections

Fig. 8.15 The motion of two bodies around a common centre of mass

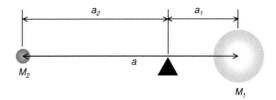

$$a_1 = a \cdot \frac{M_2}{M_1 + M_2}.$$

Now $a_2 = a - a_1$ is the semi-major axis of the second body's orbit. If the barycentre is located within the more massive body, like a star, that body will appear to 'wobble' rather than following a discernible orbit (see Chap. 6 on the methods of detection of extrasolar planets).

The third law of Kepler is expressed now as

$$\frac{G(M_1 + M_2)}{4\pi^2} = \frac{a^3}{P^2}.$$

Let us assume a body of mass M_2 orbiting the Sun of mass M_1, at a distance r, with respective velocities V_1 and V_2. The conservation of the energy is expressed as the sum of its kinetic and potential energies (Virial theorem). We have

$$E = \frac{1}{2}M_1 V_1^2 + \frac{1}{2}M_2 V_2^2 - \frac{GM_1 M_2}{r} \sim -G\frac{M_1 M_2}{2a}.$$

Expressing the orbit in polar coordinates $(r, \theta)^5$, we can define the linear momentum $p = \mu V$, where $\mu = (M_1 M_2)/(M_1 + M_2)$ is the reduced mass, and now the angular momentum

$$J = r \times p.$$

Applying the conservation of angular momentum,

$$r_1^2 \frac{d\theta}{dt} = \text{constant} = h_1 = \left[\frac{M_2}{M_1 + M_2}\right]\frac{h}{2},$$

$$r_2^2 \frac{d\theta}{dt} = \text{constant} = h_2 = \left[\frac{M_1}{M_1 + M_2}\right]\frac{h}{2}.$$

We have for the general formulation of the orbit

$$r = \frac{h^2}{\mu} \frac{1}{1 + e\cos(\theta)},$$

where h is the specific angular momentum and e the eccentricity of the orbit. The total angular momentum of the system is now given by

$$J = m_1 h_1 + m_2 h_2 = \mu h.$$

If $E < 0$, the orbits is bound (Ellipse).
If $E = 0$, the orbit is marginally bound or critical (Parabola).
If $E > 0$, the orbit is unbound (Hyperbola).

In the real world, the gravitational force upon each planet, F_i, is the sum of the forces of the Sun and all the other bodies in the system

$$F_i = m_i \frac{d^2 r_i}{dt^2} = Gm_i \sum_{j \neq i} m_j \frac{r_j - r_i}{|r_j - r_i|^3},$$

where m_i are the masses and r their positions in space. We have the n-body problem, which contains 6n variables, since each body is represented by three space and three velocity components. The resolution implies 10 independent algebraic integrals: 3 for the centre of mass, 3 for the linear momentum, 3 for the angular momentum and finally 1 for the energy.

We have an analytical solution only for the two-body case. In our Solar System, this simplification works reasonably well, since the gravitational force exerted by the Sun is dominant. Nevertheless, the better the precision of the measurements became, the more complicated the planetary movements turned out to be. The object that presented the most problems was definitely the Moon. L. Euler (1707–1783)

---

[5] r is the distance between the orbiting body and the central body, and θ the direction of the orbiting body, the true anomaly.

and A.C. Clairaut (1713–1765) developed celestial mechanics calculations, assuming that the true orbit of a body was the superposition of an average value (Two-Body Problem) and a disturbance.

8.3.3 Harmony and Chaos

8.3.3.1 Historical Background

The laws of Kepler and Newton led to a perfectly predictable harmonic world. However, I. Newton was already worried that the accumulated effects of the weak gravitational tugs between neighbouring planets would increase their orbital eccentricities, leading to collisions and finally to the destruction of the Solar System. In Book III of his book *Optiks*, first published in 1704, he made some reflections on the destiny of the Solar System:

> For while comets move in very eccentric orbs in all manner and position, blind fate could never make all the planets move one and the same way in orbs concentric, some inconsiderable irregularities excepted, which could have arisen from the mutual actions of planets upon one another, and which will be apt to increase, till this system wants a reformation.

This idea has been developed in detail until our days.

The calculations of planetary trajectories had been based on geometric methods. The first passage in the simplification was introduced by R. Descartes (1596–1650) and P. Fermat (1601–1665) with the development of analytical geometry, which allowed the description of curves with algebraic equations. Overcoming philosophical prejudices on the idea of the infinite, I. Newton and G. Leibniz (1646–1716) introduced the concepts of the derivative and integral, a great advancement in the determination of trajectories of celestial objects and the theoretical exposition of new problems. Shortly afterwards, J. Bermoulli (1654–1705) raised the basic differential equations for the movement of planets. J.L. Lagrange (1736–1813) found an analytical solution for the three-body problem when the mass of the third is negligible compared with the other two, and its movement is in the same orbital plane. It is what we have described as the 'Restricted Three-body Problem'.[6]

Still the prediction of the behavior of the Solar System reached its greatest landmark with the discovery, by means of mathematical calculations, of a new planet, Neptune. W. Herschel (1738–1822) had discovered the planet Uranus in 1781. Soon afterwards, and independently, the British astronomer J.C. Adams (1819–1892) and the French astronomer U. Leverrier (1811–1877) found anomalies in the Uranus orbit, which they interpreted as due to the gravitational action of an unknown planet

---

[6] The third body has a mass small enough that it does not influence other objects, which are usually assumed to be in circular orbits. Our Solar System can be separated into several subsystems of this kind (e.g. Sun-planet-satellite, Sun-planet-asteroid etc.), making the numerical solution more accurate.

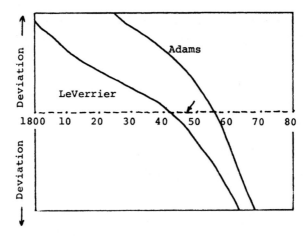

Fig. 8.16 The deviation of the predictions of Adams and Leverrier (*solid lines*) from the true position of Neptune for different years from 1800 to 1880. The *arrow* points to 1846 when the discovery was made and the predictions were incorrect by only a few degrees

located even farther from the Sun.[7] Each of them found it difficult to convince their colleagues to dedicate telescope time to observe the region where the new planet was to be found. Finally, two astronomers at the Observatory of Berlin, J.S. Galle (1812–1910) and H.L. D'Arrest (1822–1875), followed the suggestions of Leverrier and they discovered Neptune on the 23rd of September 1846 (see Fig. 8.16). Galle wrote to Le Verrier on 25th September saying: 'Monsieur, the planet of which you indicated the position really exists'. Le Verrier soon replied: 'I thank you for the alacrity with which you applied my instructions. We are thereby, thanks to you, definitely in possession of a new world'.[8]

At the beginning of nineteenth century, P.S. Laplace (1749–1827) thought that with the tools available he could predict, with total security, the future of a planetary system. An idea that fitted perfectly with the harmonic model of skies. The average movements of planets and satellites were stable, the deviations were compensated and the anomalies were periodic. His work is the basis of the casual determinism expressed in his own words in the introduction of his 'Essai Philosophique sur les Probabilités'[9]:

---

[7] There were some important errors in the calculations of Adams and Leverrier. Following the Titius–Bode law, they estimated the distance Neptune–Sun at 38 AU. Fortunately, between 1790 and 1850, Uranus and Neptune were in conjunction (aligned with the Sun on the same side of the planetary system), and this helped to minimize errors.

[8] In fact, Galileo first observed Neptune on 28 December 1612 and 28 January 1613, but he believed it to be a fixed star (Kowal and Drake 1980).

[9] It first appeared as an introduction to the third edition of his *Théorie Analytique des Probabilitiés*, which was originally published in 1812. There is an English version of the complete essay published by Dover in 1951.

8.3 Planetary Orbits

> We may regard the present state of the universe as the effect of its past and the cause of its future. An intelligence which at a certain determined moment would know all forces that set nature in motion, and all positions of all objects of which nature is composed, if this intelligence were also vast enough to submit these data to analysis, it would embrace in a single formula the movements of the greatest bodies of the universe and those of the tiniest atom; for such an intellect nothing would be uncertain and the future just like the past would be present before its eyes.

This was the expression of an absolute determinism, leading to an almost infallible science. But soon the situation became more complicated and at the same time more interesting.

N. Poincaré (1854–1912) presented a paper to an international competition held by the Swedish court in 1888, in which he demonstrated that: 'it was not possible to integrate the equations of motion of three bodies suffering mutual interactions, not being feasible to find an analytical solution to the movement of planets valid for an infinite interval of time'.[10] First published in 1890 and later developed in his monumental work *Les Méthodes Nouvelles de la Mécanique Céleste*,[11] Poincaré's work comes to mark the beginning of the theory of chaos (Poincaré, 1890). In his 1908 essay *Science et Méthode*[12] he noted:

> If we knew exactly the laws of the nature and the situation of the universe at the initial moment, we could predict exactly the situation of that same universe at a succeeding moment. But even if it were the case that the natural laws had no longer any secret for us, we could still only know the initial situation approximately. If that enabled us to predict the succeeding situation with the same approximation, that is all we require, and we should say that the phenomenon had been predicted, that it is governed by laws. But it is not always so; it may happen that small differences in the initial conditions produce great ones in the final phenomena. A small error in the former will produce an enormous error in the latter. Prediction becomes impossible.

A.N. Kolgomorov (1903–1987), V.I. Arnold and J.K. Moser showed that, for certain values of the initial conditions, it was nonetheless possible to obtain convergent series. If the masses, eccentricities and inclinations of the planets are small enough, then many initial conditions lead to quasi-periodic orbits. This is known as the KAM theory (Tabor 1989). Unfortunately, the masses of the Solar System bodies are much too large and its application to planetary motions is strongly limited.

The dynamic systems are divided into stochastic, with random movements, and deterministic when these are due to some cause. In the latter case, if the relationship cause–effect is non-linear, a temporary limit in the precision of the predictions, and often chaos, appears. Already in the twentieth century, E.N. Lorenz (1917–2008) defined chaos as the sensitivity of the system to the initial conditions and applied the concept to the climate system (Lorenz 1963, 1993).

---

[10] The jury members for the competition were K. Weierstrass (1815–1897), G. Mittag-Leffler (1846–1927) and C. Hermite (1822–1901), who awarded him the prize.

[11] English version by D. Goroff (1993) published by the American Institute of Physics and reprinted by Springer in 2007.

[12] English translation: *Science and Method*, Lancaster, PA: Science Press 1913.

8.3.4 Relevant Parameters of Dynamical Stability

The *Roche limit*, d, is the distance within which a planetary body held together only by its own gravity will disintegrate due to the tidal forces of the star or a second planet, exceeding the gravitational self-attraction:

$$d = R \left(2 \frac{\rho_M}{\rho_m} \right)^{1/3},$$

where R is the stellar radius, ρ_M the stellar density and ρ_m the planetary density. For a fluid planet, tidal forces produce elongation, causing it to break apart more readily. Planetary rings are located inside the Roche limit.

A *Hill sphere* is the gravitational sphere of influence of an astronomical body in the face of perturbations from another heavier body around its orbit (Gladman 1993). The radius of the Hill sphere was defined by G.W. Hill (1838–1914) and is given by

$$R_H = a \left(\frac{m}{3M} \right)^{1/3},$$

where m is the mass of the smaller body, orbiting around the heavier body of mass M at a distance a. Orbits at or just within the Hill sphere are not stable in the long term. In the Solar System the planet with the largest Hill sphere is Neptune (0.775 AU).

Barnes and Greenberg (2006) have applied the Hill stability criteria to different extrasolar systems. The stability of a dynamic system, composed by a central star (Mass M_S) and two non-resonant planets (Masses M_1 and M_2), is achieved when it fulfills the following inequality

$$-\frac{2(M_S + M_1 + M_2)}{G^2 M_*} J^2 E > 1 + 3^{4/3} \frac{M_1 M_2}{M_S^{2/3} (M_1 + M_2)^{4/3}} - \frac{M_1 M_2 (11 M_1 + 7 M_2)}{3 M_S (M_1 + M_2)^2},$$

where J is the total angular momentum and E the energy. Note that the left hand side of the inequality, usually named β, is a function of the positions and velocities of the system, and the right hand side, β_{crit}, is purely a function of the masses. We have stability when $\beta/\beta_{crit} > 1$.

The tidal forces between two objects tend to synchronize the orbital and rotational periods in order to preserve the angular momentum of the system. For an Earth-like planet in a circular orbit, this distance, R_{TL}, was estimated by Peale (1977) as

$$R_{TL}(t) = 0.027 (P_0 t/Q)^{1/6} M^{1/3},$$

where P_0 is the original rotation period and Q the solid plus ocean dissipation rate. For the Earth $P_0 = 13.5$ h and $Q = 100$ (Burns 1986).

8.3.4.1 Uncertainty

The harmonic world of Kepler and Laplace results from a short-term vision of our System. In his 1993 book, I. Peterson commented: *Was it an accident of celestial mechanics that the Solar System happens to be simple enough to have permitted the formulation of Kepler's laws and to ensure predictability on a human time scale? Or could we have evolved and pondered the skies only in a Solar System afflicted with a mild case of chaos?*.

Let us consider the distance, d, between two possible trajectories in a dynamical system based on slightly different initial condition and its time evolution (Fig. 8.17). It illustrates on the transition between two kinds of dynamical systems: (1) A regular and highly predictable one, where small differences grow linearly – Δx, $\Delta v \propto t$, and (2) A chaotic difficult-to-predict one, where small differences grow exponentially – Δx, $\Delta v \propto \exp(-t/t_L)$. This zone of transition between the two regimes occurs at the so-called Lyapunov time, T_L, which for the terrestrial planets will be of the order of five million years, increasing to ten for giant planets (Lecar et al. 2001). A second characteristic time scale is the escape time, the time for a major change in the orbit

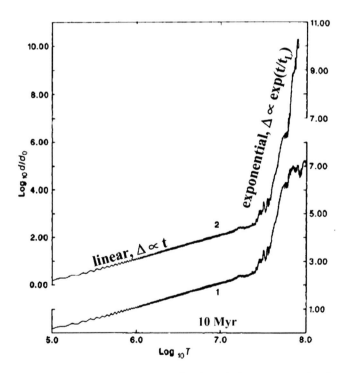

Fig. 8.17 Transition from a regular behaviour of a dynamical system to a chaotic one for the distance between two nearby trajectories. Adapted from Laskar (1989). Reprinted by permission from Macmillan Publishers Ltd: Nature Vol. 338, p. 237. Copyright (1989)

as an expulsion from the system. Finally, we could mention the time to first close encounter, T_{CE}, the time it takes to the first close encounter between any pair of planets, usually considered as $3R_H$.

Chaotic trajectories can be better visualized in the phase space, where we see the set of all possible states of a physical system, described by its position, q_i, and momentum, p_i, variables. Every degree of freedom or parameter of the dynamical system is represented as an axis of a multidimensional space. Each state corresponds to a point in the phase space.

Following these studies we can say that there are two types of trajectories in dynamical systems: 'Nice' deterministic trajectories and 'random' trajectories associated with resonances. When we increase the energy of the system, we also increase the random regions. For some critical value of energy, chaos appears.

For monographs and reviews on chaos in the Solar System, see Peterson (1993), Diacu and Holmes (1996), Malhotra et al. (2001) and Laskar (2008).

8.3.5 Resonances in Planetary Systems

A resonance occurs when some of the quantities characterizing the motion of two or more celestial bodies can be considered as commensurable, that is their ratio is close to an integer fraction:

$$\frac{T_1}{T_2} = \frac{p_1}{p_2},$$

where T_1 and T_2 are parameters of the corresponding orbits of the two bodies and P_i the positive integers. Mean motion resonances occur when two bodies have orbital periods that fulfill this condition.

A resonance assures the repetition of some geometrical configurations, implying periodic recurrence of gravitational interactions. Overlapping resonances, that is multiple gravitational resonances in close proximity provide the route to chaos in the Solar System (Malhotra et al. 2001).

8.3.5.1 Laplace Resonances

The best example of mean motion resonance, also called Laplace resonance, involves Jupiter and three of its satellites: Io (1:1), Europa (2:1) and Ganymede (4:1). Stabilization of the orbits occur in this case, because the two implicated bodies move in such a synchronized way that they never approach closely.

The orbital periods, T, of the satellites follows the relation

$$n_{Io} - 3n_{Europa} + 2n_{Ganym} = 0,$$

where $n_{Io} = 360/T$ (degrees/day) $= 203.488992435$. Io is in a 2:1 resonance with Europa, which is itself in a 2:1 resonance with Ganymede (Fig. 8.18).

8.3 Planetary Orbits

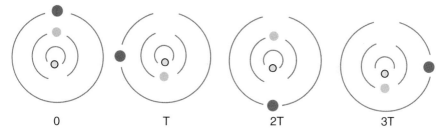

Fig. 8.18 Laplace resonance in the Jupiter system of satellites

This resonance prevents the occurrence of triple conjunction (the three satellites aligned on the same side of the planet), which would give rise to a peak in their mutual perturbations.

8.3.5.2 Kirkwood Gaps

Orbital resonances can also destabilize the orbits when small bodies are implicated.

In 1867, D. Kirkwood (1814–1895) found that the positions that agreed with the resonances of the greater Jupiter semi-axis lacked asteroids. Models of their dynamic behaviour showed that, on scales of hundreds of thousands of years, abrupt variations in the eccentricity took place, which sent asteroids to the interior of the Solar System, where they could collide with the terrestrial planets (Kirkwood 1867). The best known example is the impact that caused the extinction of the dinosaurs, and many other species, 65 million years ago (see Chap. 2). These gaps are like holes, through which the asteroid population is slowly draining away. In our days the collision danger persists; therefore, the dynamic instability of the Solar System can lead to a real threat for life on our planet.

The most prominent Kirkwood gaps (see Fig. 8.19) are located at the following mean orbital radii:

- 2.06 AU (4:1 resonance)
- 2.5 AU (3:1 resonance)
- 2.82 AU (5:2 resonance)
- 2.95 AU (7:3 resonance)
- 3.27 AU (2:1 resonance)

The 3:1 resonance is of great importance for the future of Earth, because it is the source of the Near-Earth Objects (NEOs) (Wisdom 1985). In the 2:1 resonance, the reality has been shown to be very different, and the trials to extend the Wisdom model to this resonance have shown that the diffusion times are of the order of Ga. The dynamics of the asteroids in the 3:2 resonance (the Hildas) is very similar to that of the asteroids in the 2:1 resonance, but the chaotic diffusion is much slower, allowing the Hildas to remain in the resonance for periods of time much larger than the age of the Solar System, thus explaining the important population observed there.

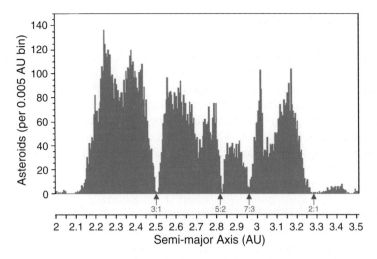

Fig. 8.19 The Kirkwood gaps in the distribution of asteroids

Mean motion resonances provide a basic mechanism for long-term stability, but overlapping of resonant orbits leads to a chaotic behaviour. Chaos among the Jovian planets results from the overlap of the components of a mean motion resonance between Jupiter, Saturn and Uranus (Murray and Holman 1999). Numerical simulations by Adams et al. (2008) suggest that the presence of turbulence in circumstellar disks, after the formation of the planets, strongly limits the fraction of planets that could survive in a mean motion resonance.

Minton and Malhotra (2009) have simulated the 4 Ga evolution of the asteroid population finding not only the Kirkwood gaps but also the track[13] of the migration event of Jupiter and Saturn, which gave rise to the 'Late Heavy Bombardment' 3.9 Ga ago.

8.3.5.3 Spin–Orbit Resonance

Other resonances exist between different orbital parameters, but their effects on stability are smaller. A good example of spin–orbit resonance is the Moon leading to a synchronous orbit. As a consequence the Moon is tidally locked to the Earth.

Mercury is also in a 3:2 spin–orbit resonance. The chaotic evolution of its orbit brought its eccentricity beyond 0.32, causing it to be captured in this resonance (Correia and Laskar 2004).

---

[13] Excess depletion of asteroids both on the inner edge of the belt and in the outer edges of each Kirkwood gap.

8.3.6 Lagrangian Points

The N-body problem could lead to stable situations for some specific geometries, such as the Lagrangian Points, first described by J.L. Lagrange (1736–1837). Two configurations are adequate for this purpose: (1) three bodies lying along the same line or (2) three bodies located at the vertices of an equilateral triangle (Fig. 8.20).

The Lagrangian points (also called libration points) are the five positions in an orbital configuration where a small object affected only by gravity can theoretically be stationary relative to two larger objects (such as a satellite with respect to the Earth and Moon).

A first observational test was the discovery of the Trojan group of asteroids, which are divided into two groups located along the orbital path of Jupiter, 60° ahead and behind the planet (L4 and L5 points).

The possibility of moving back and forth from L4 to L5, passing through L3, identifies a stable 'horseshoe' orbit. A good example is given by the asteroids Cruithne[14], 2002aa29[15] (Wajer 2009) and 54509 YORP, secret companions of the Earth (Fig. 8.21). They are in a normal elliptic orbit around the Sun. However, because its period of revolution around the Sun is almost exactly equal to that of the Earth, they appear to 'follow' each other in their paths around the Sun. Figure 8.21 shows the variable distance of one of these objects to the Earth while describing its horseshoe orbit.

The Sun–Earth L1 is ideal for making observations of the Sun, because objects here are never shadowed by the Earth or the Moon.[16]

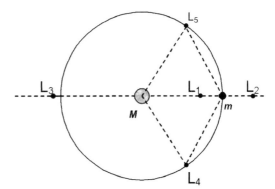

Fig. 8.20 Location of the Lagrangian Points

---

[14] Cruithne was discovered on 10 October 1986 by Duncan Waldron on a photographic plate taken with the UK Schmidt Telescope at Siding Spring Observatory, Australia. It is approximately 5 km in diameter.

[15] Discovered on 9 January 2002 has a diameter of about 40–100 m.

[16] Such is the case with the ESA/NASA satellite SOHO, observing the Sun since 1995.

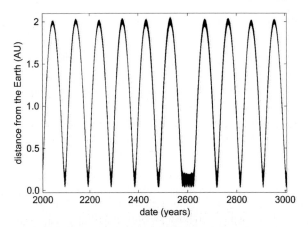

Fig. 8.21 Distance of 2002aa29 from the Earth as a function of time. For several hundred years this object moves in a horseshoe orbit and regularly approaches the Earth. The horseshoe motion is interrupted occasionally, transforming into the quasi-satellite motion. In this state, the asteroid stays within 0.2 AU from the Earth for several decades. Credit: Wajer (2001), reprinted with permission of Elsevier

Schwarz et al. (2005) have investigated the stability and habitability of fictitious Trojan planets situated in a 1:1 resonance around a giant planet, solving a three body problem with different mass ratios of the primary bodies.

8.4 The Dynamically Habitable Zone

Previously we have defined the habitable zone (HZ) of a planetary system as the region where water can exist in liquid phase on the surface of a planet by virtue of a combination of stellar and planetary parameters.

Following this idea, its dynamically habitable zone (DHZ) is defined as the region of the planetary system where a terrestrial planet can survive without suffering major perturbations in its orbit due to the influence of giant planets. The inner (R_{int}) and outer (R_{out}) boundaries for stable orbits of a planet with orbital parameters (a,e) are defined by

$$R_{int} = a(1-e) - n_{int}(e)R_H,$$
$$R_{ext} = a(1+e) + n_{ext}(e)R_H,$$

where n_{int} should range in the interval [2–3], with [3–16] for n_{ext}.

If hot giant planets are present in the system, and simplifying the problem to circular and co-planar orbits, we can consider three main types of stable orbits for terrestrial planets (Fig. 8.22).

8.4 The Dynamically Habitable Zone

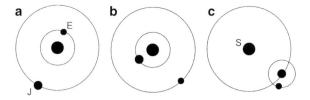

Fig. 8.22 Three main types of stable orbits for terrestrial planets: (**a**) a Solar system configuration, (**b**) a hot giant planet orbiting close to the star and (**c**) terrestrial planet orbiting a giant planet

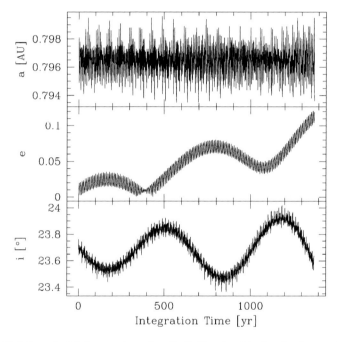

Fig. 8.23 Variation of orbital parameters of an Earth-like planet in the planetary system around HD 114783, a K2V star. In a relatively short time, 10^6 years, the eccentricity experiences significant secular variations, until the planet first crosses the limits of the habitable zone. From Fig. 4 of Menou et al. (2003). Reprinted with permission of the American Astronomical Society

Menou and Tabachnik (2003) have quantified the dynamical habitability of 85 known extrasolar planetary systems via simulations of their orbital dynamics in the presence of potentially habitable terrestrial planets. Their results indicate that more than half the known extrasolar planetary systems (mostly those with distant, eccentric giant planets) are unlikely to harbour habitable terrestrial planets (Fig. 8.23). About one-fourth of the systems (mostly those with close-in giant planets) appear as dynamically habitable as our own solar system. The influence of yet undetected giant planets on these systems, however, could compromise their dynamical habitability.

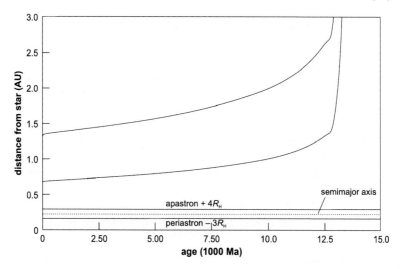

Fig. 8.24 Habitable zone of ρCrB (*curved lines*) and the giant planet's gravitational reach (1.3 times minimum giant mass). The HZ is a safe haven throughout the main sequence. From Fig. 6 of Jones et al. (2005). Reprinted with permission of the American Astronomical Society

In Chap. 5 we discussed the meaning of the Habitable Zone (HZ), the region around the star where liquid water can exist on the surface of a terrestrial planet. This zone is not static but evolves with time. Here we discuss its relation with the DHZ.

Stable orbits of terrestrial planets inside the DHZ exist only if the orbits of the giant planets are located sufficiently far away from either the inner or the outer edge of the habitable zone (Noble 2002). Figure 8.24 illustrates the possible conditions of habitability of terrestrial planets in the system ρCrB (Jones et al. 2005), showing how an Earth-like planet can remain safe with a giant planet close to the star.

A final aspect to comment is the influence of eccentric orbits on habitability. Even if the extremes in stellar insolation near periastron and apoastron are compromising for liquid–water environments, such worlds might still be habitable (cf. Williams and Pollard 2002), if they receive a stellar flux that, when averaged over a complete orbit, is not too different from the nearly constant flux received by the Earth from the Sun. The time-averaged flux over an eccentric orbit is given by

$$<F> = \frac{L}{4\pi a^2 (1-e^2)^{1/2}},$$

where L is the star's luminosity, e is the eccentricity of the planet and a is the distance between the planet and the star.

8.5 Architecture of Planetary Systems

According to the most recent research, different types of planetary systems can be expected. Figure 8.25 shows a composite of different possible architectures around Sun-like stars. HZ and Snow Lines are parameters for differentiating parts of a planetary system under distinct criteria.

It would be highly desirable to know the decisive parameter that determines the fate of a protoplanetary disk and its transformation in a set of planets and debris rotating around a star. The initial mass and dust content of the disk seem to be the genetic component that determines the future of the planetary system. After the birth of the system, gravitational interactions between the components will condition the final configuration of the system, much as education conditions a human being.

Greaves et al. (2007) proposed four likely system configurations, listed in order of their expected frequency:

1. Systems with detectable debris but no detectable planets. The very old star (10 Ga) Tau Ceti is a good example, with a debris disk located at 55 AU, similar to our Kuiper belt, but one order of magnitude more massive, 1.2 M_E (Greaves et al. 2004).
2. Systems with both planets and debris. The young (0.5–1.0 Ga) star Epsilon Eridani belongs to this class. The dust content of its disk is 1,000 times larger than that of the Solar System. However, it is likely that at this age the Solar System looked very similar.
3. Systems with gas giants orbiting beyond 0.1 AU ('Cool Jupiters') but not debris. Our Solar System is an example of the third class, since neither of our debris belts is dense and massive enough to be observed from another star.
4. Systems with gas giants orbiting inside 0.1 AU ('Hot Jupiters') and no debris.

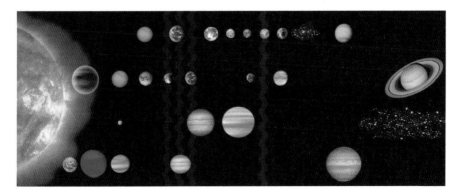

Fig. 8.25 Possible architectures of planetary systems. *Top to bottom*: (**a**) a Uranus of 10M_E orbits at 5.2 AU and seven terrestrial planets form (Raymond et al. 2004), (**b**) A giant planet migrates to 0.05 AU, a Saturn analogue orbits at 9.5 AU, and six terrestrial planets form between them (Mandell et al. 2007), (**c**) based on the actual configuration of the HD 128311 system, in which two massive gas giants orbit between 1 and 2 AU and a debris belt occupies the outer system (the Mercury analogue is speculative) and (**d**) based on the planetary configuration of planets around 55 Cancri. Copyright: Raymond Harris

Meyer et al. (2008) found that the frequency of systems with evidence for dust debris, around stars with masses between 0.7 and 2.2 M_{Sun}, ranges from 8.5–19% at ages <300 Ma to <4% for older stars. They finally suggest that many, perhaps most, Sun-like stars might form terrestrial planets. The mass and density of the protoplanetary disk is an important parameter for the evolution of the system. Kokubo and Ida (2002) have modelled the density of the solid component of the disk by a function of the form

$$\Sigma = \Sigma_1 \left(\frac{a}{1\text{AU}}\right)^{-\alpha} \text{g cm}^{-2},$$

where Σ_1 is the reference density at 1 AU and α a factor that defines how flat or steep the density distribution is across the disk. He found that the condition for the formation of gas giants is when the lifetimes fulfill the inequalities $T_{grow} < T_{disk}$ and $T_{cont} < T_{disk}$, indicating that the accretion of gas onto a rocky core starts. We have already mentioned that the lifetime of a disk is in the range of 10^6–10^7 years, T_{grow} is the growth time of a protoplanet and T_{cont} the contraction time.

The diversity of planetary systems as a function of the disk mass (density) is summarized in Fig. 8.26.

- *Light disks* ($\Sigma_1 < 3$): Gas giants will not be formed.
- *Medium disks* ($\Sigma_1 \sim 10$): A system similar to our own will be formed, with gas giants and ice and rocky planets.
- *Massive disks* ($\Sigma_1 > 30$): Gas giants can form in practically all the regions of the disk, leaving only space at the outer edge for ice planets.

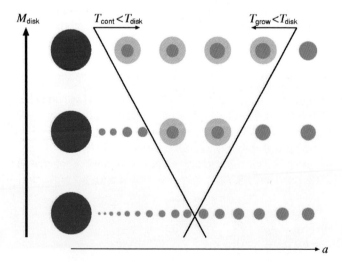

Fig. 8.26 Scheme of the diversity of planetary systems against the initial mass disk. The *left circle* stands for central stars. The *double circles* (cores with envelopes) are Jupiter-like gas giants, and the others are Terrestrial and Neptunian planets. Reprinted from Fig. 13 of Kokubo and Ida (2002) with permission of the American Astronomical Society

Kokubo et al. (2006) investigated the final stage of terrestrial planet formation for different protoplanetary disks characterized for distinct surface density (Σ_1, α), orbital separation of the initial protoplanet system and the bulk density of protoplanets, while the initial system radial range was fixed at 0.5–1.5 AU. For the standard disk model,[17] typically two Earth-sized planets formed in the terrestrial planet region. The number of planets slowly decreases as the surface density of the initial protoplanets increases, while the masses of individual planets increase almost linearly. For a steeper surface density profile, large planets tend to form closer to the star.

Planets that form early in the lifetime of a disk are likely to be lost, whereas late-forming planets will survive (Chambers 2009). For details on the peculiarity of the Solar System see next chapter.

8.5.1 Systems with Hot Jupiters: The Planetary Migration

Giant planets are necessarily formed in the outer parts of the planetary systems (see Chap. 7). As some of them are detected very close to their parent star, a mechanism is needed to explain their migration and simultaneously their high eccentricities. A pioneering idea was proposed by Goldreich and Tremaine (1980) based on the angular momentum transfer between a disk and a satellite orbiting a parent star. See Papaloizou and Terquem (2006), Armitage (2008) and Chambers (2009) for reviews on the different mechanisms suggested.

However, it is important to remark that when biased trends are removed, there is a deficit of massive planets orbiting closer than 1 AU (Cumming et al. 2008). Simply, they were the easiest to be detected.

8.5.1.1 Planetesimal-driven Migration

Planets can migrate via the interaction with smaller bodies in their vicinity. A planet that ejects a planetesimal from the planetary system must give up energy, moving closer toward the star. Conversely, a planet that scatters planetesimals into shorter period gains energy, moving outward (Levison et al. 2007).

The distribution of Trans-Neptunian objects is a good proof that this occurred during the LHB event and the subsequent migration outward of Neptune (Malhotra 1995).

8.5.1.2 Planet–Planet Scattering

After the initial phase of planetary formation, mutual gravitational interactions between the formed planets bring the system into a phase of chaotic evolution during

---

[17] Valid for the Solar System, corresponds to a disk with $\Sigma_1 = 10$ and $\alpha = 3/2$.

which the orbits become highly irregular (Ward 1988). This mechanism reproduced quite well the observed eccentricity distribution (Ford and Rasio 2008) and dynamical configurations (β/β_{crit} distribution, recall Sect. 8.3.4) of observed extrasolar planets (Raymond et al. 2009).

An initially formed planetary system can evolve via one of the following options:

- Ejection of one or more planets
- An increase in the orbital separation of the planets, toward a more stable configuration
- Collisions between planets

Many extrasolar systems formed by two giant planets lie near instability, and then at least one additional planet must exist in the stable regions of well-separated extrasolar planetary systems to push these systems to the edge of stability (Barnes and Raymond 2004; Raymond and Barnes 2005). These ideas form part of the theory of Packed Planetary Systems (see Soter 2007 for a popular version), which predicts that when there are gaps large enough for a planet to be on a stable orbit, there is a planet there. Following these criteria, the planet HD 74156d was discovered by Bean et al. (2008). In summary, all planetary systems are packed as tightly as possible and large spaces between planets are rare.

8.5.1.3 The LHB Event and the Nice Model

As we have already mentioned, the Late Heavy Bombardment, described in Chap. 2, was probably produced by a reorganization of the giant planets of our System. Gomes et al. (2005), Morbidelli et al. (2005) and Tsiganis et al. (2005) proposed the so-called *Nice model* to explain the event. After the dissipation of the gas and dust of the primordial disk, the four giant planets were on near-circular orbits between ∼5.5 and ∼17 astronomical units (AU), much more closely spaced and more compact than in the present. A large, dense disk of planetesimals, totalling about 35 M_E, extended from the orbit of the outermost giant planet to some 35 AU. Planetesimals at this disk's inner edge suffered gravitational encounters with the outermost giant planet, which changed the planetesimals' orbits. The planets scattered inwards the majority of the small icy bodies that they encountered, exchanging angular momentum with the scattered objects so that the planets moved outwards in response, preserving the angular momentum of the system.

This process continued until the planetesimals interacted with Jupiter, whose immense gravity sent them into highly elliptical orbits or even ejected them outright from the Solar System. This, in contrast, caused Jupiter to move slightly inward. After several hundreds of millions of years of slow migration, Jupiter and Saturn, the two innermost giant planets, crossed their mutual 1:2 mean-motion resonance. This resonance increased their orbital eccentricities and destabilized the entire planetary system. After that, the arrangement of the giant planets altered quickly and dramatically. Jupiter shifted Saturn out towards its present position, and this relocation caused mutual gravitational encounters between Saturn and the two ice giants,

8.5 Architecture of Planetary Systems

which propelled Neptune and Uranus onto much more eccentric orbits. These ice giants then ploughed into the planetesimal disk, scattering tens of thousands of planetesimals from their formerly stable orbits in the outer Solar System. This disruption almost entirely scattered the primordial disk, removing 99% of its mass, a scenario that explains the modern-day absence of a dense trans-Neptunian population. Some of the planetesimals were thrown into the inner Solar System, resulting in the Late Heavy Bombardment. Figure 8.27 illustrates the main phases of this process.

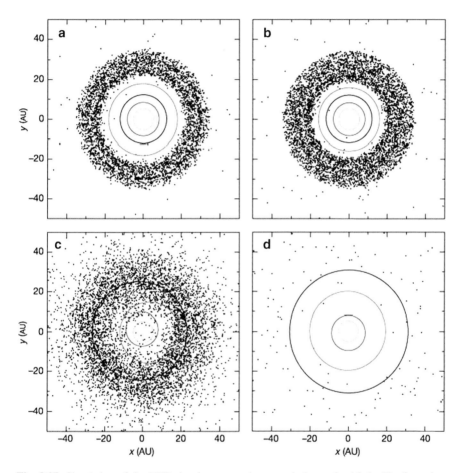

Fig. 8.27 Simulation of the LHB showing outer planets and planetesimal belt. The four giant planets were initially on nearly circular, co-planar orbits with semimajor axes of 5.45, 8.18, 11.5 and 14.2 AU. The dynamically cold planetesimal disk was 35 M_E, with an inner edge at 15.5 AU and an outer edge at 34 AU. Each panel represents the state of the planetary system at four different epochs: (**a**) the beginning of planetary migration (100 Ma); (**b**) just before the beginning of LHB (879 Ma); (**c**) just after the LHB has started (882 Ma); and (**d**) 200 Ma later, when only 3% of the initial mass of the disk is left and the planets have achieved their final orbits. Adapted from Gomes et al. (2005) Fig. 2. Reprinted by permission from Macmillan Publishers Ltd: Nature Vol. 435, p. 466. Copyright (2005)

8.5.1.4 Interaction with a Distant Companion Star

A binary companion in an orbit highly inclined to the planetary orbit could also cause a large fluctuation in the eccentricity of the planet (Malmberg et al. 2007). It is the so-called Kozai oscillation (Kozai 1962), a very long-range effect, where its amplitude is purely dependent on the relative orbital inclination.

8.5.1.5 Gas Disk Migration

A planet embedded in a gas protoplanetary disk at a radius r_p exerts a perturbation in the disk, modifying the distribution of gas in the planet's vicinity. Gravitational interactions between the planet and the non-uniform gas generates torques[18] that alter the planetary orbit. If the angular momentum exchange with the inner parts ($r < r_p$) is cancelled by the interaction with the outer parts of the disk ($r > r_p$), the planet remains stationary. However, if the angular momentum exchange is not balanced, the planet will drift radially. Three types of migration are possible (De Val-Borro 2006):

Type I. Weak perturbation, ($M_p \ll M_J$): the surface density profile of the disk is only weakly altered by the planet and the migration rate is proportional to the planetary mass. The planet remains entirely embedded within the gas. As the mass of the giant planet start to grow, its gravitational field produces spiral structures in the disk that destroy the original symmetry. A icy-rocky kernel is formed and the spiral structure in the interior of the disk tends to push it outwards, while the same effect in the exterior part tends to push it inwards. The rate of migration is proportional to the mass of the disk and the kernel. The migration time for an Earth-like planet at 5 AU turns out to be 1 Ma for a typical disk model (Tanaka et al. 2002).

Type II. Strong perturbation, for larger planetary masses. An annular gap (hole) is created in which the surface density of gas is low. The migration continues until the kernel grows to more than ten Earth masses. At this value, a gap opens in the disk, dramatically slowing the migration but without stopping it. If the migration is slow, the circumstellar disk disappears rapidly and embryos of Earth-like planets can be formed. However, this interesting and tricky process is not able to reproduce the observed eccentricities. Type II migration is typically slower than type I.

Type III. Also called runaway migration, this is a new model developed for Jupiter–Saturn sized planets. The planet clears a partial gap in the disk. The remaining gas close to the planet exerts a co-rotation torque that grows in proportion to the migration speed (Peplinski et al. 2008a–c).

Trilling et al. (1998) discussed the different fate of giant planets considering three classes: (1) planets that migrate inward too rapidly and lose all their mass; (2) planets that migrate inward, lose some but not all of their mass and survive in orbits

---

[18] A torque is a measure of how much a force acting on an object causes the object to rotate.

close to the star; and (3) planets that do not lose any mass. Some planets in the latter class do not migrate very far from their formation locations.

In summary, the diversity of planetary systems could be the result of the following:

- Disk–Planet Interaction
- Planet–Planet Interaction
- Star–Planet Interaction
- Disk Breakup

8.5.1.6 Stopping the Migration

The migration tends to accelerate with decreasing distance from the star. Therefore, we need a mechanism to stop the migration, although perhaps most of the Hot Jupiters end their lives on colliding with their star (Ida and Lin 2004). HD 82943 probably had a catastrophic event in the past with the fall of one planet on the star (Israelian et al. 2001). Figure 8.28 shows an artistic view of the process. However, in some cases the migration stops. Halting a giant planet's inward migration at small semi-major axes is even less well understood than initiating and maintaining the migration. Various braking mechanisms have been proposed:

- Outflowing stellar winds and EUV radiation from the young star.
- Interaction with the magnetosphere of the star (Terquem 2003). The observed enhancement in the flare strength of solar-like stars with respect to the Sun could be caused by the presence of Hot Jupiters. In such conditions, a reconnection of the magnetic field of the stellar corona with those of the planetary magnetosphere

Fig. 8.28 Artistic view of a planet engulfed by the star. Credit: G. Perez (IAC)

could give rise to the strong release of energy (Rubenstein and Schaefer 2000). A different approach has been used by Rice et al. (2008) who suggest that the entry of the planet into the stellar magnetosphere results in a rapid growth of its orbital eccentricity, which finally leads to its destruction.
- Gravitational Tides (Lin et al. 1996; Trilling et al. 1998).
- Scattering by multiple planetesimals (Weidenschilling and Marzari 1996).
- Resonant interactions by two (or more) protoplanets (Kley 2000).
- Mass transfer from the planet onto the star (Trilling et al. 1998).
- Low viscosity regions (Dead Zones) in the protoplanetary disk (Matsumura et al. 2007).

8.5.1.7 Survival of Terrestrial Planets

Fogg and Nelson (2007) remark that the migration of giant planets takes place in the first few Ma of the life of the planetary system, whereas $\sim$10–100 Ma are required to complete the formation of a terrestrial planet. Therefore, most of the material of inner solid disks is not destroyed by the intrusion of a migrating giant planet, allowing in principle the formation of terrestrial planets in habitable zones.

A similar behaviour results from the simulations by Raymond et al. (2006) and Mandell et al. (2007), with the mass and water content decreasing with the presence of an outer giant planet. They model the formation of terrestrial planets consisting in the accumulation of 1,000 km embryos and a swarm of billions of 1–10 km planetesimals. Several Earth-mass planets form inside the orbit of the migrating Jovian planet, analogous to the discovered 'Hot Super-Earths'. Very water-rich Earth-mass planets can form from surviving material outside the giant planet's orbit, often in the habitable zone and with low orbital eccentricities. As a reference, Fig. 8.29 shows the simulation for an analogue of the Solar System with a fixed Jupiter (Raymond et al. 2006) .

The circular orbit of Jupiter in our Solar System promotes the stability of circular orbits among the other planets. Probably, the Solar System has kept its present configuration because there is not enough mass in the vicinity of the giant planets to produce gravitational interactions, giving rise to the migration of the planet inwards (Thommes 2005).

Ito and Tanikawa (1999) have studied various kinds of TPs with equal dynamical separations, d_d, and determined their instability scale under the distortion from the Jovian planets. They find that TP with $d_d < 18R_H$ are unstable in a short time scale ($< 10^7$ yr). The present value of $d_d > 26R_H$ is therefore a significant condition for maintaining the stability of the Solar System.

Sándor et al. (2007) have compiled a stability catalogue for a dynamical model consisting of a star, a giant planet and an Earth-like planet, establishing also the stability properties of known planetary systems. Süli et al. (2005) have made an interesting simulation by increasing the mass of the Earth by a factor of k_E and testing

8.5 Architecture of Planetary Systems

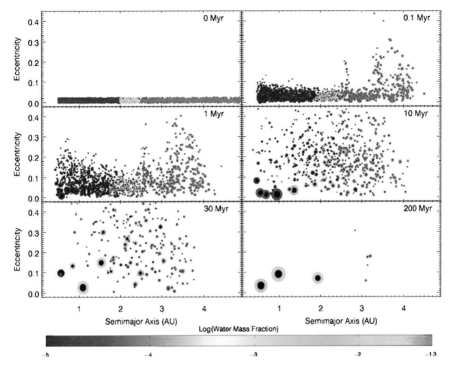

Fig. 8.29 Snapshots of the growth, on a 200 Ma scale, of terrestrial planets and water delivery for a situation similar to the Solar System. Each plot shows the orbital semimajor axis vs. eccentricity for the orbit of each object. *Red*, Dry and *Blue*, Wet (see scale on the colour bar) Credit: S. Raymond For more details see http://www.obs.u-bordeaux1.fr/e3arths/raymond/

the stability of the Solar System under such conditions. The only effect would seem to be that the motion of Mars becomes more chaotic, leading to its ejection, while Venus is well protected.

Jones et al. (2005) have investigated whether Earth-mass planets could remain confined inside the HZ of 111 discovered extrasolar systems. The HZ migrates outward during the main-sequence lifetime, and they find that in about two-thirds of the systems, an Earth-mass planet could be confined inside the HZ for at least 1,000 Ma sometime during the main-sequence lifetime. Later, Jones et al. (2006) extended this work to 152 systems. For those with Hot Jupiters, 60% offer DHZ greater than 20% of the HZ width and in 50% of the cases habitability is possible 1 Ga in the past. Von Bloh et al. (2007) studied the likelihood of finding habitable Earth-like planets on stable orbits for 86 selected extrasolar systems. Almost 60% of the investigated systems could harbour habitable Earth-like planets on stable orbits and for 18 of these extrasolar systems, they find even better prerequisites for dynamic habitability than in our own Solar System.

8.5.2 Binary Systems

Stellar systems are born in the collapse of molecular clouds. Modelling by Machida (2008) found that fast rotation and low metallicity favoured cloud fragmentation and therefore the formation of multiple stellar systems. Two thirds of the stars in the Milky Way are members of binary or multiple stellar systems, which surely complicates the stability of the planets encircling one of the members of the system. However, planetary systems are at least as abundant in twin-star systems as they are in those like our own, with only one star.

Two types of stable orbits can be considered for Earth-like planets around a binary system: (1) S-type with planets encircling one component of the binary system (Eggenberger 2004) and (2) P-type with planets encircling both components of the binary stellar system. No exoplanet with a P-type orbit has been detected as yet, but it is also true that close binary systems are not included in the current surveys.

Trilling et al. (2007) observations seem to indicate that debris disks are more abundant around the tightest and widest binary stars (Fig. 8.30).

Holman and Wiegert (1999) have investigated the orbital parameters of a binary system, where planets can persist for long times.[19] They obtain the best fit for a critical semi-major axis[20] of the orbit given by

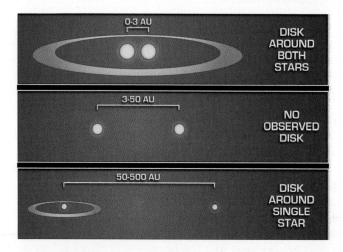

Fig. 8.30 Observations with the Spitzer telescope indicate more debris disks around binary stars that are 0–3 AU apart (*top panel*) and 50–500 AU apart (*bottom panel*) than binary stellar systems that are in the range 3–50 AU (*middle panel*). Credit: NASA/JPL - Caltech/T. Pyle (SSC)

---

[19] This study investigates orbital stability numerically within the elliptic restricted three-body problem. That is, the planets are modelled as test particles moving in the gravitational field of a pair of stars on fixed eccentric orbits about each other.

[20] Semimajor axis at which the test particles at all initial longitudes survived the full integration time

$$a_c = (1.60 \pm 0.04) + (5.10 \pm 0.05)e + (-2.22 \pm 0.11)e^2 + (4.12 \pm 0.09)\mu,$$

where m_1 is the mass of the star the test particle is orbiting, m_2 is the mass of the perturbing star, $\mu = m_2/(m_1 + m_2)$ and e is the eccentricity of the orbit.

Musielak et al. (2005) simulations indicate that the stability of Jupiter-type planets depends on both the distance ratio between the star and the planet and the mass ratio of the possible stellar companion(s). Lissauer et al. (2004) have studied the formation of terrestrial planets in binary star systems. If the orbit of a wide binary system is inclined with respect to the planet orbit by more than 40°, it can cause an increase in the planetary eccentricity (Ford et al. 2000). Once in such an orbit, the planet may plunge into the star and be destroyed or its orbit may circularize through tides.

Recently, Takeda et al. (2008) have modelled the dynamical influence of a secondary stellar companion on the stability of a two-planet system orbiting the primary star. They find three distinct classes: completely decoupled systems in which planetary orbits are independently affected by the binary perturbation; Weakly coupled systems in which a large mutual inclination angle grows; and Strongly coupled, dynamically rigid systems.

8.5.3 Multiple Planetary Systems

In the previous chapter, we described the main statistical features of the giant exoplanets detected and we also discussed hypothetical terrestrial exoplanets within the realm of future surveys. Now, it is time to study the collective properties of the planets forming a planetary system.

At this moment, 25 planetary systems with more than one planet have been discovered. Probably about half of the stars with planets have multiple systems (Wright et al. 2007). These systems can be divided in two broad categories: hierarchical systems with well separated planets and Resonant systems with rational period ratios. Kley et al. (2004) predicted that planets in a 2:1 MMR resonance should have a larger mass and less eccentricity for the outer planet, result that must be confronted with future observations.

These planetary systems will be the starting point to place our own planetary system in a universal perspective. In the following sections we discuss in more detail some of the most relevant systems.

8.5.3.1 Gliese 581

Gliese 581 is a low-mass M2 red dwarf located 20 light-years away from Earth. Table 8.2 lists the orbital parameters of the exoplanets discovered so far around this star: a Hot-Neptune and two Super-Earths. The age of the system is estimated in 4.3 Ga.

Table 8.2 Main orbital parameters of the exoplanets discovered around Gliese 581

Planet	Mass (sin i) [$M_J(M_E)$]	Orbital period (Days)	Semimajor axis (AU)	Eccentricity
b	0.049	5.4	0.041	0.02 ± 0.01
c	0.016	12.9	0.073	0.16 ± 0.07
d	0.024 (7.7)	83.6	0.25	0.20 ± 0.10

See Table 8.1 for comparison with the Solar System and http://exoplanet.eu/catalog.hhp for more details

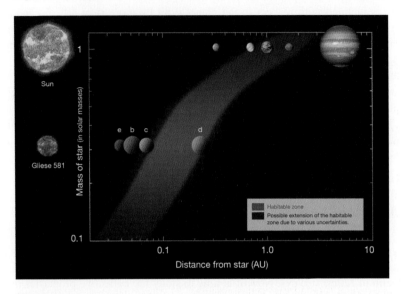

Fig. 8.31 This diagram shows the distances of the planets in the Solar System (*upper row*) and in the Gliese 581 system (*lower row*) from their respective stars (*left*). The habitable zone is indicated as the *blue* area, showing that Gliese 581 d is located inside the habitable zone around its low-mass red star. Based on a diagram by Franck Selsis, Univ. of Bordeaux. Credit: ESO

The system is dynamically stable (Beust et al. 2008) and the planets could be included within the habitable zone (Fig. 8.31). Von Bloh et al. (2007) and Selsis et al. (2007) have studied the conditions of habitability of these exoplanets. Recently, Mayor et al. (2009b) have announced the discovery of Gliese 581e, a planet of only 1.9 M_E orbiting the red dwarf in just 3.15 days.

8.5.3.2 Gliese 876

Gliese 876 is a M4-old dwarf (8.9 Ga) located at a distance of 4.72 pc with three planets. Shankland (2006) investigated the possibility of transits of the two outer components of the system. Using radio observations in the millimetre range, Shankland et al. (2008) ruled out the presence of a dust disk with either a mass greater than 0.0006 M_E or less than $\sim$250 AU across (Table 8.3).

8.5 Architecture of Planetary Systems

Table 8.3 Main orbital parameters of the exoplanets discovered around Gliese 876

Planet	Mass (sin i) [$M_J (M_E)$]	Orbital period (Days)	Semimajor axis (AU)	Eccentricity
d	0.018 (7.5)	1.937	0.020	0.00
c	0.56	30.1	0.13	0.27
b	1.935	60.94	0.208	0.025

See Table 8.1 for comparison with the Solar System and http://exoplanet.eu/catalog.hhp for more details

Table 8.4 Main orbital parameters of the exoplanets discovered around υ Andromeda

Planet	Mass (sin i) [$M_J (M_E)$]	Orbital period (Days)	Semimajor axis (AU)	Eccentricity
b	>0.687 ± 0.058	4.617	0.059	0.023 ± 0.018
c	>1.970 ± 0.170	241.23	0.830	0.262 ± 0.021
d	>3.930 ± 0.330	1290.1	2.540	0.258 ± 0.032

See Table 8.1 for comparison with the Solar System and http://exoplanet.eu/catalog.hhp for more details

Kinoshita and Nakai (2001) suggested that the planetary system is stabilized by a 2:1 mean motion resonance of the two outer planets. The third inner planet is a Hot Super-Earth (Rivera et al. 2005).

8.5.3.3 Upsilon Andromeda (HD 9826)

υ And is an S-type binary star composed of a yellow solar-like star, F8, and a red dwarf, separated 750 AU from each other and located at 43.9 light-years (13.5 parsecs) from the Earth. The age of the system is estimated to be 3.8 (±1) Ga. Around the primary star three exoplanets (Table 8.4) have been detected (Butler et al. 1999).

Ford et al. (2005) report that the current configuration of the three giant exoplanets probably results from a close dynamical interaction with another planet, now ejected from the system. The planets started on nearly circular orbits, but chaotic evolution caused the outer planet (υ And d) to be perturbed suddenly into a higher-eccentric orbit.

Lissauer and Rivera (2001) and Michtchenko and Malhotra (2004) have studied the dynamical stability of the system. In general, the system is stable for a large domain of eccentricities, although the inner planet can suffer strong oscillations.

8.5.3.4 55 Cancri (HD 75732)

55 Cancri is also a binary stellar system located 41 light-years away from Earth. The primary star is a G8 yellow star, classified as 'super metal-rich' and with an age of

Table 8.5 Main orbital parameters of the exoplanets discovered around 55 Cancri A

Planet	Mass (sin i) $M_J(M_E)$	Orbital period (Days)	Semimajor axis (AU)	Eccentricity
e	>0.034	2.817	0.038	0.07
b	>0.824	14.651	0.115	0.014
c	>0.169	43.93	0.240	0.086
f	>0.144	260	0.781	0.200
d	3.835	5218	577	0.025

See Table 8.1 for comparison with the Solar System and http://exoplanet.eu/catalog.hhp for more details

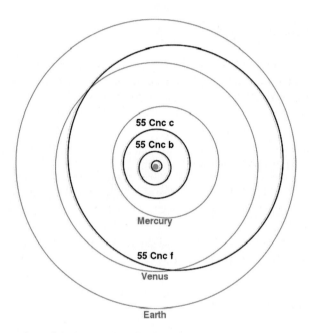

Fig. 8.32 Comparison of the inner orbits of 55 Cancri (*black*) and the Solar System

5.5 Ga. The characteristics of the five detected exoplanets rotating around this star are listed in Table 8.5 and their orbits plotted in Fig. 8.32. The system seems to be dynamically stable (Fischer et al. 2008). Raymond et al. (2008) show that the stable zone may contain two to three additional planets; if they are there then they are ∼50 M_E each.

Far IR observations by Jayawardhana et al. (2000) seem to suggest the presence of a dust cloud beyond the space occupied by the planets, equivalent to our Kuiper belt.

8.5.3.5 47 UMa (HD 95128)

The planetary system around 47 UMa, a G0 dwarf star located 46 light-years away, is probably one of the best analogues of the Solar System observed to date with two giant planets similar to Jupiter/Saturn, with masses of 2.6 and 0.46 M_J. Ji et al. (2005) found that the most likely candidates for habitable environments are Earth-like planets with orbits in the ranges of 0.8 AU < a < 1.3 AU.

However, the presence of a giant planet within 2.5 AU of the star may have disrupted planet formation in the inner system, and reduced the amount of water delivered to the inner planets during accretion (Raymond 2006).

8.5.3.6 HD 69830

HD 69830 is an orange K0 dwarf star, located at approximately 41 light-years from Earth. The star has no gas giant planets, but three Neptune-mass planets have been detected (Lovis et al. 2006 and Table 8.6). Alibert et al. (2006) presented a scenario for the formation of this system, but Payne et al. (2009) were not able to explain the observed eccentricities of the planets as the result of planetary perturbations during migration, unless the planetary system is aligned nearly face-on (Fig. 8.33).

Based on infrared observations, Beichman et al. (2005) and Lisse et al. (2007) presented evidence of the existence of an asteroid belt, 20 times more massive than the solar one and located at around 1.0 AU from the central source, coincident with the 2:1 and 5:2 mean motion resonances of the outermost of the three Neptune-sized planets. The asteroids were detected by a ring of warm dust that is probably generated by asteroidal collisions. The dust appears to lie within the equivalent of Venus' orbital distance in the Solar System (see Fig. 7.8).

8.5.3.7 HD160691 (μ Arae)

This is a G3 dwarf with an estimated age of 6.4 Ga and is located 15.3 pc (49.9 l.y) away. Its planetary components are listed in Table 8.7. The habitable zone spans the range 0.69–2.16 AU and the configuration recalls that of the Solar System with a Super-Earth and almost circular planetary orbits.

Table 8.6 Main orbital parameters of the exoplanets discovered around HD69830

Planet	Mass (M_J)	Orbital period (Days)	Semi-major axis (AU)	Eccentricity
b	>0.033	8.667± 0.003	0.0785	0.10
c	>0.038	31.56 ± 0.04	0.186	0.13
d	>0.058	197 ± 3	0.63	0.07

See Table 8.1 for comparison with the Solar System and http://exoplanet.eu/catalog.hhp for more details

Fig. 8.33 Artistic view of the Zodiacal light produced by the dust disk around HD 69830. Credit: NASA/JPL-Caltech/R. Hurt (SSC)

Table 8.7 Main orbital parameters of the exoplanets discovered around HD160691 (Data: Pepe et al. 2007)

Planet	Mass (M_J)	Orbital period (Days)	Semi-major axis (AU)	Eccentricity
c	0.033	9.639	0.091	0.172 ± 0.040
d	0.522	310.55	0.921	0.067 ± 0.001
b	1.676	643.25	1.497	0.128 ± 0.017
e	1.814	4205.8	5.235	0.098 ± 0.063

See Table 8.1 for comparison with the Solar System

Table 8.8 Main orbital parameters of the exoplanets discovered around HD 40307

Planet	Mass (M_T)	Orbital period (Days)	Semi-major axis (AU)	Eccentricity
b	>4.2	4.3115 ± 0.0006	0.047	0.0
c	>6.8	9.620 ± 0.002	0.081	0.0
d	>9.2	20.46 ± 0.01	0.134	0.0

See Table 8.1 for comparison with the Solar System

8.5.3.8 HD 40307

Recently, Mayor et al. (2009a, b) detected three Super-Earths orbiting very close to a K2V star, HD 40307 (see Table 8.8), located approximately 42 light-years away. Barnes et al. (2009) have proposed that they are probably mini-Neptunes, which have migrated into their current orbits closer to the star.

8.6 Violence and Harmony

In this chapter, we began by summarizing the existing theories on the Origin of the Solar System. According to current knowledge, all planetary systems seem to have a common process for their formation. The genetic print of each primordial nebula (rotation, mass, metallicity) configures the general architecture of the corresponding planetary system. The location where the planets form determines the density profile of the nebula, which in turn determines the migration path of the planets, which drives the planets to a new position after disk dissipation (Crida 2009).

After the formation of the first protoplanetary bodies, the gravitational interactions between them will leave only the winners in a process where collisions, migrations and different kinds of perturbations are dominant. Yet some members were, and can still be, fired from the system. The peaceful aspect of our Solar System is mainly a question of its age. Just as life seems to require conditions at the edge of chaos (Lewin 2000), planetary systems seem to like to be close to dynamical instability.

A great deal of observational work remains to be done in the coming decades to fill in the gaps in our knowledge about different types of planets and planetary systems. Our Solar System seems to be just one member of a large community with some peculiarities determined by the position and masses of the different components (star, planets, satellites and minor bodies). We dedicate the last chapter to a topic well worth studying: the question of how common or how unusual our environment actually is.

References

Adams, F.C., Hollenbach, D., Laughlin, G., Gorti, U.: Photoevaporation of circumstellar disks due to external far-ultraviolet radiation in stellar aggregates. Astrophys. J. **611**, 360–379 (2004)

Adams, F.C., Laughlin, G., Bloch, A.M.: Turbulence in extrasolar planetary systems implies that mean motion resonances are rare. Astrophys. J. **683**, 1117–1128 (2008)

Aiton, E.J.: The vortex theory of planetary motions. Mac Donald, London (1972)

Alibert, Y., Baraffe, I., Benz, W., Chabrier, G., Mordasini, C., Lovis, C., Mayor, M., Pepe, F., Bouchy, F., Queloz, D., Udry, S.: Formation and structure of the three Neptune-mass planets system around HD 69830. Astron. Astrophys. **455**, L25–L28 (2006)

Armitage, P.: Planetary formation and migration. Scholarpedia **3**, 4479 (2008)

Aumann, H.H., Beichman, C.A., Gillett, F.C., de Jong, T., Houck, J.R., Low, F.J., Neugebauer, G., Walker, R.G., Wesselius, P.R.: Discovery of a shell around Alpha Lyrae. Astrophys. J. **278**, L23–L27 (1984)

Barnes, R., Greenberg, R.: Stability limits in extrasolar planetary systems. Astrophys. J. **647**, L163–L166 (2006)

Barnes, R., Quinn, T.: The (in)stability of planetary systems. Astrophys. J. **611**, 494–516 (2004)

Barnes, R., Raymond, S.N.: Predicting planets in known extrasolar planetary systems I. Test particle simulation, Astrophys. J. **617**, 569–574 (2004)

Barnes, R., Jackson, B., Raymond, S.N., West, A.A., Greenberg, R.: The HD 40307 planetary system: Super-Earths or mini-Neptunes? Astrophys. J. **695**, 1006–1011 (2009)

Bean, J.L., McArthur, B.E., Benedict, G.F., Armstrong, A.: Detection of a third planet in the HD 74156 system using the Hobby-Eberly telescope. Astrophys. J. **672**, 1202–1208 (2008)

Beichman, C.A., Bryden, G., Gautier, T.N., Stapelfeldt, K.R., Werner, M.W., Misselt, K., Rieke, G., Stansberry, J., Trilling, D.: An excess due to small grains around the nearby K0 V star HD 69830: Asteroid or cometary debris? Astrophys. J. **626**, 1061–1069 (2005)

Bertout, C.: T Tauri stars – Wild as dust. Ann. Rev. Astron. Astrophys. **27**, 351–395 (1989)

Beust, H., Bonfils, X., Delfosse, X., Udry, S.: Dynamical evolution of the Gliese 581 planetary system. Astron. Astrophys. **479**, 277–282 (2008)

Boss, A.P.: Giant planet formation by gravitational instability. Science **276**, 1836–1839 (1997)

Boss, A.P.: The origin of protoplanetary disks. In: Woodward, C.E., Shull, J.M., Thronson, H.A., Jr. (eds.) Origins, Astronomical Society of the Pacific Conference Series, vol. 148, pp. 314–326 (1998)

Boss, A.P.: Collapse and fragmentation of molecular cloud cores. VII. Magnetic fields and multiple protostar formation. Astrophys. J. **568**, 743–753 (2002)

Boss, A.P.: Rapid formation of outer giant planets by disk instability. Astrophys. J. **599**, 577–581 (2003)

Boss, A.P.: Rapid formation of super-Earths around M dwarf stars. Astrophys. J. **644**, L79–L82 (2006)

Boss, A.P.: Testing disk instability models for giant planet formation. Astrophys. J. **661**, L73–L76 (2007)

Brahic, A.: Theories of the origin of the solar system – Some historical remarks. In: Brahic, A. (ed.) Formation of Planetary Systems, Toulouse, Cepadues-Editions, p. 15, 17–58 (1982)

Brush, S.: Meteorites and the origin of the solar system. Geol. Soc. Lond. Spec. Publ. **256**, 417–441 (2006)

Brush, S.G.: A history of modern planetary physics: Nebulous Earth. Cambridge University Press, London (1996); reprinted 2009

Bryden, G., Beichman, C.A., Trilling, D.E., Rieke, G.H., Holmes, E.K., Lawler, S.M., Stapelfeldt, K.R., Werner, M.W., Gautier, T.N., Blaylock, M., Gordon, K.D., Stansberry, J.A., Su, K.Y.L.: Frequency of debris disks around solar-type stars: First results from a Spitzer MIPS survey. Astrophys. J. **636**, 1098–1113 (2006)

Burns, J.A.: The evolution of satellite orbits. In: Burns, J.A., Matthews, M.S. (eds.) Satellites, pp. 117–158. University of Arizona Press, AZ (1986)

Butler, R.P., Marcy, G.W., Fischer, D.A., Brown, T.M., Contos, A.R., Korzennik, S.G., Nisenson, P., Noyes, R.W.: Evidence for multiple companions to υ Andromedae. Astrophys. J. **526**, 916–927 (1999)

Caballero, J.A.: Stars and brown dwarfs in the σ Orionis cluster: The Mayrit catalogue. Astron. Astrophys. **478**, 667–674 (2008)

Cameron, A.G.W.: The early evolution of the solar system. In: Hemenway, C.L., Millman, P.M., Cook, A.F. (eds.) IAU Colloq. 13: Evolutionary and physical properties of Meteoroids, pp. 347–354 (1973)

Celletti, A., Perozzi, E.: Celestial mechanics: The Waltz of the planets. Springer, Heidelberg (2007)

Chambers, J.E.: Terrestrial planet formation. In: Holt, S.S., Deming, D. (eds.) The search for other worlds, American Institute of Physics Conference Series, vol. 713, pp. 203–212 (2004)

Chambers, J.E.: Planetary migration: What does it mean for planet formation? Ann. Rev. Earth Planet. Sci. **37**, 321–344 (2009)

Correia, A.C.M., Laskar, J.: Mercury's capture into the 3/2 spin-orbit resonance as a result of its chaotic dynamics. Nature **429**, 848–850 (2004)

Crida, A.: Minimum mass solar nebulae and planetary migration. Astrophys. J. **698**, 606–614 (2009)

Cumming, A., Butler, R.P., Marcy, G.W., Vogt, S.S., Wright, J.T., Fischer, D.A.: The Keck planet search: Detectability and the minimum mass and orbital period distribution of extrasolar planets. Publ. Astron. Soc. Pac. **120**, 531–554 (2008)

de Val-Borro, M., Edgar, R.G., Artymowicz, P., Ciecielag, P., Cresswell, P., D'Angelo, G., Delgado-Donate, E.J., Dirksen, G., Fromang, S., Gawryszczak, A., Klahr, H., Kley, W.,

Lyra, W., Masset, F., Mellema, G., Nelson, R.P., Paardekooper, S.J., Peplinski, A., Pierens, A., Plewa, T., Rice, K., Schäfer, C., Speith, R.: A comparative study of disc-planet interaction. Mon. Not. Roy. Astron. Soc. **370**, 529–558 (2006)

Diacu, F., Holmes, P.: Celestial encounters. The origins of chaos and stability. Princeton University Press, NJ(1996)

Eggenberger, A., Udry, S., Mayor, M.: Statistical properties of exoplanets. III. Planet properties and stellar multiplicity. Astron. Astrophys. **417**, 353–360 (2004)

Fischer, D.A., Marcy, G.W., Butler, R.P., Vogt, S.S., Laughlin, G., Henry, G.W., Abouav, D., Peek, K.M.G., Wright, J.T., Johnson, J.A., McCarthy, C., Isaacson, H.: Five planets orbiting 55 Cancri. Astrophys. J. **675**, 790–801 (2008)

Fixler, J.B., Foster, G.T., Mc Guirk, J.M., Kasevich, M.A.: Atom interferometer measurement of the newtonian constant of gravity. Science **315**, 74–77 (2007)

Fogg, M.J., Nelson, R.P.: On the formation of terrestrial planets in hot-Jupiter systems. Astron. Astrophys. **461**, 1195–1208 (2007)

Ford, E.B., Lystad, V., Rasio, F.A.: Planet-planet scattering in the upsilon Andromedae system. Nature **434**, 873–876 (2005)

Ford, E.B., Rasio, F.A.: Origins of eccentric extrasolar planets: Testing the planet-planet scattering model. Astrophys. J. **686**, 621–636 (2008)

Forrest, W.J., Sargent, B., Furlan, E., D'Alessio, P., Calvet, N., Hartmann, L., Uchida, K.I., Green, J.D., Watson, D.M., Chen, C.H., Kemper, F., Keller, L.D., Sloan, G.C., Herter, T.L., Brandl, B.R., Houck, J.R., Barry, D.J., Hall, P., Morris, P.W., Najita, J., Myers, P.C.: Mid-infrared spectroscopy of disks around classical T Tauri stars. Astrophys. J. Suppl. **154**, 443–447 (2004)

Freistetter, F., Krivov, A.V., Löhne, T.: Planets of β Pictoris revisited. Astron. Astrophys. **466**, 389–393 (2007)

Gladman, B.: Dynamics of systems of two close planets. Icarus **106**, 247–263 (1993)

Goldreich, P., Tremaine, S.: Disk-satellite interactions. Astrophys. J. **241**, 425–441 (1980)

Goldreich, P., Ward, W.R.: The formation of planetesimals. Astrophys. J. **183**, 1051–1062 (1973)

Golimowski, D.A., Ardila, D.R., Krist, J.E., Clampin, M., Ford, H.C., Illingworth, G.D., Bartko, F., Benítez, N., Blakeslee, J.P., Bouwens, R.J., Bradley, L.D., Broadhurst, T.J., Brown, R.A., Burrows, C.J., Cheng, E.S., Cross, N.J.G., Demarco, R., Feldman, P.D., Franx, M., Goto, T., Gronwall, C., Hartig, G.F., Holden, B.P., Homeier, N.L., Infante, L., Jee, M.J., Kimble, R.A., Lesser, M.P., Martel, A.R., Mei, S., Menanteau, F., Meurer, G.R., Miley, G.K., Motta, V., Postman, M., Rosati, P., Sirianni, M., Sparks, W.B., Tran, H.D., Tsvetanov, Z.I., White, R.L., Zheng, W., Zirm, A.W.: Hubble space telescope ACS multiband coronagraphic imaging of the debris disk around β Pictoris. Astron. J. **131**, 3109–3130 (2006)

Gomes, R., Levison, H.F., Tsiganis, K., Morbidelli, A.: Origin of the cataclysmic late heavy bombardment period of the terrestrial planets. Nature **435**, 466–469 (2005)

Gónzalez Martínez-Pais, J.I.: Introducción a la Mecánica Celeste (formulación newtoniana. Servicio de Publicaciones, Universidad de La Laguna (2003)

Greaves, J.S., Wyatt, M.C., Holland, W.S., Dent, W.R.F.: The debris disc around τ Ceti: A massive analogue to the Kuiper Belt. Mon. Not. Roy. Astron. Soc. **351**, L54–L58 (2004)

Greaves, J.S., Fischer, D.A., Wyatt, M.C., Beichman, C.A., Bryden, G.: Predicting the frequencies of diverse exo-planetary systems. Mon. Not. Roy. Astron. Soc. **378**, L1–L5 (2007)

Hayashi, C.: Stellar evolution in early phases of gravitational contraction. Publ. Astron. Soc. Pac. **13**, 450–452 (1961)

Hillenbrand, L.A., Carpenter, J.M., Kim, J.S., Meyer, M.R., Backman, D.E., Moro-Martín, A., Hollenbach, D.J., Hines, D.C., Pascucci, I., Bouwman, J.: The complete census of 70 μm-bright debris disks within "the formation and evolution of planetary systems" Spitzer legacy survey of sun-like stars. Astrophys. J. **677**, 630–656 (2008)

Holman, M.J., Wiegert, P.A.: Long-term stability of planets in binary systems. Astron. J. **117**, 621–628 (1999)

Ida, S., Lin, D.N.C.: Toward a deterministic model of planetary formation. I. A desert in the mass and semimajor axis distributions of extrasolar planets. Astrophys. J. **604**, 388–413 (2004)

Inaba, S., Ikoma, M.: Enhanced collisional growth of a protoplanet that has an atmosphere. Astron. Astrophys. **410**, 711–723 (2003)

Inaba, S., Wetherill, G.W., Ikoma, M.: Formation of gas giant planets: Core accretion models with fragmentation and planetary envelope. Icarus **166**, 46–62 (2003)

Israelian, G., Santos, N.C., Mayor, M., Rebolo, R.: Evidence for planet engulfment by the star HD82943. Nature **411**, 163–166 (2001)

Ito, T., Tanikawa, K.: Stability and instability of the terrestrial protoplanet system and their possible roles in the final stage of planet formation. Icarus **139**, 336–349 (1999)

Ivanova, N., Taam, R.E.: Magnetic braking revisited. Astrophys. J. **599**, 516–521 (2003)

Jaki, S.L.: Planets and planetarians: A history of theories of the origin of planetary systems. Wiley, NY (1977)

Jayawardhana, R., Holland, W.S., Greaves, J.S., Dent, W.R.F., Marcy, G.W., Hartmann, L.W., Fazio, G.G.: Dust in the 55 Cancri planetary system. Astrophys. J. **536**, 425–428 (2000)

Ji, J., Liu, L., Kinoshita, H., Li, G.: Could the 47 Ursae Majoris planetary system be a second solar system? Predicting the Earth-like planets. Astrophys. J. **631**, 1191–1197 (2005)

Jones, B.W., Underwood, D.R., Sleep, P.N.: Prospects for habitable "Earths" in known exoplanetary systems. Astrophys. J. **622**, 1091–1101 (2005)

Jones, B.W., Sleep, P.N., Underwood, D.R.: Which exoplanetary systems could harbour habitable planets? Int. J. Astrobiol. **5**, 251–259 (2006)

Jura, M., Chen, C.H., Furlan, E., Green, J., Sargent, B., Forrest, W.J., Watson, D.M., Barry, D.J., Hall, P., Herter, T.L., Houck, J.R., Sloan, G.C., Uchida, K., D'Alessio, P., Brandl, B.R., Keller, L.D., Kemper, F., Morris, P., Najita, J., Calvet, N., Hartmann, L., Myers, P.C.: Mid-infrared spectra of dust debris around main-sequence stars. Astrophys. J. Suppl. **154**, 453–457 (2004)

Kalas, P., Graham, J.R., Chiang, E., Fitzgerald, M.P., Clampin, M., Kite, E.S., Stapelfeldt, K., Marois, C., Krist, J.: Optical images of an exosolar planet 25 light-years from earth. Science **322**, 1345–1348 (2008)

Kennedy, G.M., Kenyon, S.J.: Planet formation around stars of various masses: The snow line and the frequency of giant planets. Astrophys. J. **673**, 502–512 (2008)

Kinoshita, H., Nakai, H.: Stability of the GJ 876 planetary system. Publ. Astron. Soc. Jpn. **53**, L25–L26 (2001)

Kirkwood, D.: Meteoritic astronomy. Lipincott, Philadelphia (1867)

Kleine, T., Münker, C., Mezger, K., Palme, H.: Rapid accretion and early core formation on asteroids and the terrestrial planets from Hf-W chronometry. Nature **418**, 952–955 (2002)

Kley, W.: On the migration of a system of protoplanets. Mon. Not. Roy. Astron. Soc. **313**, L47–L51 (2000)

Kley, W., Peitz, J., Bryden, G.: Evolution of planetary systems in resonance. Astron. Astrophys. **414**, 735–747 (2004)

Kokubo, E., Ida, S.: Formation of protoplanet systems and diversity of planetary systems. Astrophys. J. **581**, 666–680 (2002)

Kokubo, E., Kominami, J., Ida, S.: Formation of terrestrial planets from protoplanets. I. Statistics of basic dynamical properties. Astrophys. J. **642**, 1131–1139 (2006)

Kowal, C.T., Drake, S.: Galileo's observations of Neptune. Nature **287**, 311–313 (1980)

Kozai, Y.: Secular perturbations of asteroids with high inclination and eccentricity. Astron. J. **67**, 591–598 (1962)

Kuiper, G.P.: On the origin of the solar system. Proc. Natl. Acad. Sci. **37**, 1–14 (1951)

Larson, R.B.: The physics of star formation. Rep. Progr. Phys. **66**, 1651–1697 (2003)

Laskar, J.: A numerical experiment on the chaotic behaviour of the solar system. Nature **338**, 237–238 (1989)

Laskar, J.: Chaotic diffusion in the Solar System. Icarus **196**, 1–15 (2008)

Laughlin, G., Bodenheimer, P., Adams, F.C.: The core accretion model predicts few Jovian-mass planets orbiting red dwarfs. Astrophys. J. Lett. **612**, L73–L76 (2004)

Lecar, M., Franklin, F.A., Holman, M.J., Murray, N.J.: Chaos in the solar system. Ann. Rev. Astron. Astrophys. **39**, 581–631 (2001)

Levison, H.F., Lissauer, J.J., Duncan, M.J.: Modeling the diversity of outer planetary systems. Astron. J. **116**, 1998–2014 (1998)

Levison, H.F., Morbidelli, A., Gomes, R., Backman, D.: Planet migration in planetesimal disks. In: Reipurth, B., Jewitt, D., Keil, K. (eds.) Protostars and Planets V, University of Arizona Press, Tucson, pp. 669–684 (2007)

Lewin, R.: Complexity: Life at the edge of chaos. University of Chicago Press, IL (2000)

Lin, D.N.C., Bodenheimer, P., Richardson, D.C.: Orbital migration of the planetary companion of 51 Pegasi to its present location. Nature **380**, 606–607 (1996)

Lissauer, J.J., Rivera, E.J.: Stability analysis of the planetary system orbiting υ Andromedae. II. Simulations using new lick observatory fits. Astrophys. J. **554**, 1141–1150 (2001)

Lissauer, J.J., Quintana, E.V., Chambers, J.E., Duncan, M.J., Adams, F.C.: Terrestrial planet formation in binary star systems. In: Garcia-Segura, G., Tenorio-Tagle, G., Franco, J., Yorke, H.W. (eds.) Revista Mexicana de Astronomia y Astrofisica Conference Series, pp. 99–103 (2004)

Lisse, C.M., Beichman, C.A., Bryden, G., Wyatt, M.C.: On the nature of the dust in the debris disk around HD 69830. Astrophys. J. **658**, 584–592 (2007)

Lorenz, E.N.: Deterministic nonperiodic flow. J. Atmos. Sci. **20**, 130–141 (1963)

Lorenz, E.: The essence of chaos. University of Washington Press, WA (1993)

Lovis, C., Mayor, M., Pepe, F., Alibert, Y., Benz, W., Bouchy, F., Correia, A.C.M., Laskar, J., Mordasini, C., Queloz, D., Santos, N.C., Udry, S., Bertaux, J.L., Sivan, J.P.: An extrasolar planetary system with three Neptune-mass planets. Nature **441**, 305–309 (2006)

Lyttleton, R.A.: An accretion hypothesis for the origin of the solar system. Mon. Not. Roy. Astron. Soc. **122**, 399–407 (1961)

Machida, M.N.: Binary formation in star-forming clouds with various metallicities. Astrophys. J. **682**, L1–L4 (2008)

Malhotra, R.: The origin of Pluto's orbit: Implications for the solar system beyond Neptune. Astron. J. **110**, 420–429 (1995)

Malhotra, R., Holman, M., Ito, T.: Chaos and stability of the solar system. Proc. Natl. Acad. Sci. **98**, 12,342–12,343 (2001)

Malmberg, D., Davies, M.B., Chambers, J.E.: The instability of planetary systems in binaries: how the Kozai mechanism leads to strong planet-planet interactions. Mon. Not. Roy. Astron. Soc. **377**, L1–L4 (2007)

Mandell, A.M., Raymond, S.N., Sigurdsson, S.: Formation of Earth-like planets during and after giant planet migration. Astrophys. J. **660**, 823–844 (2007)

Matsumura, S., Pudritz, R.E., Thommes, E.W.: Saving planetary systems: Dead zones and planetary migration. Astrophys. J. **660**, 1609–1623 (2007)

Mayor, M., Udry, S., Lovis, C., Pepe, F., Queloz, D., Benz, W., Bertaux, J.L., Bouchy, F., Mordasini, C., Segransan, D.: The HARPS search for southern extra-solar planets. XIII. A planetary system with 3 super-Earths (4.2, 6.9, and 9.2 M_T). Astron. Astrophys. **493**, 639–644 (2009a)

Mayor, M., Bonfils, X., Forveille, T., Delfosse, X., Udry, S., Bertaux, J.L., Beust, H., Bouchy, F., Lovis, C.L., Pepe, F., Perrier, C., Queloz, D., Santos, N.C.: The HARPS search for southern extra-solar planets. XVIII. An Earth-mass planet in the GJ 581 planetary system. Astron. Astrophys. **507**, 487–494 (2009b)

McCrea, W.H.: The origin of the solar system. Proceedings of the Royal Society of London, Series A, **256**, 245–266 (1960)

McCrea, W.: Formation of the solar system – Brief review and revised protoplanet theory. In: Runcorn, S.K. (ed.) The Physics of the Planets: Their Origin, Evolution and Structure, pp. 421–439. Wiley, NY (1988)

Menou, K., Tabachnik, S.: Dynamical habitability of known extrasolar planetary systems. Astrophys. J. **583**, 473–488 (2003)

Meyer, M.R., Hillenbrand, L.A., Backman, D., Beckwith, S., Bouwman, J., Brooke, T., Carpenter, J., Cohen, M., Cortes, S., Crockett, N., Gorti, U., Henning, T., Hines, D., Hollenbach, D., Kim, J.S., Lunine, J., Malhotra, R., Mamajek, E., Metchev, S., Moro-Martin, A., Morris, P., Najita, J., Padgett, D., Pascucci, I., Rodmann, J., Schlingman, W., Silverstone, M., Soderblom, D.,

Stauffer, J., Stobie, E., Strom, S., Watson, D., Weidenschilling, S., Wolf, S., Young, E.: The formation and evolution of planetary systems: Placing our solar system in context with Spitzer. Publ. Astron. Soc. Pac. **118**, 1690–1710 (2006)

Meyer, M.R., Carpenter, J.M., Mamajek, E.E., Hillenbrand, L.A., Hollenbach, D., Moro-Martin, A., Kim, J.S., Silverstone, M.D., Najita, J., Hines, D.C., Pascucci, I., Stauffer, J.R., Bouwman, J., Backman, D.E.: Evolution of mid-infrared excess around sun-like stars: Constraints on models of terrestrial planet formation. Astrophys. J. **673**, L181–L184 (2008)

Michtchenko, T.A., Malhotra, R.: Secular dynamics of the three-body problem: application to the υ Andromedae planetary system. Icarus **168**, 237–248 (2004)

Militzer, B., Hubbard, W.B., Vorberger, J., Tamblyn, I., Bonev, S.A.: A massive core in Jupiter predicted from first-principles simulations. Astrophys. J. **688**, L45–L48 (2008)

Minton, D.A., Malhotra, R.: A record of planet migration in the main asteroid belt. Nature **457**, 1109–1111 (2009)

Morbidelli, A., Levison, H.F., Tsiganis, K., Gomes, R.: Chaotic capture of Jupiter's Trojan asteroids in the early solar system. Nature **435**, 462–465 (2005)

Moulton, F.R.: On the evolution of the solar system. Astrophys. J. **22**, 165–181 (1905)

Moulton, F.R.: An introduction to celestial mechanics. Second edition revised, Dover Publications (1970)

Murray, N., Holman, M.: The origin of chaos in the outer solar system. Science **283**, 1877–1881 (1999)

Musielak, Z.E., Cuntz, M., Marshall, E.A., Stuit, T.D.: Stability of planetary orbits in binary systems. Astron. Astrophys. **434**, 355–364 (2005)

Natta, A., Testi, L., Calvet, N., Henning, T., Waters, R., Wilner, D.: Dust in protoplanetary disks: Properties and evolution. In: Reipurth, B., Jewitt, D., Keil, K. (eds.) Protostars and Planets V, University of Arizona Press, Tucson, pp. 767–781 (2007)

Noble, M., Musielak, Z.E., Cuntz, M.: Orbital stability of terrestrial planets inside the habitable zones of extrasolar planetary systems. Astrophys. J. **572**, 1024–1030 (2002)

Papaloizou, J.C.B., Terquem, C.: Planet formation and migration. Rep. Progr. Phys. **69**, 119–180 (2006)

Pascucci, I., Gorti, U., Hollenbach, D., Najita, J., Meyer, M.R., Carpenter, J.M., Hillenbrand, L.A., Herczeg, G.J., Padgett, D.L., Mamajek, E.E., Silverstone, M.D., Schlingman, W.M., Kim, J.S., Stobie, E.B., Bouwman, J., Wolf, S., Rodmann, J., Hines, D.C., Lunine, J., Malhotra, R.: Formation and evolution of planetary systems: Upper limits to the gas mass in disks around sun-like stars. Astrophys. J. **651**, 1177–1193 (2006)

Payne, M.J., Ford, E.B., Wyatt, M.C., Booth, M.: Dynamical simulations of the planetary system HD69830. Mon. Not. Roy. Astron. Soc. **393**, 1219–1234 (2009)

Peale, S.J.: Rotation histories of the natural satellites, pp. 87–111. IAU Colloq. 28: Planetary satellites (1977)

Pepe, F., Correia, A.C.M., Mayor, M., Tamuz, O., Couetdic, J., Benz, W., Bertaux, J.L., Bouchy, F., Laskar, J., Lovis, C., Naef, D., Queloz, D., Santos, N.C., Sivan, J.P., Sosnowska, D., Udry, S.: The HARPS search for southern extra-solar planets. VIII. μ Arae, a system with four planets. Astron. Astrophys. **462**, 769–776 (2007)

Pepliński, A., Artymowicz, P., Mellema, G.: Numerical simulations of type III planetary migration - I. Disc model and convergence tests. Mon. Not. Roy. Astron. Soc. **386**, 164–178 (2008a)

Pepliński, A., Artymowicz, P., Mellema, G.: Numerical simulations of type III planetary migration - II. Inward migration of massive planets. Mon. Not. Roy. Astron. Soc. **386**, 179–198 (2008b)

Pepliński, A., Artymowicz, P., Mellema, G.: Numerical simulations of type III planetary migration - III. Outward migration of massive planets. Mon. Not. Roy. Astron. Soc. **387**, 1063–1079 (2008c)

Peterson, I.: Newton's clock: Chaos in the solar system. Freeman, New York (1993)

Poincaré, N.: Sur le probleme de tres corps et les équations de la dynamique. Acta Math. **13**, 1–270 (1890)

Poincaré, N.: Les Méthodes nouvelles de la Mécanique Céleste. 3 volumes, Gauthier-Villars, Paris (1892)

Pontoppidan, K.M., Blake, G.A., van Dishoeck, E.F., Smette, A., Ireland, M.J., Brown, J.: Spectroastrometric imaging of molecular gas within protoplanetary disk gaps. Astrophys. J. **684**, 1323–1329 (2008)
Raymond, S.N.: The search for other earths: Limits on the giant planet orbits that allow habitable terrestrial planets to form. Astrophys. J. **643**, L131–L134 (2006)
Raymond, S.N., Barnes, R.: Predicting planets in known extrasolar planetary systems. II. Testing for saturn mass planets. Astrophys. J. **619**, 549–557 (2005)
Raymond, S.N., Quinn, T., Lunine, J.I.: Terrestrial planet formation in disks with varying surface density profiles. Astrophys. J. **632**, 670–676 (2005)
Raymond, S.N., Quinn, T., Lunine, J.I.: High-resolution simulations of the final assembly of Earth-like planets I. Terrestrial accretion and dynamics. Icarus **183**, 265–282 (2006)
Raymond, S.N., Scalo, J., Meadows, V.S.: A decreased probability of habitable planet formation around low-mass stars. Astrophys. J. **669**, 606–614 (2007)
Raymond, S.N., Barnes, R., Gorelick, N.: A dynamical perspective on additional planets in 55 Cancri. Astrophys. J. **808** (2008)
Raymond, S.N., Barnes, R., Veras, D., Armitage, P.J., Gorelick, N., Greenberg, R.: Planet-planet scattering leads to tightly packed planetary systems. Astrophys. J. **696**, L98–L101 (2009)
Rice, W.K.M., Armitage, P.J., Hogg, D.F.: Why are there so few hot Jupiters? Mon. Not. Roy. Astron. Soc. **384**, 1242–1248 (2008)
Rieke, G.H., Su, K.Y.L., Stansberry, J.A., Trilling, D., Bryden, G., Muzerolle, J., White, B., Gorlova, N., Young, E.T., Beichman, C.A., Stapelfeldt, K.R., Hines, D.C.: Decay of planetary debris disks. Astrophys. J. **620**, 1010–1026 (2005)
Rivera, E.J., Lissauer, J.J., Butler, R.P., Marcy, G.W., Vogt, S.S., Fischer, D.A., Brown, T.M., Laughlin, G., Henry, G.W.: A 7.5 M_E planet orbiting the nearby star, GJ 876. Astrophys. J. **634**, 625–640 (2005)
Rubenstein, E.P., Schaefer, B.E.: Are superflares on solar analogues caused by extrasolar planets? Astrophys. J. **529**, 1031–1033 (2000)
Russell, H.N.: The solar system and its origin. The Macmillan Company, New York (1935)
Sándor, Z., Süli, Á., Érdi, B., Pilat-Lohinger, E., Dvorak, R.: A stability catalogue of the habitable zones in extrasolar planetary systems. Mon. Not. Roy. Astron. Soc. **375**, 1495–1502 (2007)
Saumon, D., Guillot, T.: Shock compression of deuterium and the interiors of Jupiter and Saturn. Astrophys. J. **609**, 1170–1180 (2004)
Schmidt, O.Y.: Dokl. Akad. Nauk. USSR 45, No. **6**, 229–233 (1944)
Schwarz, R., Pilat-Lohinger, E., Dvorak, R., Érdi, B., Sándor, Z.: Trojans in habitable zones. Astrobiology **5**, 579–586 (2005)
Selsis, F., Kasting, J.F., Levrard, B., Paillet, J., Ribas, I., Delfosse, X.: Habitable planets around the star Gliese 581? Astron. Astrophys. **476**, 1373–1387 (2007)
Shankland, P.D., Blank, D.L., Boboltz, D.A., Lazio, T.J.W., White, G.: Further Constraints on the presence of a debris disk in the multiplanet system Gliese 876. Astron. J. **135**, 2194–2198 (2008)
Shu, F.H., Adams, F.C., Lizano, S.: Star formation in molecular clouds - Observation and theory. Annu. Rev. Astron. Astrophys. **25**, 23–81 (1987)
Smith, B.A., Terrile, R.J.: A circumstellar disk around Beta Pictoris. Science **226**, 1421–1424 (1984)
Soter, S.: Are planetary systems filled to capacity? Am. Sci. **95**, 414–421 (2007)
Su, K.Y.L., Rieke, G.H., Misselt, K.A., Stansberry, J.A., Moro-Martin, A., Stapelfeldt, K.R., Werner, M.W., Trilling, D.E., Bendo, G.J., Gordon, K.D., Hines, D.C., Wyatt, M.C., Holland, W.S., Marengo, M., Megeath, S.T., Fazio, G.G.: The vega debris disk: A surprise from Spitzer. Astrophys. J. **628**, 487–500 (2005)
Süli, Á., Dvorak, R., Freistetter, F.: The stability of the terrestrial planets with a more massive Earth. Mon. Not. Roy. Astron. Soc. **363**, 241–250 (2005)
Tabor, M.: Non-linear dynamics: An introduction. Wiley, New York (1989)
Takeda, G., Kita, R., Rasio, F.A.: Planetary systems in binaries. I. dynamical classification. ArXiv e-prints **802** (2008)

Tanaka, H., Takeuchi, T., Ward, W.R.: Three-dimensional interaction between a planet and an isothermal gaseous disk. I. Corotation and lindblad torques and planet migration. Astrophys. J. **565**, 1257–1274 (2002)

Ter Haar, D.: Further studies on the origin of the solar system. Astrophys. J. **111**, 179–190 (1950)

Terquem, C.E.J.M.L.J.: Stopping inward planetary migration by a toroidal magnetic field. Mon. Not. Roy. Astron. Soc. **341**, 1157–1173 (2003)

Thommes, E.W.: A safety net for fast migrators: Interactions between gap-opening and sub-gap-opening bodies in a protoplanetary disk. Astrophys. J. **626**, 1033–1044 (2005)

Trilling, D.E., Benz, W., Guillot, T., Lunine, J.I., Hubbard, W.B., Burrows, A.: Orbital evolution and migration of giant planets: Modeling extrasolar planets. Astrophys. J. **500**, 428–439 (1998)

Trilling, D.E., Stansberry, J.A., Stapelfeldt, K.R., Rieke, G.H., Su, K.Y.L., Gray, R.O., Corbally, C.J., Bryden, G., Chen, C.H., Boden, A., Beichman, C.A.: Debris disks in main-sequence binary systems. Astrophys. J. **658**, 1289–1311 (2007)

Trilling, D.E., Bryden, G., Beichman, C.A., Rieke, G.H., Su, K.Y.L., Stansberry, J.A., Blaylock, M., Stapelfeldt, K.R., Beeman, J.W., Haller, E.E.: Debris disks around sun-like stars. Astrophys. J. **674**, 1086–1105 (2008)

Tsiganis, K., Gomes, R., Morbidelli, A., Levison, H.F.: Origin of the orbital architecture of the giant planets of the solar system. Nature**435**, 459–461 (2005)

Van Boekel, R., Min, M., Leinert, C., Waters, L.B.F.M., Richichi, A., Chesneau, O., Dominik, C., Jaffe, W., Dutrey, A., Graser, U., Henning, T., de Jong, J., Köhler, R., de Koter, A., Lopez, B., Malbet, F., Morel, S., Paresce, F., Perrin, G., Preibisch, T., Przygodda, F., Schöller, M., Wittkowski, M.: The building blocks of planets within the terrestrial region of protoplanetary disks. Nature **432**, 479–482 (2004)

von Bloh, W., Bounama, C., Franck, S.: Dynamic habitability for Earth-like planets in 86 extrasolar planetary systems. Planet. Space Sci. **55**, 651–660 (2007a)

von Bloh, W., Bounama, C., Cuntz, M., Franck, S.: The habitability of super-Earths in Gliese 581. Astron. Astrophys. **476**, 1365–1371 (2007b)

Wajer, P.: 2002 AA29: Earth's recurrent quasi-satellite? Icarus **200**, 147–153 (2009)

Ward, W.R.: On disk-planet interactions and orbital eccentricities. Icarus **73**, 330–348 (1988)

Weidenschilling, S.J., Marzari, F.: Gravitational scattering as a possible origin for giant planets at small stellar distances. Nature **384**, 619–621 (1996)

Weizsäcker, C.F.V.: Über die Entstehung des Planetensystems. Zeitschrift fur Astrophysik **22**, 319–355 (1943)

Williams, D.M., Pollard, D.: Habitable Planets on Eccentric Orbits. In: Montesinos, B., Gimenez, A., Guinan, E.F. (eds.) The Evolving Sun and its Influence on Planetary Environments, Astronomical Society of the Pacific Conference Series, vol. 269, pp. 201–213 (2002)

Wisdom, J.: Meteorites may follow a chaotic route to Earth. Nature **315**, 731–733 (1985)

Woolfson, M.M.: A capture theory of the origin of the solar system. Proc. Roy. Soc. Lond. A **282**, 485–507 (1964)

Woolfson, M.M.: The capture theory and the origin of the solar system, pp. 179–198. In: Origin of the Solar System (1978)

Wright, J.T., Marcy, G.W., Fischer, D.A., Butler, R.P., Vogt, S.S., Tinney, C.G., Jones, H.R.A., Carter, B.D., Johnson, J.A., McCarthy, C., Apps, K.: Four new exoplanets and hints of additional substellar companions to exoplanet host stars. Astrophys. J. **657**, 533–545 (2007)

Wyatt, M.C.: Evolution of debris disks. Annu. Rev. Astron. Astrophys. **46**, 339–383 (2008)

Wyatt, M.C., Holland, W.S., Greaves, J.S., Dent, W.R.F.: Extrasolar analogues to the Kuiper belt. Earth Moon Planet **92**, 423–434 (2003)

Yin, Q., Jacobsen, S.B., Yamashita, K., Blichert-Toft, J., Télouk, P., Albarède, F.: A short timescale for terrestrial planet formation from Hf-W chronometry of meteorites. Nature **418**, 949–952 (2002)

Zapatero Osorio, M.R., Caballero, J.A., Béjar, V.J.S., Rebolo, R., Barrado Y Navascués, D., Bihain, G., Eislöffel, J., Martín, E.L., Bailer-Jones, C.A.L., Mundt, R., Forveille, T., Bouy, H.: Discs of planetary-mass objects in σ Orionis. Astron. Astrophys. **472**, L9–L12 (2007)

Chapter 9
Is Our Environment Special?

In this book we have tried to established the Earth–Exoplanets connection. However, to validate this approach we must discuss how special is our planet, and its nearby environment, compared with the rest of exoplanets detected and to be discovered. In the first chapter, we described this environment from our local neighbourhood to our location in the Galaxy.

With this background in mind, we are now prepared to discuss three classical questions: 'Is the Sun something special?', 'Is the Earth a rare or unique planet?' and 'Is the solar system something rare or unique?'. As in other branches of Science, only future observations and careful data interpretation will provide a more definite answer. In formulating such questions, semantics also plays a role: anomalous, rare or unique are often used to describe results from studies on this topic with ambiguous meanings. Finally, we can also ask ourselves: special, for what? The answer is obvious: for the habitability of the planetary system and the existence of life.

A few comments are relevant at this point. To place our planet in an universal context is a difficult task depending on the emphasis given to the similarities and differences found:

- As commented by Gonzalez (1999) in a footnote, an anomaly may be due to a simple low-probability statistical fluctuation, or it may be an indication of a physical process or selection bias.
- Gustafsson (2008) remarked: The key issue here is *how similar* the comparison stars have to be relative to the Sun and *how similar* the methods have to be in order for systematic errors to be reduced to an insignificant level.
- Studying the planetary habitability, it is important to avoid a circular reasoning, where the conclusion is in the assumption, and the abuse of mere correlations without a physical background. Our reference point is limited to one, the Earth and its current environment.

We also want to stress that it is clearly outside our purposes to derive any teleological implications from these purely natural facts. Let us start with our own star.

9.1 Is the Sun Anomalous?

This question was studied in detail by Gustafsson (1998) and Gonzalez (1999), leading to different conclusions. For the first, the Sun is a normal star for its mass, if only a little iron rich, whereas for the latter we are living in a privileged environment with regards to the habitability of our planet. The availability of an unbiased stellar sample for comparison is probably the most important problem that needs to be solved in order to answer this question. The most complete work to date has been carried out by Robles et al. (2008),[1] who identified 11 properties that have some plausible association with habitability. We discuss separately some of these parameters.

9.1.1 Singularity

The Sun is a yellow dwarf star, a common stellar type in our Galaxy. However, there are much more criteria than the spectral type which can be used to classify a star as a solar twin and evaluate how common our Sun is. In principle, the first distinguishing characteristic point out is its solitude. Between 60 and 80% of solar-like stars are part of binary or multiple stellar systems (Duquennoy and Mayor 1991). However, Lada (2006) claims a lower binary fraction for low-mass stars, concluding that most of these stars are solitary red dwarfs.

The explanation for the fact that there is only one Sun in our planeatry system lies in the process of stellar formation via the contraction of a molecular cloud (Tohline 2002). Fragmentation of the original molecular cloud during the formation of a protostar is an acceptable explanation[2] for the formation of a binary or a multiple star system (Boss 1992). It is now pertinent to look for factors that favour such fragmentation.

Cloud rotation promotes fragmentation, while the process tends to be suppressed in clouds with higher metallicity (Machida 2008). This latter relationship is supported by the observations of Lucatello et al. (2005), who found that 100% of the stars in the galactic halo (metal-poor) are in binary/multiple systems. Grether and Lineweaver (2007) also found that the frequency of stellar companions tends to be more abundant around low metallicity hosts.

Attwood et al. (2009) have simulated star formation for different levels of turbulence.[3] They found that essentially all the statistical parameters of the stars

---

[1] See a short erratum in The Astrophysical Journal (2008) 689:2, 1457, which does not affect the conclusions of the paper.

[2] Other mechanisms have been proposed such as fission (a star splits in two, a rather unlikely process), capture (requiring high stellar densities) and disk fragmentation.

[3] Physical process in which kinetic energy cascades from large to small spatial scales.

formed are independent of the turbulence. Magnetic fields tend to suppress cloud fragmentation and therefore the formation of binary and multiple stellar systems (Price and Bate 2007).

9.1.2 Mass

Mass is one of the most relevant parameters of a star: it determines the spectral type and its lifetime in the main sequence. Figure 9.1 (Fig. 1 from Robles et al. 2008) compares the solar mass to the stellar mass distribution of the 125 nearest stars. Also overplotted is the initial mass function[4] by Kroupa (2002). We can see clearly that the Sun is more massive than 95% ± 2% of the stars.

What is the relevance of this expected result? In Chap. 5 we explained the concept of the circumstellar habitable zone and the role played by stellar luminosity, intrinsically related to the mass. Although low-mass stars are more numerous and stay longer in the main sequence, intermediate-mass stars are perhaps the best compromise between the stellar lifetime and the distance at which habitable planets can be formed.

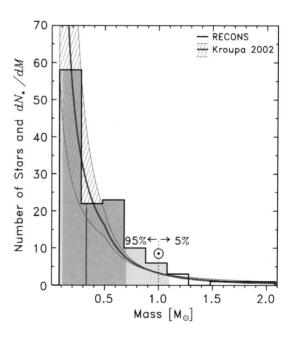

Fig. 9.1 Mass histogram of the 125 nearest stars (see RECONS database, http://www.recons.org/). The median of the distribution is indicated by the vertical *gray* line. The 68 and 95% bands around the median are indicated, respectively, by the vertical *dark gray* and *light gray* bands. The *solid curve* and hatched area around it represents the initial mass function and its associated uncertainty (Kroupa 2002). The Sun is indicated by the symbol (☉), Courtesy: J.A. Robles. Reproduced by permission of the American Astronomical Society

---

[4] It specifies the mass distribution of a newly formed stellar population and it is frequently assumed to be a simple power law (Miller and Scalo 1979).

9.1.3 Location

The relevant question is: What role has the birth site of our Sun played in its subsequent evolution? The first step is to locate the original site.

The Sun is presently located in the Galaxy close to the inner rim of the Orion arm. At 8.5 ± 1.1 kpc from the centre (Kerr and Lynden-Bell 1986), it is inside the co-rotation distance,[5] minimising the crossing of the spiral arms to only one every one billion years (Mishurov et al. 1997; Mishurov and Zenina 1999). The crossing of a spiral arm could be associated to an enhancement in Supernova explosions and in the impact rate of giant comets. We should remark that a star placed at exactly the co-rotation distance would experience a mean motion resonance (see last chapter), probably resulting in a destabilization of its orbit.

The Sun describes an almost circular orbit, which is less eccentric than $93\% \pm 1\%$ of FGK stars within 40 parsecs (Robles et al. 2008).

The Sun likely formed in a cluster with more than 100 members, some of them very massive. The presence of short-lived isotopes in primitive meteorites seem to indicate that the 1.8 Ma young Sun was polluted by a supernova explosion of one of its massive siblings located at a distance of 0.02–1.6 pc (Looney et al. 2006). The local environment of the forming solar system was therefore likely reminiscent of a high-mass star-forming region like the Orion region. The young Sun and its circumstellar disk may have resided in a HII region for a considerable amount of time (Hester et al. 2004).

Based on observations of young stellar clusters, Adams et al. (2006) found a correlation between the number of stars, N, and the initial size of the cluster:

$$R(N) = R_{300} \sqrt{(N/300)},$$

where $R_{300} = 1\text{–}2$ pc. However, the Sun has long lost the connection with its cradle.[6] Based in its apparent higher metallicity, Wielen et al. (1996) proposed that the Sun was formed at a galactocentric distance of 6.6 ± 0.9 kpc and migrated outwards to its present position during its lifetime. Another approach has been taken by Roskar et al. (2008) in their numerical simulations. They show that spiral arms are created and destroyed during the evolution of a spiral galaxy, such as the Milky Way, launching stars from their birth position while still preserving their circular orbits.

However, all hopes of identifying the solar siblings and the characteristics of the cradle may not be lost. Portegies Zwart (2009) estimated that still 1% of these stars are within 100 pc of our current location in the Galaxy, increasing to 10% within 1 kpc. Future astrometric projects could detect many of them based on their

---

[5] The distance from the core where stars orbit the galaxy at the same rate as its spiral arm structure does.

[6] Since its formation, the Sun has made 16 revolutions around the Galaxy with billions of gravitational interactions.

kinematical properties. Based on the data on the nearby supernova explosion, he was also able to estimate between 500 and 3,000 solar masses the mass of the initial cluster with a radius in the range of 1–3 pc.

9.1.4 Age

Life needs elements heavier than hydrogen and helium, and time is needed to produce them in the interiors of the stars. Carbon is essential at least for the type of life we know. Livio (1999) suggest that the emergence of the first civilizations is related to a peak in the production of carbon in the Galaxy. Figure 9.2 uses the planetary nebulae and the rate of stellar formation as parameters related to the carbon production and the red-shift,[7] z, as the descriptor of the Universe age. We see a peak at z ∼1, which corresponds to 5.6 Ga. If we now add the time necessary for the development of complex life (∼3.5 Ga), we obtain a figure close to the age of the Galaxy disk (∼10 Ga). In this approach we would be one of the first civilizations in our Galaxy.

Another aspect is the intrinsic age of the Sun and its relation with the neighbourhood. Robles et al. (2008) conclude that the Sun is younger than 53% ± 2% of the stars in the thin disk of our Galaxy. By comparing with the cosmic star formation rate[8] of Hopkins and Beacom (2006), with the median at 9.15 Ga, they suggest that the Sun was born after 86% ± 5% of the stars that have ever been born (see also Fig. 2 of Robles et al. 2008).

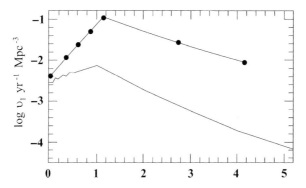

Fig. 9.2 Dependence of the planetary nebula formation rate on redshift (x-axis). *Dashed curve*, the star formation rate; *solid line*, planetary nebula formation rate. Adapted from Livio (1999) Fig. 2. Reproduced by permission of the American Astronomical Society

---

[7] Dimensionless quantity given by $z = (\lambda - \lambda_{ref})/\lambda_{ref}$, where $(\lambda - \lambda_{ref})$ express the difference between the observed and emitted wavelengths of an object. It is related to the expansion of the Universe.

[8] It is dominated by bulges and elliptical galaxies, where most of the mass of the Universe resides.

The main implication of this age is that the planetary system is already dynamically stable (see Chap. 8), and that in planets around the Sun enough time elapsed for complex life to develop.

9.1.5 Chemical Composition: Metallicity

The level of metallicity is clearly a function of the stellar age and galactocentric distance. Giant planet formation is favoured with increasing stellar metallicities (see Chaps. 7 and 8), but little is known for Earth-like planets (see, however, Fig. 9.3). The Sun is more metal-rich than ∼2/3 of local solar-like stars and less metal-rich than ∼2/3 of the stars hosting Hot Jupiters (Gonzalez 1999; Lineweaver 2001; Grether and Lineweaver 2007; Robles et al. 2008).

Supernova explosions are less frequent going away from the galactic centre, although their effect on the habitability of a planet is far from clear (Ruderman 1974 and Chap. 7 of Vázquez and Hanslmeier 2005). These two factors (metallicity and Supernova explosions) have led to the formation of the hypothesis of galactic habitable zone (Gonzalez et al. 2001; Linewaver et al. 2003), an annular region between 7 and 9 kparsec from the Galactic centre, where 75% of the stars are older than the Sun (see Chap. 5). Therefore, it would seem we are in the safe region of the Galaxy. However, if the mechanism proposed by Roskar et al. (2008) works, this zone would lose much of its relevance. See also the criticisms of this concept raised by Prantzos (2008).

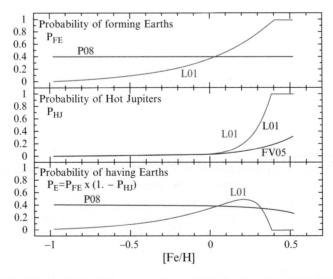

Fig. 9.3 Role of the metallicity of the proto-stellar nebula in the formation and presence of Earth-like planets around solar-like stars. L01 (Lineweaver 2001); FV05 (Fischer and Valenti 2005); P08 (Prantzos 2008). The [Fe/H] value for the Sun is taken as 0. Adapted from Fig. 3 of Prantzos (2008). Copyright: Springer

The composition and chemistry of the planets is determined by the initial elemental composition of the protoplanetary disk and this depends on the environment where the star was born. We have already stressed in Chap. 7 the importance of the C/O ratio. Edvardsson et al. (1993) and Gustafsson et al. (1999) propose that the Sun has a low [C/O] ratio relative to solar-like stars at similar galactocentric radii. However, the available stellar carbon and oxygen data are probably not accurate enough to evaluate the possible anomaly of the Sun. Also, the oxygen abundance of the Sun is under debate (Centeno and Socas-Navarro 2008; Scott et al. 2009).

Allende-Prieto (2008) studied a sample of 80 nearby stars with solar-like colours and luminosity, concluding that objects with solar iron abundances also exhibit solar abundances of carbon, silicon, calcium, titanium and nickel.

9.1.6 Magnetic Activity

The Sun exhibits a series of phenomena related with the emergence of magnetic fields produced at the bottom of the convection zone via a dynamo process. The outer layers are heated by a process also related to the magnetic fields, emitting XUV radiation. Stars show all the phenomena associated with solar magnetic activity, the base of the solar–stellar connection, but under different conditions of luminosity, mass, age, chemical composition, etc. Magnetic activity decays with the stellar age (Skumanich 1972) and is proportional to the rotation period, P_{rot}, and the depth of the convection zone, usually expressed by convective turnover time (Noyes et al. 1984). The Mount Wilson survey has monitored in the last decades the flux at the CaII line as a proxy of stellar activity (Baliunas et al. 1998). A similar program is being carried out with data from the Lick and Keck observatories (Wright 2004). Stars show different average lengths of the main activity cycle, P_{cyc}.

Vaughan and Preston (1980) first noticed a gap in the distribution of CaII flux with the spectral type. Further analysis confirmed this gap, especially visible by plotting P_{cyc}/P_{rot} (Brandenburg 1998; Saar and Brandenburg 1999 and Fig. 9.4). We have two sequences, the active A sequence and the inactive (I) with cooler and more slowly rotating stars. Böhm-Vitense (2007) suggest that different kinds of dynamos are working for stars in the two sequences. This behaviour characterized the differences between young and old stars, with the transition taking pace at a P_{rot} of 21 days (2–3 Ga). Curiously, the Sun lies squarely between the two sequences. It is very tempting to relate this transition with one of the main phases of life evolution on the Earth (Whitehouse 1983), but it is still premature and the subject needs more detailed study.

Israelian et al. (2004) and Gonzalez (2008) have found that exoplanet host stars are lithium-depleted compared to solar-type stars without detected massive planets. Bouvier (2008) suggests that, for solar-like stars, this dichotomy may result from their rotational history. Long-lasting protoplanetary disks, braking the stellar rotation, should be associated to massive planet formation becoming slow rotators, developing a high degree of differential rotation between the radiative core and

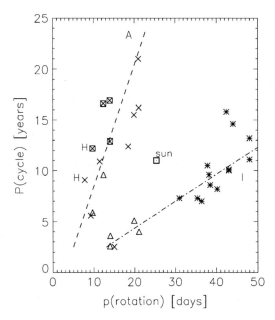

Fig. 9.4 Period of the activity cycles, P_{cyc} in years, plotted as a function of the rotation periods, P_{rot}, in days. The data follow two sequences, the relatively young, active A sequence (*dashed line*) and the generally older, less active I sequences (*dash–dotted line*). The letter H indicates Hyades group stars, *crosses* indicate stars on the A sequence, and *asterisks* stars on the I sequences. *Squares* around the crosses show stars brighter than the Sun. The solar point is plotted as a *square* with a *dot* inside. Adapted from Fig. 1 of Böhm-Vitense (2007). Reproduced by permission of the American Astronomical Society

the convective envelope[9] (Fig. 9.5). He finally proposes that the strong differential rotation at the base of the convection zone is responsible for the enhanced Li depletion in slow rotators. The two curves converge at an age of about 1 Ga and therefore the memory of our old Sun is lost.

The Sun in Time program has studied the evolution of a selected group of stars, similar to the Sun but with different ages (Güdel 2007). They confirm the dispersion of values for young stars, supporting an early Sun more active than in our days (see also Chap. 4).

It is also worth considering whether our Sun is exceptional or normal according to the strength of transitory events associated with the magnetic activity. Giant flares seem to have been detected in solar-like stars (Schaefer et al. 2000). Although the sample is not very large, the Sun seems to have an anomalous behaviour, at least in this aspect. It has been suggested (Rubenstein and Schaefer 2000) that the enhancement in flare strength could be caused by the presence of giant planets close to the star. In such conditions, a reconnection of the magnetic field lines of the stellar

---

[9] The respective time scales for core-surface coupling would be 100 Ma for slow-rotators and only 10 Ma for fast-rotators.

9.1 Is the Sun Anomalous?

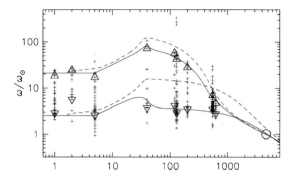

Fig. 9.5 Rotational models for slow and fast rotators. The modelled evolution of surface rotation for slow and fast rotators is shown by the *solid lines*. For both models, the rotation of the radiative core is shown by the *dashed lines*. Individual measurements of angular velocities are also shown. Time is indicated in years and the angular rotation is referred to the Sun. Reprinted with permission from the Astronomy and Astrophysics Editorial Office

corona with those of the planetary magnetosphere could be produced, giving rise to an important release of energy. A recent analysis by Kashyap et al. (2008) reveals that stars with Hot Jupiters have on average more X-ray activity, by a factor of ∼4, than those with more distant planets.

Along a typical cycle, variations in brightness occur at the photosphere, produced by the energy balance between different magnetic structures (sunspots, plages and network). Lockwood et al. (2007) have recently stated, contradicting previous claims, that the amplitude of solar variations related to the magnetic activity is typical, compared to those of sun-like stars of similar age.

9.1.7 Solar Analogs

Solar-type, solar-analog and solar-twin are three expressions meaning increasing similarity with our Sun. Temperature, metallicity, level of magnetic activity and absence of stellar companion are some of the criteria used in this classification. Pioneering attempts to find solar analogs were made by Cayrel de Strobel et al. (1981, 1990, 1996) and Neckel (1986). Pasquini et al. (2008) have identified five solar twins in M67, a stellar cluster with an age similar to the Sun's.

Gaidos (1998) has presented a list of 38 young (0.2–0.8 Ga) stars analog to our Sun. More recently Soubiran and Triaud (2004) have established their own top ten list of solar analogs. The best candidates to be stars analog to the Sun are summarized in Table 9.1

Robles et al. (2008) conclude that the Sun is a typical star and that there are no special requirements for a star to host a planet with life. This team also mentioned in a newsletter (www.phys.org May 21, 2008) that: 'it [the Sun] seems to be a random star that was blindly pulled out of the bag of all stars'.

Table 9.1 List of solar analogs in our neighbourhood

Star	Spectral type	Mass (M_S)	Age (Ga)	Reference
Alpha Centaury A	G2V	1.100	5–6	Quintana et al. (2002)
18 Scorpii (HD 146233)	G2V	1.01 ± 0.03	4.2	Porto de Mello and da Silva (1997)
37 Gem (HD 50692)	G0V	1.1	5.5	Soderblom (1986)
51 Pegasi (HD 217014)	G2.5V	1.06	7.5–8.5	
β Canum Venaticorum	G0V	1.08	4.05	Porto de Mello et al. (2006)
HD 143436 (HIP 78399)	G0V	1.01 ± 0.02	3.8 ± 2.9	King et al. (2005)
HD 98618	G5V	1.02 ± 0.03	4.21 ± 0.90	Meléndez et al. (2006)
HIP 100963	G5V			Takeda and Tajitsu (2009)
HIP 56948 (HD101364)	G5V	1.0	5.5	Meléndez and Ramírez (2007)

9.2 Is the Solar System Unique?

We have previously discussed that stars of solar mass are relatively rare in our environment. Now we would like to know how many of them have planets orbiting around them. The number of exoplanets is still too small to conclude anything in this respect. Marcy et al. (2005) estimated that 12% of the stars have a gas giant within 20 AU, and this value has recently increased to 17–19% by Cumming et al. (2008). It would be crucial to know empirically in the future how many of them have Earth-like planets.

We have seen that other planetary systems do not look like the solar system. It is therefore pertinent to ask whether the solar system is special in some way compared to the majority of planetary systems to be found in the Galaxy (Beer et al. 2004). Only after more intensive observational work can this question be answered.

9.2.1 Nature vs. Nurture

Two main factors have brought the solar system to the present situation. On the one hand, there are the initial conditions of the nebular cloud and, on the other hand, the varying environmental conditions the components (star, planets, satellites and minor bodies) have undergone. Planetary systems can be defined by how they were born and how the different components were located (heritage) and by the interactions the components have been subjected to (mainly gravitational instabilities) during the childhood of the system. This debate between heritage and environment is typical of other astronomical fields, such as the debate between monolithic and hierarchical models in the formation of galaxies (e.g. Van der Wel 2008) and also in psychology with the debate nature vs. nurture[10] (Meaney 2001). This dichotomy would seem to be too naive and both approaches are probably complementary as they are the distinction between random (stochastic) and deterministic factors. Our solar system seems to be unique in the same sense that every human being is unique but at the same time part of a common biological species.

9.2.1.1 Formation

One remarkable feature of our solar system is the presence of gas giants, which have not migrated very far from the initial location and show circular and stable orbits. How common are these bodies in planetary systems?

Marchi (2007) and Marchi and Ortolani (2008) have classified the giant exoplanets into five different families based on parameters such as stellar metallicity and the orbital parameters of the planet. They show the importance of including environmental factors, such as the stellar mass, to discriminate between otherwise similar exoplanets. The first two classes correspond to Hot Jupiters differentiated by stellar mass and metallicity.

Thommes et al. (2008a) have carried out simulations of planetary formation varying the disk mass ($0.01 M_S < M_{disk} < 0.1 M_S$) and the gas viscosity. The emerging systems result from a set of planet–disk and planet–planet interactions. Gas giant planets form within 1–10 Ma, during the time that their parent stars possess a gas disk (see Haisch et al. 2001). Two parameters seem to be fundamental for the destiny of a planetary system: τ_{disk}, gas disk depletion time, and τ_{giant}, time to form the first gas giant (see Kokubo and Ida 2002). In cases with $\tau_{giant} > \tau_{disk}$ (less massive disks and high viscosity), we have systems without gas giants consisting solely of rocky-ice bodies, the barren scenario. If $\tau_{giant} < \tau_{disk}$ (large massive disks and low viscosity), planet formation is a sort of anarchic mess, Hot Jupiters are produced and eccentric orbits are originated. Finally, we can expect a relative small number of solar-like systems with $\tau_{giant} \sim \tau_{disk}$, where gas giants undergo only a modest migration and eccentricity growth (see Fig. 9.6 and 8.26). The solar system would be the exception rather than the rule, and would be formed for a narrow range of disk masses around $0.05\ M_S$. However, the authors remark that scaled-down versions of the solar system (see the one recently detected by Gaudi et al. 2008) are likely to be more common. Nevertheless, in all the simulated cases terrestrial planets can be formed and remain stable.

How important are gas giants for the development of life in a planetary system? In Chap. 2 we described the biological extinctions that occurred on our planet during the last 500 Ma. Some of them have been attributed to the impact of objects from our debris disk, which nowadays is not very massive ($\sim 0.1 M_E$). The role that the giant planets play in these events is a that of double-edged sword (Horner and Jones 2008). On the one hand, they shield the inner solar system from the impacts by

---

[10] Defined as the sum of the environmental factors influencing the behaviour and traits expressed by an organism.

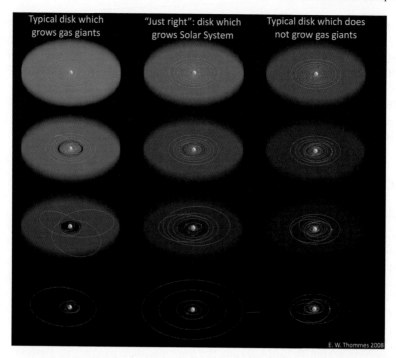

Fig. 9.6 Three different cases of planets forming in a gas disk: the violent version (*left*), the 'barren' version (*right*) and in between these two, where conditions are just right (solar system). Courtesy: E. Thommes

comets of the Oort cloud (Wetherill 1994), and on the other hand, they favour the formation of the asteroid belt and therefore the impacts of these bodies, and also perturb the bodies of the Kuiper belt (Horner and Jones 2009). In this debate the supply of volatiles to the terrestrial planets must also be considered (see Lunine 2001).

The metallicity of the primordial nebula plays an important role in the formation of planets. Prantzos (2008) summarizes the question, plotting separately the fraction of Earth-like planets (P_{FE}) and Hot Jupiters (P_{HJ}) as a function of metallicity, Z. Under the assumption that Hot Jupiters destroy the formed Earth-like planets during the process of migration (see however Raymond et al. (2006) and Fogg and Nelson (2009)), he obtains the probability of having Earths for a planetary system of a given metallicity.

9.2.1.2 Stellar Encounters

Some of the formed systems emerged from the protoplanetary phase intrinsically unstable, but others, which emerged stable, became unstable through perturbations from close encounters with other stars. This is specially relevant in stellar clusters. Malmberg et al. (2007a, b, 2008) define a singleton as a star which is formed single and has never suffered close encounters[11] with other stars or spent time within a binary system. Close fly-by encounters may result in the direct ejection of planets or that these remain bound but on more eccentric orbits. They also remark that even small perturbations can sometimes lead to significant instabilities via planet–planet interactions within the planetary system.

They suggest that planetary systems like our own solar system can exist only around singleton stars and estimate the singleton frequency with masses similar to that of the Sun to be between 90 and 95% in the solar neighbourhood.

9.2.1.3 Gravitational Interactions: LHB Events

Laskar (1996) and Lecar et al. (2001) remark that the solar system is not stable, it is just old. Objects, with shorter instability times, have long since been ejected or collided with others. The system is now full, dynamically dense, that is if you tried to squeeze another planet in between the existing ones, the resulting gravitational disturbances would dynamically excite the system, leading to a collision or ejection before the system could settle down again (see Barnes and Quinn (2004) and their theory of 'packed planetary systems'). The solar system has increased its internal order[12] by exporting disorder (entropy) to the rest of the Galaxy, which receives the chaotically ejected objects[13]: a process of self-organization or 'natural selection of the fittest'. In fact, the definition of planet given in the past chapter implies this kind of final stable configuration. On the other hand, systems that have experienced any important migration must appear dynamically underdense.

Formation of the Moon was one of the events belonging to the early violent phase of the solar system. The number and position of terrestrial planets in the solar system could easily have been different at the beginning. Tidal interactions with nebular gas may have caused early-formed inner terrestrial planets to migrate inward while they were forming, and several planets may have been lost this way into the Sun before the gas dispersed. The existing four terrestrial planets are simply the survivors of the process (Mc Neil et al. 2005). However, it was in the outer parts of the system where the last crucial event took place, leading to the present harmonic system.

As we have mentioned previously in the book, lunar records indicate that a spike in the cratering occurred 700 Ma after the formation of the solar system, the late heavy bombardment (LHB). This event probably added a substantial amount of water and organics to the crust of the Earth and the other terrestrial planets, favouring the conditions for life to emerge. This was produced by the entrance in a 1:2 resonance of the migrating Jupiter and Saturn (see Chap. 8 for more details).

A migration of only 1–2 AU was probably enough to produce this effect (Malhotra 2007). Morbidelli and Crida (2007) suggest that the stronger inner torque

---

[11] The authors consider a close encounter as when two objects pass within 1,000 AU of each other.
[12] Reducing the number of planets and increasing the spacing between them.
[13] This process, called dynamical relaxation, also operates in star clusters and in entire galaxies.

of Jupiter was able to slow down or even stop the type II migration, leading to the present configuration of our solar system. Why was the migration so long delayed from the formation of the system? Gomes et al. (2005) point out that the time at which Jupiter and Saturn reached their 1:2 resonance depends on (1) their distance from their resonance position, (2) the mass of materials in the planetesimal disk, particularly near its inner edge, and (3) the relative location of the inner edge of the disk and the outermost ice giant. LHB events are probably a common evolutionary step in many planetary systems (Thommes et el. 2008b).

The resonances between planets of a multiple system can decide the final configuration. Beaugé et al. (2006) remark that, although migration does not always lead to resonance trapping, the existence of massive bodies in exact mean-motion resonance (MMR) can be explained via a migration mechanism. The fact that the planets of the solar system did not experience significant migration is consistent with the fact that none of them are trapped in resonance. About a quarter of discovered systems contain planet pairs locked into MMR (Udry et al. 2007).

Pilat-Lohinger et al. (2008) have simulated the stability of a planetary system, a clone of the solar system, for different mass ratios of the two gas giants of the system. The dynamical map of Fig. 9.7 shows the chaotic regions and their associated resonances. The highest degree of chaos around the 2:1 MMR, especially for high Saturn masses, is visible near 8 AU.

Using terrestrial planets as test particles (bodies without mass), these authors studied the influence in the habitable zone. They found (1) an increase of Venus's eccentricity for the real Jupiter and Saturn masses and the actual semimajor axis of Saturn; (2) an increase of the eccentricity to nearly 0.3 of a test planet at Earth's

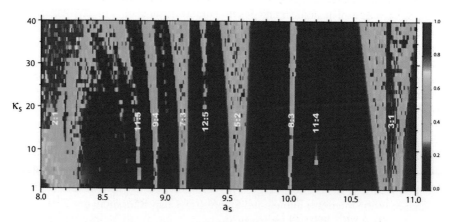

Fig. 9.7 MMRs appearing between Jupiter and Saturn in the region between 8 and 11 AU (x-axis) for various masses of Saturn (y-axis). The colours measure the closeness of the considered trajectory to a quasi-periodic one: *dark blue* shows the orbits that cannot be distinguished from a quasi-periodic one; *blue to yellow* shows the region where the instability is increased, but the disruption probably does not occur on a Ga timescale; *orange to red* areas indicate that the planetary system is chaotic and its destruction is possible. Adapted from Fig. 1 of Pilat-Lohinger et al. (2008). Reproduced by permission of the American Astronomical Society

9.2 Is the Solar System Unique?

position when Saturn's mass was increased by a factor of 3 or more; and (3) if the two giant planets are in 2:1 resonance, a strong influence on the outer region of the habitable zone is observed.

9.2.1.4 Mercury: The Achilles Heel

Our old solar system seems to be protected from 'catastrophes' by its hierarchy and the associated stabilizing resonances. However, some planetary events could still happen. Laskar's (1994) simulations of planetary orbits of the solar system in the past, and into the future, showed dramatic variations in the orbits of Mercury ($0.1 < e < 0.5; 8° < i < 21$). Shifting the initial positions of the Earth by only 150 m, he found that Mercury could reach $e \sim 1$ (i.e. near hyperbolic trajectory) 3.5 Ga in the future, leading to its ejection from the solar system. A three-body resonance between Jupiter, Saturn and Uranus may also lead to an ejection of Uranus, but the escape time (10^{18} years) is longer than the lifetime of the solar system (10^{10} years). Long-term simulations by Batygin and Laughlin (2008) indicate that the instability of Mercury will lead to the ejection of Mars from the solar system $\sim$822 Ma from now, the collision of Venus and Mercury at $\sim$862 Ma, and finally, the fall of Mercury onto the Sun at $\sim$1.261 Ma.

Recent simulations by Laskar and Gastineau (2009) have included contributions from the Moon and general relativity. They found, as in previous studies, that 1% of the solutions lead to a large increase in Mercury's eccentricity, large enough to allow collisions with Venus or the Sun. Surprisingly, in one of these solutions Mercury destabilizes all the terrestrial planets $\sim$3.34 Ga from now, with possible collisions of Mercury, Mars or Venus with the Earth.

9.2.2 Debris Disks

Kornet et al. (2001) found that disks with low values of specific angular momentum are bled out of solids and do not form planetary systems. Disks with high and intermediate values of specific angular momentum form diverse planetary systems. Solar-like planetary systems form from disks with initial masses of 0.02 solar masses and angular momenta $J = 3 \times 10^{52}$ g cm^2 s^{-1}. Our solar system still contains two major debris belts: the Kuiper and the asteroid belts, generating dust through mutual collisions of larger parent bodies. These 'debris disks' share a similar evolution, but with a wide range of initial masses.

The solar system was probably dustier in the past, but interactions between the giant planets and encounters with nearby stars have configured the present situation, as we have described previously. Morbidelli and Levison (2008) remark the importance of the mass of the original dust disk and the position of the planets in the system.

Images of debris disks at different evolutionary stages (ages) could be equivalent to a solar system 'time machine' (Moro Martín, 2008). Gaspar et al. (2009) found that the fraction of debris decreases with the age of the planetary system. They derive that during the first Ga of their evolution, up to 15–30% of solar-type stars might undergo an orbital realignment of giant planets such as the one thought to have led to the late heavy bombardment. This has been evidenced by the observation of hot dust around some solar-like stars (Wyatt et al. 2007; Rhee et al. 2008; Fujiwara et al. 2009).

Recently, the Spitzer Space Telescope has detected a prebiotic molecule, hydrogen cyanide (HCN),[14] in the planet forming disks around yellow stars, but not in the disks around cooler, reddish (M5-M9) stars (Pascucci et al. 2009).

9.2.3 The Energetic Environment

The birth of our planetary system was probably connected with the death (Supernova explosion) of one nearby massive star (Bizzarro et al. 2007). Observations with the Spitzer Space Telescope indicate that the radiation environment of a planetary system can play a crucial role in its evolution. The protoplanetary disk of a sun-like star can be ripped away by the powerful winds of a nearby hot O-type star through a process called photoevapouration (Fig. 9.8). Our own Sun and its suite of planets might have grown up on the edge of an O-star's danger zone before migrating to its current, peaceful home. However, we know that our young sun did not linger for too long in any hazardous territory, or our planets, and life would not be here today (Balog et al. 2007).

A danger zone can be established around these hot stars, where no planetary systems can be formed and where, therefore, no Earth-like planets are expected. Figure 9.9 shows such regions in the Rosetta nebula, as spheres with a radius of approximately 1.6 light-years.

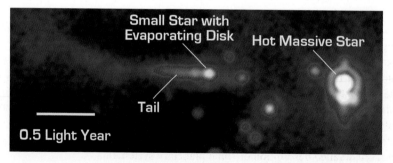

Fig. 9.8 The potential planet-forming disk (or 'protoplanetary disk') of a sun-like star is being violently ripped away by the powerful winds of a nearby hot O-type star in this image from NASA's Spitzer Space Telescope. Credit: NASA/JPL-Caltech/Z. Balog (Univ. of Arizona/Univ. of Szeged)

---

[14] Five HCN molecules can link up to form adenine, one of the four chemical bases of DNA.

9.2 Is the Solar System Unique?

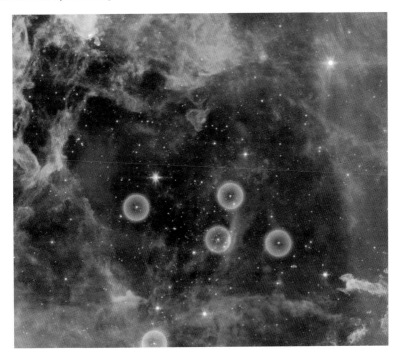

Fig. 9.9 This infrared image from NASA's Spitzer Space Telescope shows the Rosetta nebula, a pretty star-forming region more than 5,000 light-years away in the constellation Monoceros. The danger zones around massive stars are shown with circles. Credit: NASA/JPL-Caltech/Univ.of Arizona

9.2.4 Solar System Analogs

The correct answer to the question raised in this section depends, of course, on future observational results, and also on the kind of detail we need for the comparison. The main objects characterising a planetary system, planets and debris clouds, have been already discovered around other stars. There are so many dynamical parameters in a planetary system that it will be almost impossible to find an exact clone of the solar system in our Galaxy. We have seen how numerical simulations predict that planetary systems with gas giants should be rather rare.

Still awaiting the detection of Earth-like planets, a few systems can be pinpointed (Table 9.2), where the position and eccentricity of a giant planet, similar in mass to Jupiter, is adequate for a similar distribution to that of the solar system (see Chap. 8 for a more detailed description of some planetary systems).

Using the microlensing technique, Gaudi et al. (2008) have discovered a planetary system, located at 1.5 kpc, with two giants of ~ 0.7 and $0.27\ M_J$ orbiting a primary star of $0.5\ M_S$ at respective distances of ~ 2.3 and 4.6 AU. They infer equilibrium temperatures of 82 ± 12 K and 59 ± 7 K, a little cooler than our Jupiter and Saturn.

Table 9.2 List of planetary systems with a Jupiter-like planet located between 2.5 and 6 AU and showing almost circular orbits. Data Source: J. Schneider Encyclopedia of Exoplanets

Star	Mass (M_J)	a (AU)	Eccentricity	Reference
HD 70642	>2 (±0.06)	3.3	0.1 ± 0.06	Carter et al. (2003)
				Hinse et al. (2008)
HD 216425	1.23	2.6	0.14 ± 0.07	Jones et al. (2003)
HD 154345	1.0	4.19	0.044 ± 0.046	Wright et al. (2008)
Sun	1.0	5.20	0.049	

9.3 Is the Earth Something Special?

This question was broadly formulated by the philosophers in the context of the plurality of worlds. At the present time, we possess the first data that can give us insights to the answer. The final query is related to the presence or not of life in other exoplanets.

Cumming et al. (2008) have extrapolated the results of their survey to low planetary mass, finding that 11% of all the stars will have an Earth-mass planet within 1 AU. However, this is a field yet to be explored.

9.3.1 Habitability

We have seen in previous chapters how the Earth is located in the habitable zone of the Sun. Its mass and internal structure have allowed the greenhouse gases to recycle and to be in balance with the increase of solar luminosity and the climatic effects of life. These conditions resulted in the continuous presence of liquid water on its surface along the last 4 Ga, a feature that does not seem easy to be repeated often at a cosmic scale. The low eccentricity of the Earth's orbit was also an essential feature of the stability of its climate and therefore its aptitude to be habitable during such a long period of time. The different evolution of our siblings, Venus and Mars, support this assumption.

Lineweaver (2001) calculated, as a function of the metallicity, the probability of a stellar system to harbour an Earth-like planet. Combined with an estimate of the stellar formation rate, he was able to obtain an estimate of the age distribution of Earth-like planets in the Universe. The final analysis indicated that three-quarters of the Earth-like planets in the Universe are older than the Earth and that their average age is 1.8 ± 0.9 Ga older than the Earth. Franck et al. (2007) have estimated that the number of stellar systems with Earth-like habitable planets reached a maximum at the time of the Earth's formation. Bounama et al. (2007) estimated that primitive life-bearing planets were more numerous at 3.4 Ga before present and that the maximum number of planets inhabited by complex life occurred 1.8 Ga ago and subsequently declined. The best times for the biological efficiency in our Galaxy are probably over.

9.3 Is the Earth Something Special?

After the LHB event the Earth has remained relatively dynamically stable during its lifetime, with the main threat coming from collisions with minor bodies. Here, we are interested in whether the Earth is special from the dynamic point of view compared to the other rocky planets of the system and to other similar exoplanets of the same kind. Shortly after its birth, a violent event conditioned its stability, marking an important difference with other terrestrial planets.

9.3.2 Variations of Orbital Parameters

Laskar (1994, 2008) has studied in depth the very long-term (Ga) stability of the solar system. The maximum eccentricity of the Earth reached through chaotic diffusion reaches about 0.08, while its current variations are approximately 0.06. In summary, our planet is well protected in the inner part of our planetary system. However, other smaller perturbations at shorter time scales can now be considered.

Following the pioneering work of James Croll (1821–1890), Milutin Milankovitch (1879–1958) calculated variations in solar insolation driven by changes in the orbital parameters of the Earth at the scale of the last millions of years (see Berger 1980; Berger et al. 1984). The implicated parameters are the following:

- Obliquity, i, of the Earth rotation axis with an amplitude of $\pm 2, 4°$ and a period of 41,000 years. In addition, we have short-term (18.6 yr) variations, known as nutation.
- Eccentricity, e, of the orbit with two main periods of variation (413,000 and 100,000 years). It varies from an orbit nearly circular ($e = 0.005$) to one mildly elliptical ($e = 0.058$). The changes are primarily due to gravitational interactions with Jupiter and Saturn.
- The lunar and solar tidal torques on Earth's ellipticity give rise to the 26,000 years' precession with the gravitational pull of other planets (mainly Saturn and Jupiter) slowly perturbing the orientation of the ecliptic in space. This leads to changes in the longitude of the perihelion. The climatic effect is expressed as $\delta e \sin \omega$ with periods of 23,000 and 19,000 years.

Because of the present configuration of the continents, the amount of insolation at 65° of latitude in the northern hemisphere is especially critical. Cooler summers trigger the start of an ice-age due to their incomplete melting of the previous winter's ice and snow.

Figure 9.10 shows the temporal variations of these parameters compared to the glaciation cycles in the last million years.

Berger and Loutre (1991) give analytical expressions for the variations of orbital parameters. Recently Laskar et al. (2004) have calculated insolation values spanning from −250 to 250 Ma. Calculations indicate that the insolation should increase gradually over the next 25,000 years and no ice ages are expected to occur in the next 50,000 years (Berger and Loutre 2002).

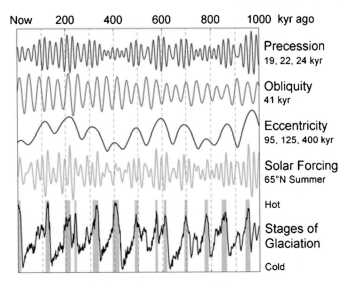

Fig. 9.10 Variation of the orbital parameters of the Earth, the insolation at 65° North and the different glacial and interglacial periods

9.3.3 Presence of a Large Satellite

Laskar et al. (1993) showed mathematically that if Earth did not have a large moon, and if it was spinning at the same rate as it is now, its obliquity would vary chaotically, with variations extending from 0 to 85°. This behaviour would induce dramatic changes in climate.

We have seen in Chap. 2 that, as a result of its gravitational interaction with Earth, the Moon is gradually being accelerated to a higher orbit; it is receding from us at a rate of about 1 in. per year. A billion years from now, its pull will be so weak that Earth's obliquity will begin to fluctuate chaotically.

The stabilizing effect of the Moon is possible, thanks to its relatively large size. Comparing the mass of our satellite with other secondary bodies of the solar system, we see that the Earth has a very massive Moon evidenced both by an extrapolation of the sizes of the satellites of the giant planets and still more dramatically by the Martian satellites (Alfven and Arrehnius 1972 and Fig. 9.11). Only Triton shares this condition in our planetary system.

The Moon is also responsible for an important part of the ocean tides. At the early Earth, the Moon was ∼10 times closer to our planet (see Chap. 2) and tides were ∼1,000 times larger. If life originated around deep ocean hydrothermal vents or at a *warm little pond*[15] close to the ocean shores, probably tides played a decisive role (Comins 1993; Benn 2001; Lathe 2004).

---

[15] We quote here a famous phrase of Charles Darwin (1809–1882) about the origin of life, contained in a letter to the botanist Joseph Hooker.

9.3 Is the Earth Something Special?

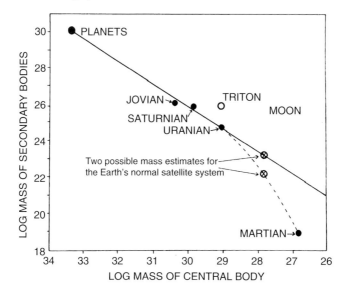

Fig. 9.11 Total mass of secondary body system as a function of central body mass. Both Triton and the Moon have much larger mass than expected of normal satellites. Two possible mass estimates for a standard Earth satellite are shown

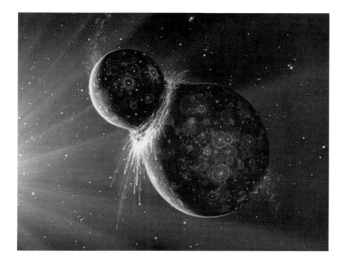

Fig. 9.12 Artist's conception of the giant impact event forming the Moon. Credit: NASA

The formation of the Moon was originated by the collision with a Mars-sized object, called Theia, in the first phase of the solar system (4.4 Ga ago), an apparently stochastic event (Fig. 9.12). What happened with the rest of the terrestrial planets? This invites us to consider the existence of similar impacts on other terrestrial planets during the early phase of the solar system. In fact, this seems to have been the case.

In Chap. 7, we commented on the high density of Mercury and its relatively large core. The explanation lies in an impact that vapourized an important part of its original crust. The impactor had approximately 1/6 the Mercury's mass and was several hundred kilometres across (Benz et al. 1988).

Davies (2008) suggests that Venus was formed by a near head-on collision of two large planetary embryos. The event explains the slow[16] retrograde rotation of the planet and its relatively dry interior. Alemi and Stevenson (2006) proposed a two-impacts hypothesis. In the first, a moon was created by the impact of a large body. This satellite was destroyed by a second megaimpact that produced the slow retrograde rotation of Venus.

The hemispheric dichotomy of Mars has been explained in terms of an impact (Wilhems and Squyres 1984; Nimmo et al. 2008; Andrews-Hanna et al. 2008). The observed eccentricity of the resulting basin can be explained by an oblique impact angle (30–60°) (Marinova et al. 2008). Other possible explanations are erosion by an ancient ocean and plate tectonics (Sleep 1994). See Watters et al. (2007) for a recent review on the subject.

In summary, we have the same process but different consequences depending on the details of the impact. Only the Earth was fortunate in having a large Moon.

9.4 The Ultimate Factor: Life

The formation of a star and its planetary system is a consequence of a process of collapse in a molecular cloud. Detailed interactions within the molecular cloud gives rise to a myriad of objects of different masses, positions and chemical conditions, ranging from very massive stars to small grains. We can say that, at least partially, our destiny was written in our cradle.

All the structures in the Universe evolve with time, and in this evolution the environmental conditions also play a decisive role. The LHB event was decisive for the final arrangement of planets and perhaps for the emergence of life in the third planet. Our solar system is already old enough to have lost most of the imprints of the processes that have led to its present status. The violent first phases of the evolution are now over. The Earth has a set of properties that as the genetic print of a living being makes it something special, and also similar to other members of its family: the rocky planets.

Evidently, what makes our environment so special is the presence of life in the third planet rotating around the star Sol. Gustafsson (2008) concluded that, 'The Sun is odd in certain aspects since a habitable planetary system has to be there too'. Nowadays, we have started to explore the Universe looking for life, but our attempts are still very timid. We are lacking an universal definition of life, to allow its non-ambiguous detection in an extraterrestrial body.

---

[16] A Venusian sidereal day (243 Earth days) lasts more than a Venusian year (224.7 Earth days).

At the time this book went into print, several space missions and ground-based projects are starting to give data on Earth-like planets. Many topics discussed in the last chapters will undoubtedly have to be rewritten.

We do not yet have the necessary understanding to test empirically how rare or common our planet and its surroundings are. Only a detailed cartography of Earth-like planets in our Galaxy will give us the necessary empirical data to have an answer to this fundamental question. Only then will we complete the second branch of the Earth–Exoplanets connection. Nevertheless, the Earth will continue to be our reference, our Rosetta stone for understanding other worlds.

References

Adams, F.C., Proszkow, E.M., Fatuzzo, M., Myers, P.C.: Early evolution of stellar groups and clusters: Environmental effects on forming planetary systems. Astrophys. J. **641**, 504–525 (2006)
Alemi, A., Stevenson, D.: Why venus has no moon. Bull. Am Astron. Soc. **38**, 491 (2006)
Alfvén, H., Arrhenius, G.: Origin and evolution of the earth-moon system. Moon **5**, 210–230 (1972)
Allende Prieto, C.: Solar chemical peculiarities? In: Israelian, G., Meynet, G. (eds.) The Metal-Rich Universe, pp. 36–40. Cambridge University Press, London (2008)
Andrews-Hanna, J.C., Zuber, M.T., Banerdt, W.B.: The Borealis basin and the origin of the martian crustal dichotomy. Nature **453**, 1212–1215 (2008)
Attwood, R.E., Goodwin, S.P., Stamatellos, D., Whitworth, A.P.: Simulating star formation in molecular cloud cores. IV. The role of turbulence and thermodynamics. Astron. Astrophys. **495**, 201–215 (2009)
Baliunas, S.L., Donahue, R.A., Soon, W., Henry, G.W.: Activity cycles in lower main sequence and POST main sequence stars: The HK project. In: Donahue, R.A., Bookbinder, J.A. (eds.) Cool Stars, Stellar Systems, and the Sun, Astronomical Society of the Pacific Conference Series, vol. 154, pp. 153–172 (1998)
Balog, Z., Muzerolle, J., Rieke, G.H., Su, K.Y.L., Young, E.T., Megeath, S.T.: Spitzer/IRAC-MIPS survey of NGC 2244: Protostellar disk survival in the vicinity of hot stars. Astrophys. J. **660**, 1532–1540 (2007)
Barnes, R., Quinn, T.: The (in)stability of planetary systems. Astrophys. J. **611**, 494–516 (2004)
Batygin, K., Laughlin, G.: On the dynamical stability of the solar system. Astrophys. J. **683**, 1207–1216 (2008)
Beaugé, C., Michtchenko, T.A., Ferraz-Mello, S.: Planetary migration and extrasolar planets in the 2/1 mean-motion resonance. Mon. Not. Roy. Astron. Soc. **365**, 1160–1170 (2006)
Beer, M.E., King, A.R., Livio, M., Pringle, J.E.: How special is the solar system? Mon. Not. Roy. Astron. Soc. **354**, 763–768 (2004)
Benn, C.R.: The moon and the origin of life. Earth Moon Planets **85**, 61–66 (2001)
Benz, W., Slattery, W.L., Cameron, A.G.W.: Collisional stripping of Mercury's mantle. Icarus **74**, 516–528 (1988)
Berger, A.: The Milankovitch astronomical theory of paleoclimates: A modern review. Vistas Astron. **24**, 103–122 (1980)
Berger, A., Loutre, M.: Insolation values for the climate of the last 10 million years. Quaternary Sci. Rev. **10**, 297–317 (1991)
Berger, A., Loutre, H.: An exceptionally long interglacial ahead? Science **297**, 1287–1288 (2002)
Berger, A., Imbrie, J., Hays, J., Kukla, G., Saltzman, B. (eds.): Milankovitch and climate: Understanding the response to astronomical forcing, Dordrecht, D. Reidel (1984)

Bizzarro, M., Ulfbeck, D., Trinquier, A., Thrane, K., Connelly, J.N., Meyer, B.S.: Evidence for a late supernova injection of 60Fe into the protoplanetary disk. Science **316**, 1178–1181 (2007)

Böhm-Vitense, E.: Chromospheric activity in G and K main-sequence stars, and what it tells us about stellar dynamos. Astrophys. J. **657**, 486–493 (2007)

Boss, A.P.: Formation of binary stars. In: Sahade, J., McCluskey, G.E., Kondo, Y. (eds.) The realm of interacting binary stars, Astrophysics and Space Science Library, vol. 177, pp. 355–380 (1992)

Bounama, C., Von Bloh, W., Franck, S.: How rare is complex life in the milky way? Astrobiology **7**, 745–756 (2007)

Bouvier, J.: Lithium depletion and the rotational history of exoplanet host stars. Astron. Astrophys. **408**, L53–L56 (2008)

Brandenburg, A., Saar, S.H., Turpin, C.R.: Time evolution of the magnetic activity cycle period. Astrophys. J. **498**, L51–L54 (1998)

Carter, B.D., Butler, R.P., Tinney, C.G., Jones, H.R.A., Marcy, G.W., McCarthy, C., Fischer, D.A., Penny, A.J.: A planet in a circular orbit with a 6 year period. Astrophys. J. **593**, L43–L46 (2003)

Cayrel de Strobel, G.: Solar analogs seen at high spectral resolution and very high S/N ratios. In: Sánchez, F., Vázquez, M. (eds.) New Windows to the Universe, pp. 195–212. Cambridge University Press, London (1990)

Cayrel de Strobel, G.: Stars resembling the Sun. The Astron. Astrophys. Rev. **7**, 243–288 (1996)

Cayrel de Strobel, G., Knowles, N., Hernandez, G., Bentolila, C.: In search of real solar twins. Astron. Astrophys. **94**, 1–11 (1981)

Centeno, R., Socas-Navarro, H.: A new approach to the solar oxygen abundance problem. Astrophys. J. **682**, L61–L64 (2008)

Comins, N.F.: What if the moon didn't exist, voyages to Earths that might have been. Harper Collins (1993)

Cumming, A., Butler, R.P., Marcy, G.W., Vogt, S.S., Wright, J.T., Fischer, D.A.: The keck planet search: Detectability and the minimum mass and orbital period distribution of extrasolar planets. Publ. Astron. Soc. Pac. **120**, 531–554 (2008)

Davies, J.H.: Did a mega-collision dry Venus' interior? Earth Planet. Sci. Lett. **268**, 376–383 (2008)

Duquennoy, A., Mayor, M.: Multiplicity among solar-type stars in the solar neighbourhood. II – Distribution of the orbital elements in an unbiased sample. Astron. Astrophys. **248**, 485–524 (1991)

Edvardsson, B., Andersen, J., Gustafsson, B., Lambert, D.L., Nissen, P.E., Tomkin, J.: The chemical evolution of the galactic disk – Part one – Analysis and results. Astron. Astrophys. **275**, 101–152 (1993)

Fischer, D.A., Valenti, J.: The planet-metallicity correlation. Astrophys. J. **622**, 1102–1117 (2005)

Fogg, M.J., Nelson, R.P.: Terrestrial planet formation in low eccentricity warm Jupiter systems. Astron. Astrophys. **498**, 575–589 (2009)

Franck, S., von Bloh, W., Bounama, C.: Maximum number of habitable planets at the time of Earth's origin: new hints for panspermia and the mediocrity principle. Int. J. Astrobiol. **6**, 153–157 (2007)

Fujiwara, H., Yamashita, T., Ishihara, D., Onaka, T., Kataza, H., Ootsubo, T., Fukagawa, M., Marshall, J.P., Murakami, H., Nakagawa, T., Hirao, T., Enya, K., White, G.J.: Hot debris dust around HD 106797. Astrophys. J. **695**, L88–L91 (2009)

Gaidos, E.J.: Nearby young solar analogs. I. Catalog and stellar characteristics. Publ. Astron. Soc. Pac. **110**, 1259–1276 (1998)

Gaspar, A., Rieke, G.H., Su, K.Y.L., Balog, Z., Trilling, D., Muzzerole, J., Apai, D., Kelly, B.C.: The low level of debris disk activity at the time of the Late Heavy Bombardment: A Spitzer study of Praesepe. Astrophys. J. **697**, 1578–1596 (2009)

Gaudi, B.S., et al.: Discovery of a Jupiter/Saturn analog with gravitational microlensing. Science **319**, 927–930 (2008)

Gonzalez, G.: Are stars with planets anomalous? Mon. Not. Roy. Astron. Soc. **308**, 447–458 (1999)

Gonzalez, G.: Parent stars of extrasolar planets - IX. Lithium abundances. Mon. Not. Roy. Astron. Soc. **386**, 928–934 (2008)

Gonzalez, G., Brownlee, D., Ward, P.: The galactic habitable zone: Galactic chemical evolution. Icarus **152**, 185–200 (2001)

Grether, D., Lineweaver, C.H.: The metallicity of stars with close companions. Astrophys. J. **669**, 1220–1234 (2007)

Güdel, M.: The sun in time: Activity and environment. Living Rev. Sol. Phys. **4**, 3 (2007)

Gustafsson, B.: Is the sun a sun-like star? Space Sci. Rev. **85**, 419–428 (1998)

Gustafsson, B.: Is the sun unique as a star – and if so, why? Phys. Scripta T **130**(1), 014,036 (2008)

Haisch, K.E., Jr., Lada, E.A., Lada, C.J.: Disk frequencies and lifetimes in young clusters. Astrophys. J. **553**, L153–L156 (2001)

Hester, J.J., Desch, S.J., Healy, K.R., Leshin, L.A.: The cradle of the solar system. Science **304**, 1116–1117 (2004)

Hinse, T.C., Michelsen, R., Jørgensen, U.G., Goździewski, K., Mikkola, S.: Dynamics and stability of telluric planets within the habitable zone of extrasolar planetary systems. Numerical simulations of test particles within the HD 4208 and HD 70642 systems. Astron. Astrophys. **488**, 1133–1147 (2008)

Hopkins, A.M., Beacom, J.F.: On the normalization of the cosmic star formation history. Astrophys. J. **651**, 142–154 (2006)

Horner, J., Jones, B.W.: Jupiter friend or foe? I: The asteroids. Int. J. Astrobiol. **7**, 251–261 (2008)

Horner, J., Jones, B.W.: Jupiter – friend or foe? II: the Centaurs. Int. J. Astrobiol. **8**, 75–80 (2009)

Israelian, G., Santos, N.C., Mayor, M., Rebolo, R.: Lithium in stars with exoplanets. Astron. Astrophys. **414**, 601–611 (2004)

Jones, H.R.A., Butler, R.P., Tinney, C.G., Marcy, G.W., Penny, A.J., McCarthy, C., Carter, B.D.: An exoplanet in orbit around τ Gruis. Mon. Not. Roy. Astron. Soc. **341**, 948–952 (2003)

Kashyap, V.L., Drake, J.J., Saar, S.H.: Extrasolar giant planets and x-ray activity. Astrophys. J. **687**, 1339–1354 (2008)

Kerr, F.J., Lynden-Bell, D.: Review of galactic constants. Mon. Not. Roy. Astron. Soc. **221**, 1023–1038 (1986)

King, J.R., Boesgaard, A.M., Schuler, S.C.: Keck HIRES spectroscopy of four candidate solar twins. Astron. J. **130**, 2318–2325 (2005)

Kokubo, E., Ida, S.: Formation of protoplanet systems and diversity of planetary systems. Astrophys. J. **581**, 666–680 (2002)

Kornet, K., Stepinski, T.F., Różyczka, M.: Diversity of planetary systems from evolution of solids in protoplanetary disks. Astron. Astrophys. **378**, 180–191 (2001)

Kroupa, P.: The initial mass function of stars: Evidence for uniformity in variable systems. Science **295**, 82–91 (2002)

Lada, C.J.: Stellar multiplicity and the initial mass function: Most stars are single. Astrophys. J. **640**, L63–L66 (2006)

Laskar, J.: Large-scale chaos in the solar system. Astron. Astrophys. **287**, L9–L12 (1994)

Laskar, J.: Large scale chaos and marginal stability in the solar system. Celestial Mech. Dyn. Astron. **64**, 115–162 (1996)

Laskar, J.: Chaotic diffusion in the solar system. Icarus **196**, 1–15 (2008)

Laskar, J., Gastineau, M.: Existence of collisional trajectories of mercury, mars and venus with the earth. Nature **459**, 817–819 (2009)

Laskar, J., Joutel, F., Robutel, P.: Stabilization of the earth's obliquity by the moon. Nature **361**, 615–617 (1993)

Laskar, J., Robutel, P., Joutel, F., Gastineau, M., Correia, A.C.M., Levrard, B.: A long-term numerical solution for the insolation quantities of the earth. Astron. Astrophys. **428**, 261–285 (2004)

Lathe, R.: Fast tidal cycling and the origin of life. Icarus **168**, 18–22 (2004)

Lineweaver, C.H., Fenner, Y., Gibson, B.K.: The galactic habitable zone and the age distribution of complex life in the milky way. Science **303**, 59–62 (2004)

Livio, M.: How rare are extraterrestrial civilizations, and when did they emerge? Astrophys. J. **511**, 429–431 (1999)

Lockwood, G.W., Skiff, B.A., Henry, G.W., Henry, S., Radick, R.R., Baliunas, S.L., Donahue, R.A., Soon, W.: Patterns of photometric and chromospheric variation among sun-like stars: A 20 year perspective. Astrophys. J. Suppl. **171**, 260–303 (2007)

Looney, L.W., Tobin, J.J., Fields, B.D.: Radioactive probes of the supernova-contaminated solar Nebula: Evidence that the sun was born in a cluster. Astrophys. J. **652**, 1755–1762 (2006)

Lucatello, S., Tsangarides, S., Beers, T.C., Carretta, E., Gratton, R.G., Ryan, S.G.: The binary frequency among carbon-enhanced, s-process-rich, metal-poor stars. Astrophys. J. **625**, 825–832 (2005)

Lunine, J.I.: The occurrence of Jovian planets and the habitability of planetary systems. Proc. Natl. Acad. Sci. **98**, 809–814 (2001)

Machida, M.N.: Binary formation in star-forming clouds with various metallicities. Astrophys. J. **682**, L1–L4 (2008)

Malhotra, R.: Dynamical cause of the late heavy bombardment. In: Lunar and Planetary Institute Science Conference Abstracts, vol. 38, pp. 2373–2374 (2007)

Malmberg, D., Davies, M.B., Chambers, J.E.: The instability of planetary systems in binaries: How the Kozai mechanism leads to strong planet-planet interactions. Mon. Not. Roy. Astron. Soc. **377**, L1–L4 (2007a)

Malmberg, D., de Angeli, F., Davies, M.B., Church, R.P., Mackey, D., Wilkinson, M.I.: Close encounters in young stellar clusters: Implications for planetary systems in the solar neighbourhood. Mon. Not. Roy. Astron. Soc. **378**, 1207–1216 (2007b)

Malmberg, D., Davies, M.B., Chambers, J.E., Church, R.P., DeAngeli, F., Mackey, D., Wilkinson, M.I.: Is our sun a singleton? Phys. Scripta T **130**(1), 014,030 (2008)

Marchi, S.: Extrasolar planet taxonomy: A new statistical approach. Astrophys. J. **666**, 475–485 (2007)

Marchi, S., Ortolani, S.: Unveiling exoplanet families. In: Exoplanets: Detection, Formation and Dynamics, IAU Symposium, vol. 249, pp. 123–128. Cambridge University Press, London (2008)

Marcy, G., Butler, R.P., Fischer, D., Vogt, S., Wright, J.T., Tinney, C.G., Jones, H.R.A.: Observed Properties of Exoplanets: Masses, Orbits, and Metallicities. Progress of Theoretical Physics Supplement **158**, 24–42 (2005)

Marinova, M.M., Aharonson, O., Asphaug, E.: Mega-impact formation of the Mars hemispheric dichotomy. Nature **453**, 1216–1219 (2008)

McNeil, D., Duncan, M., Levison, H.F.: Effects of type I migration on terrestrial planet formation. Astron. J. **130**, 2884–2899 (2005)

Meaney, M.J.: Nature, nurture, and the disunity of knowledge. Ann. N Y Acad. Sci. **935**, 50–61 (2001)

Meléndez, J., Dodds-Eden, K., Robles, J.A.: HD 98618: A star closely resembling our sun. Astrophys. J. **641**, L133–L136 (2006)

Meléndez, J., Ramírez, I.: HIP 56948: A solar twin with a low lithium abundance. Astrophys. J. **669**, L89–L92 (2007)

Miller, G.E., Scalo, J.M.: The initial mass function and stellar birthrate in the solar neighborhood. Astrophys. J. Suppl. **41**, 513–547 (1979)

Mishurov, Y.N., Zenina, I.A.: Yes, the sun is located near the corotation circle. Astron. Astrophys. **341**, 81–85 (1999)

Mishurov, Y.N., Zenina, I.A., Dambis, A.K., Mel'Nik, A.M., Rastorguev, A.S.: Is the sun located near the corotation circle? Astron. Astrophys. **323**, 775–780 (1997)

Morbidelli, A., Crida, A.: The dynamics of Jupiter and Saturn in the gaseous protoplanetary disk. Icarus **191**, 158–171 (2007)

Morbidelli, A., Levison, H.F.: Late evolution of planetary systems. Phys. Scripta T **130**(1), 014,028 (2008)

Moro-Martín, A.: On the solar system – debris disk connection. In: Exoplanets: Detection, Formation and Dynamics, IAU Symposium, vol. 249, pp. 347–354 (2008)

Nimmo, F., Hart, S.D., Korycansky, D.G., Agnor, C.B.: Implications of an impact origin for the martian hemispheric dichotomy. Nature **453**, 1220–1223 (2008)

Noyes, R.W., Hartmann, L.W., Baliunas, S.L., Duncan, D.K., Vaughan, A.H.: Rotation, convection, and magnetic activity in lower main-sequence stars. Astrophys. J. **279**, 763–777 (1984)

Pascucci, I., Apai, D., Luhman, K., Henning, T., Bouwman, J., Meyer, M.R., Lahuis, F., Natta, A.: The different evolution of gas and dust in disks around sun-like and cool stars. Astrophys. J. **696**, 143–159 (2009)

Pasquini, L., Biazzo, K., Bonifacio, P., Randich, S., Bedin, L.R.: Solar twins in M 67. Astron. Astrophys. **489**, 677–684 (2008)

Pilat-Lohinger, E., Süli, Á., Robutel, P., Freistetter, F.: The Influence of Giant Planets Near a Mean Motion Resonance on Earth-like Planets in the habitable zone of sun-like stars. Astrophys. J. **681**, 1639–1645 (2008)

Portegies Zwart, S.F.: The lost siblings of the sun. Astrophysical Journal **696**, L13–L16 (2009)

Porto de Mello, G.F., da Silva, L.: HR 6060: The closest ever solar twin? Astrophys. J. **482**, L89– (1997)

Porto de Mello, G., del Peloso, E.F., Ghezzi, L.: Astrobiologically interesting stars within 10 parsecs of the sun. Astrobiology **6**, 308–331 (2006)

Prantzos, N.: On the "galactic habitable zone". Space Sci. Rev. **135**, 313–322 (2008)

Price, D.J., Bate, M.R.: The impact of magnetic fields on single and binary star formation. Mon. Not. Roy. Astron. Soc. **377**, 77–90 (2007)

Quintana, E.V., Lissauer, J.J., Chambers, J.E., Duncan, M.J.: Terrestrial planet formation in the α centauri system. Astrophys. J. **576**, 982–996 (2002)

Raymond, S.N., Mandell, A.M., Sigurdsson, S.: Exotic earths: Forming habitable worlds with giant planet migration. Science **313**, 1413–1416 (2006)

Rhee, J.H., Song, I., Zuckerman, B.: Warm dust in the terrestrial planet zone of a sun-like pleiades star: Collisions between planetary embryos? Astrophys. J. **675**, 777–783 (2008)

Robles, J.A., Lineweaver, C.H., Grether, D., Flynn, C., Egan, C.A., Pracy, M.B., Holmberg, J., Gardner, E.: A comprehensive comparison of the sun to other stars: Searching for self-selection effects. Astrophys. J. **684**, 691–706 (2008)

Roškar, R., Debattista, V.P., Quinn, T.R., Stinson, G.S., Wadsley, J.: Riding the spiral waves: Implications of stellar migration for the properties of galactic disks. Astrophys. J. **684**, L79–L82 (2008)

Rubenstein, E.P., Schaefer, B.E.: Are superflares on solar analogues caused by extrasolar planets? Astrophys. J. **529**, 1031–1033 (2000)

Ruderman, M.A.: Possible consequences of nearby supernova explosions for atmospheric ozone and terrestrial life. Science **184**, 1079–1081 (1974)

Saar, S.H., Brandenburg, A.: Time evolution of the magnetic activity cycle period. II. Results for an expanded stellar sample. Astrophys. J. **524**, 295–310 (1999)

Schaefer, B.E., King, J.R., Deliyannis, C.P.: Superflares on ordinary solar-type stars. Astrophys. J. **529**, 1026–1030 (2000)

Scott, P., Asplund, M., Grevesse, N., Sauval, A.J.: On the solar nickel and oxygen abundances. Astrophys. J. **691**, L119–L122 (2009)

Skumanich, A.: Time scales for CA II emission decay, rotational braking, and lithium depletion. Astrophys. J. **171**, 565–568 (1972)

Sleep, N.H.: Martian plate tectonics. J. Geophys. Res. **99**, 5639–5655 (1994)

Soderblom, D.R.: A short list of SETI candidates. Icarus **67**, 184–186 (1986)

Soubiran, C., Triaud, A.: The top ten solar analogs in the ELODIE library. Astron. Astrophys. **418**, 1089–1100 (2004)

Takeda, Y., Tajitsu, A.: High-dispersion spectroscopic study of solar twins: HIP 56948, HIP 79672, and HIP 100963. Publ. Astron. Soc. Jpn. **61**, 471–480 (2009)

Thommes, E.W., Bryden, G., Wu, Y., Rasio, F.A.: From mean motion resonances to scattered planets: Producing the solar system, eccentric exoplanets and late heavy bombardment. Astrophys. J. **675**, 1538–1548 (2008a)

Thommes, E.W., Matsumura, S., Rasio, F.A.: Gas disks to gas giants: Simulating the birth of planetary systems. Science **321**, 814–817 (2008b)

Tohline, J.E.: The origin of binary stars. Ann. Rev. Astron. Astrophys. **40**, 349–385 (2002)

Udry, S., Fischer, D., Queloz, D.: A decade of radial-velocity discoveries in the exoplanet domain. In: Reipurth, B., Jewitt, D., Keil, K. (eds.) Protostars and Planets V, pp. 685–699. University of Arizona Press, AZ (2007)

Van der Wel, A.: The dependence of galaxy morphology and structure on environment and stellar mass. Astrophys. J. **675**, L13–L16 (2008)

Vaughan, A.H., Preston, G.W.: A survey of chromospheric CA II H and K emission in field stars of the solar neighborhood. Publ. Astron. Soc. Pac. **92**, 385–391 (1980)

Vázquez, M., Hanslmeier, A.: The ultraviolet radiation in the solar system. Springer, Heidelberg (2005)

Watters, T.R., McGovern, P.J., Irwin III, R.P.: Hemispheres apart: The crustal dichotomy on mars. Ann. Rev. Earth Planet. Sci. **35**, 621–652 (2007)

Wetherill, G.W.: Possible consequences of absence of Jupiters in planetary systems. Astrophys. Space Sci. **212**, 23–32 (1994)

Whitehouse, D.R.: The ancient sun and biogenesis. Observatory **103**, 160–162 (1983)

Wielen, R., Fuchs, B., Dettbarn, C.: On the birth-place of the sun and the places of formation of other nearby stars. Astron. Astrophys. **314**, 438–447 (1996)

Wilhelms, D.E., Squyres, S.W.: The martian hemispheric dichotomy may be due to a giant impact. Nature **309**, 138–140 (1984)

Wright, J.T., Marcy, G.W., Butler, R.P., Vogt, S.S.: Chromospheric Ca II emission in nearby F, G, K, and M stars. Astrophys. J. Suppl. **152**, 261–295 (2004)

Wright, J.T., Marcy, G.W., Butler, R.P., Vogt, S.S., Henry, G.W., Isaacson, H., Howard, A.W.: The Jupiter twin HD 154345b. Astrophys. J. **683**, L63–L66 (2008)

Wyatt, M.C., Smith, R., Greaves, J.S., Beichman, C.A., Bryden, G., Lisse, C.M.: Transience of hot dust around sun-like stars. Astrophys. J. **658**, 569–583 (2007)

Index

2002aa29, 363

Adams, J.C., 356
Albedo, 50, 112, 114, 115
Albedo, Bond, 118
Anthropocene, 92
Appleton, E., 172
Aristotle, 2
Assman, R., 156
Asteroids, 56, 296
ATP, 206
Autotrophs, 202

Barnard Star, 251
Barnard, E., 251
Bernal, J.D., 197
Blue Jets, 179
Brahe, T., 352
Brown dwarfs, 252, 253, 258, 261
Bruno, G., 2

Cambrian Explosion, 79
Carbon, 201
Carbon Dioxide, 56, 62, 65, 92
Carbonaceous Chondrites, 314
Cassini, G.D., 297
Chaos, 360
Chapman S., 165
Chlorophyll, 145
Chthonian Planet, 324
Clarke A.C., 111
Continuity Equation, 173
Copernicus, N., 3
Coreless Planets, 317
COROT mission, 272
Cruithne, 363

Cuvier, G., 85
Cyanobacteria, 73

Da Vinci, L., 115
Danjon, A., 117, 119, 143
DARWIN mission, 279, 281
Darwin, C., 85
Debris Disks, 343
Democritus, 2
Descartes, R., 337, 355
Dobson Unit, 157
Dobson, G., 157
Dubois J., 117

Early Earth, 62, 188
Earth
    Astenosphere, 41
    Exosphere, 163, 189
    Gamma Rays, 188
    Infrared, 16, 137
    Interior, 39
    Ionosphere, 142, 171, 239
    Lithosphere, 42
    Mantle, 41, 60
    Mesosphere, 160
    Radioemission, 184, 239
    Stratosphere, 156
    Thermosphere, 162
    Transmission Spectrum, 139
    Ultraviolet, 75, 163, 164, 233
    X-rays, 186
Earth Images
    Apollo 17, 14
    Apollo 8, 13
    Blue Marble Next Generation, 16
    EPOXI, 112
    Infrared, 16

Kaguya, 14
Lunar Orbiter 1, 11
Messenger, 20
V2 Rocket, 10
Earth-like Planets, 280, 302, 369
Earthquakes, 39, 46
Earthshine, 108, 136
Enceladus, 317
Energy
    Chemical, 208
    Nuclear, 61
    Solar Light, 206
Euler, L., 355
Europa, 317
Exoplanets
    2M1207-39b, 276
    51 Peg b, 252
    Barnard Star, 251
    Eccentricity, 305
    Formalhaut b, 276
    GJ 581e, 378
    HD 149026b, 274
    HD 189733b, 274, 323
    HD 209458b, 177, 271, 274, 324
    Mass, 302
    Mass-Radius, 319
    Metallicity, 305
    PSR B1257 + 12, 266
Exosphere, 50
Extinction
    Cretaceous, 87
    Permian, 87

Faint Sun Paradox, 62
Fourier, J., 53
Franklin, B., 45
Fullerenes, 87

GAIA mission, 258, 264
Galactic Center, 31
Galactic Habitable Zone, 234, 397
Galilei, G., 116, 356
Gauss, C.F., 172
Geocorona, 163, 164
Geosynchronous orbit, 110
Giant Planets, 348
Glint scattering, 128
Global Warming, 161
Gould Belt, 27
Greenhouse effect, 52

Habitable Zone, 226, 231, 232, 364, 366, 375, 378, 405
Halley, E., 352
Heaviside, O., 172
Heterotrophs, 202
Hill Sphere, 358
Homochirality, 145
Hot Earths, 308
Hot Jupiters, 252, 265, 304, 399
Hot Neptunes, 304
HR Diagram, 229
Hubble Space Telescope, 28
Hydrogen Bond, 203

Ice Ages, 90
International Space Station, 164, 183
Ionosphere
    Chapman layers, 175
    Radio transmission, 172

Jeans, J., 339
Jeffreys, H., 339
Jupiter, 185

KAM Theory, 357
Kant, I., 337
KEPLER mission, 272
Kepler, J., 352
Kuiper Belt, 298
Kuiper, G., 338

Lagrange, J.L., 355
Lagrangian Points, 363
Lambert
    surface, 50
Laplace, P., 338, 357
Late Heavy Bombardment, 59, 362, 371, 404
Leclerc, G.L., 339
Leucippus, 2
Leverrier, U., 356
Life
    Definition, 198
    Intelligent, 10, 27
    Origin, 59, 64, 204
    Self-organization, 199
Light Pollution, 238
Local Bubble, 27
Local Fluff, 27
Lorenz, E.N., 357

Index   421

Lovelock, J., 70
Lucretius, 2
Lyot, B., 277

M Stars, 231
Magnetosphere, 180
Mars, 326
Maul, A., 8
Maxwell, J.C., 338
Mercator, G., 5
Mercury, 316, 405
Mesosphere, 50, 160
Metallicity, 235, 305, 392, 397
Methane, 62
Methane Clathrates, 78, 84
Methanogenesis, 208
Milankovitch, M., 90
Milky Way, 30, 234, 395
Moon, 20, 56, 319, 410, 412
Moulton, F., 339

Neptune, 348, 356
Newton, I., 23, 352, 355

Ocean Planets, 312
Oort Cloud, 299
Oparin, A.I., 64
Opposition effect, 118
Orion Nebula, 341
Oxygen, 73
Ozone, 75, 156, 159

Pale Blue Dot, 112, 135, 142
Pale Red Dot, 142
Photoevaporation, 348, 406
Photography, 7
Photosynthesis, 206
Phytoplankton, 220
Planetary-mass objects, 258
Plankton blooms, 108
Plasmasphere, 182
Plate Tectonics, 46, 311
Plenitude Principle, 2
Poincaré, N., 357
Polar Mesospheric Clouds, 161
Polarization, 143, 145
Potocnik, H., 111
Protoplanetary Disks, 343
Pulsar planets, 252, 266
Purple Earth, 221

Radiation Belts, 181
Radioactivity, 44, 54
Rayleigh Number, 42
Rayleigh scattering, 51, 135, 140
Reber, G., 184
Red Dwarfs, 392
Red Edge, 145, 214
Red Giant, 95, 96, 313
Red Sprites, 179
RedOx, 205
Resonances, 404
Rocky Moons, 318

Sagan, C., 25, 94
Sarpi, P., 116
Satellite orbits, 110
Shadow hiding, 119
Silicon, 201
Snowball Earth, 122
Solar Wind, 156
Solar Analogs, 399
Solar Siblings, 395
Solar System, 21, 291
    Albedo, 139
    Nebular Theory, 337
    Protoplanet Theory, 339
Spacecrafts
    AIM, 161
    Apollo 11, 14
    Apollo 14, 14
    Apollo 16, 164
    Apollo 17, 14
    Apollo 8, 13
    Cassini, 25, 317
    CLUSTER, 182
    Galileo, 20, 135, 164
    GCRO, 188
    Kaguya, 14
    Mars Express, 137, 217
    Mars Global Surveyor, 135, 137
    Mars Reconnaissance Orbiter, 163
    Messenger, 112, 131
    SeaWiFS, 216
    TIMED, 161, 176
    TIROS, 18
    Venus Express, 112, 217
Spitzer Space Telescope, 341
Stars
    47 UMa, 381
    55 Cancri, 380
    Beta Pictoris, 315, 343
    Centauri, 25, 231
    Epsilon Eridani, 27

Formalhaut, 276
Gliese 876, 378
HD 114762, 252
HD 160691, 381
HD 40307, 382
HD 69830, 346, 381
HR 8799, 276
Tau Ceti, 27
Upsilon Andromeda, 379
Stellar Encounters, 403
Stevens, A.W., 8
Stratosphere, 50
Strutt, R.J., 165
Sun
   Energy Generation, 61
   Flares, 399
   Mass, 393
   Mass Loss, 61
   Rotation, 399
   Solitude, 393
Super-Earths, 265, 311
Super-Mercuries, 315
Supercontinents, 77, 80
Supernova, 27, 235

T Tauri Stars, 343
Terrestrial Planets, 348, 374

Thermodynamic Disequilibrium, 200
Thermosphere, 50, 162
Thompson, W., 43
Tidal Heating, 231, 318
Tidal Locking, 231
Tikhov, G.A., 135
Titan, 317
Titius-Bode Law, 292
Tournachon, G.F., 8
TPF mission, 281

Ultraviolet, 75, 234
Uranus, 348

V-2 Rockets, 10, 172
Van de Kamp, P., 251
Venus, 46, 94, 122, 234

Wallace, A., 86
Water, 41, 203, 205
Wegener, A., 45
Wright, W., 8

Zodiacal Light, 297

Printed in the United States of America